THE DOMESDAY GEOGRAPHY OF
MIDLAND ENGLAND

Upper portion of left-hand column of folio 165b of the Domesday Book (same size as original). Reproduced by courtesy of the Public Record Office. For extension and translation, see pp. 459–60.

THE
DOMESDAY GEOGRAPHY
OF
MIDLAND ENGLAND

EDITED BY

H. C. DARBY

*Professor of Geography in the
University of Cambridge*

AND

I. B. TERRETT

*Principal Lecturer in Geography
at Chester College*

SECOND EDITION

CAMBRIDGE
AT THE UNIVERSITY PRESS
1971

Published by the Syndics of the Cambridge University Press
Bentley House, 200 Euston Road, London N.W.1
American Branch: 32 East 57th Street, New York, N.Y.10022

© Cambridge University Press 1971

Library of Congress Catalogue Card Number: 78–134626

ISBN: 0 521 08078 9

First published 1954
Second edition 1971

Printed in Great Britain
at the University Printing House, Cambridge
(Brooke Crutchley, University Printer)

A century hence the student's materials will not be in the shape in which he finds them now. In the first place, the substance of Domesday Book will have been rearranged. Those villages and hundreds which the Norman clerks tore into shreds will have been reconstituted and pictured in maps, for many men from over all England will have come within King William's spell, will have bowed themselves to him and become that man's men.

From the concluding paragraph of F. W. MAITLAND'S
Domesday Book and Beyond (Cambridge, 1897)

CONTENTS

PREFACE

This book is the second of a number covering the whole of Domesday England, and it is built upon the same plan as *The Domesday Geography of Eastern England*, first published in 1952. The greater part of the preface to that volume is equally relevant to this one, and its argument must be repeated here. The Domesday Book has long been regarded as a unique source of information about legal and economic matters, but its bearing upon the reconstruction of the geography of England during the early Middle Ages has remained comparatively neglected. The extraction of this Geographical information is not always as simple as it might appear to be from a casual inspection of the Domesday folios. Not only are there general problems of interpretation, but almost every county has its own peculiarities. There is, moreover, the sheer difficulty of handling the vast mass of material, and of getting a general view of the whole. The original survey was made in terms of manors, villages and hundreds, but the clerks reassembled the information under the headings of the different landholders of each county. Their work must therefore be undone, and the survey set out once more upon a geographical basis.

The information that such an analysis makes available is of two kinds. In the first place, the details about plough-teams and about population enable a general picture of the relative prosperity of different areas to be obtained. In the second place, the details about such things as meadow, pasture, wood and salt-pans serve to illustrate further the local variations both in the face of the countryside and in its economic life. An attempt has been made to set out this variety of information as objectively as possible in the form of maps and tables. When all the maps have been drawn and all the tables compiled, we may begin to have a clearer idea of both the value and the limitations of the survey that has so captured the imagination of later generations.

But great though the bulk of the Domesday Book is, it is only a summary. The making of it not only omitted much, but has, too often, resulted in obscurity. No one works for long on the text before discovering how fascinating and tantalizing that obscurity is. In reflecting over many Domesday entries we have been reminded, time and again, of some remarks in Professor Trevelyan's inaugural lecture at Cambridge in 1927: 'On the shore where Time casts up its stray wreckage, we gather corks and broken planks, whence much indeed may be argued and more guessed; but what the great ship was that has gone down into the deep, that we shall never see.' The scene that King William's clerks looked upon has gone, and the most we can do is to try to obtain some rough outline of its lineaments; this chapter in the history of the English landscape can only be a very imperfect one.

The Domesday Geography of Eastern England contained an introductory chapter which is not included here. Amongst other things that chapter explained that the counties are considered separately and that the treatment of each follows a more or less standard pattern. This method inevitably involves some repetition, but, after experiments, it was chosen because of its convenience. It enables the account of each county to be read or consulted apart from the rest, and it also has the advantage of bringing out the peculiar features that characterise the text of each county. For although the Domesday Book is arranged on a more or less uniform plan, there are many differences between the counties, both in the nature of their information and in the way it is presented. The relevance of each of the items to a reconstruction of Domesday geography is examined, and any peculiar features that occur in the phrasing of the Domesday text are also noted. All the standard maps have been reproduced on the same scale to facilitate comparison between one county and another. A final chapter sums up some of the salient feaures of the Domesday geography of the Midland counties as a whole.

The drawing of the maps for this as for other volumes has been made possible through the generosity of the Trustees of the Leverhulme Research Fellowships, and of the University of Liverpool, and of University College, London. Most of the maps in this volume have been drawn by Miss Helen Freestone, to whom we are much indebted. Some are the work of Mr G. R. Versey to whom we are also indebted for a great deal of general assistance. For help from time to time we are grateful to Sir Frank Stenton and Professor V. H. Galbraith. To our six fellow contributors we owe warm thanks for their courtesy and patience. Our debt to the officials of the Cambridge University Press is also great.

H. C. DARBY

I. B. TERRETT

UNIVERSITY COLLEGE
LONDON
Ash Wednesday, 1954

NOTE TO SECOND EDITION

The text of the first edition has been considerably revised to take account of recent research and new place-name identifications. The treatment of the statistics for the boroughs has been brought into line with that of the other volumes of this series. The resulting changes in the densities of some units are not great, but they, together with the revision in general, have meant the alteration or redrawing of a large number of maps. The first edition appeared in 1954 and, in the light of the experience of editing the three subsequent volumes of the series, some facts and ideas would probably be presented a little differently today. It is true that a number of paragraphs have been re-written (e.g. those dealing with Waste), but otherwise changes in wording have been kept to a minimum.

I am greatly indebted to Mr G. R. Versey who has not only altered or redrawn the maps, but has helped at all stages in the checking and revision of the material.

H.C.D.

KING'S COLLEGE,
CAMBRIDGE
St Patrick's Day, 1971

MAPS

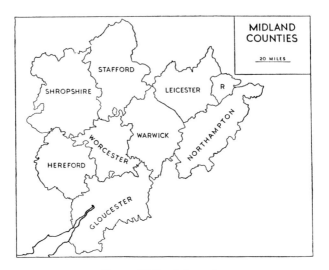

Fig. 1. Midland Counties.
R indicates the county of Rutland.

CHAPTER I

GLOUCESTERSHIRE

BY H. C. DARBY, LITT.D., F.B.A.

The Gloucestershire folios were analysed and discussed as early as 1887–9 by the Rev. C. S. Taylor, who tells us that he applied 'the methods of interpretation used by the Rev. R. W. Eyton' for Dorset and Somerset.[1] Though we cannot accept Taylor's ideas of Domesday mensuration, his analysis is still very useful, and it will always be of interest as one of the early attempts to set out the information of the Domesday Book in tabular form. This information, for Gloucestershire as for other counties, occasionally bears witness to the human element in its assembling. The account of Hawling on fo. 170, for example, speaks of '20 villeins and 5 bordars with 9 bordars' which obviously should read '9 plough-teams'. Other entries omit, or seem to omit, any reference to plough-teams or population. In an entry for Earthcott in Alveston (165) the word *silva* is followed by a blank space where measurements were apparently intended to be inserted. These are exceptions and the Gloucestershire text does not contain many such omissions or apparent omissions. The most unusual feature of the text is the account of the lands beyond the Wye that appears on the first folio, sandwiched between the description of Gloucester and that of Winchcomb. It is all too meagre an account but, at any rate, it does give us a glimpse of the peculiar economy and social structure of Wales, so different from that of the Anglo-Saxon lands.

The county of Gloucester, in terms of which this study is written, is far from being the same as the Domesday county. The main difference is in the north-east. Here, in Domesday times, portions of the counties of Gloucester, Warwick and Worcester were intermingled in a strange manner that reflected the scattered holdings of the bishop of Worcester. These complicated arrangements proved remarkably stable until the nineteenth century. Successive boundary Acts of Parliament in the nineteenth and twentieth centuries have, however, obliterated the detached

[1] C. S. Taylor, 'An Analysis of the Domesday Survey of Gloucestershire', *Bristol and Gloucs. Archaeol. Soc.* (Bristol, 1887–9).

portions of Worcestershire, and so given territorial continuity to each of the three counties.[1] There have, moreover, been some other adjustments along the Gloucestershire–Worcestershire boundary (Fig. 2). Elsewhere, a number of minor changes have also taken place. Along the east of the

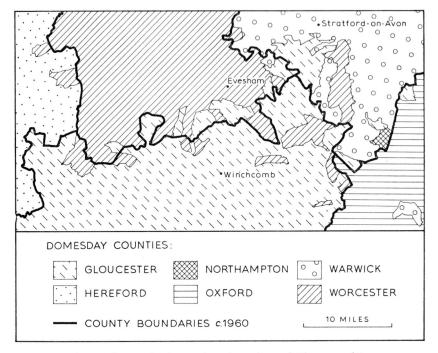

Fig. 2. Changes in the northern boundary of Gloucestershire.
The Northamptonshire area was Whichford – see p. 385 below.

county, Shenington and Widford are now in Oxfordshire. Along the south, three areas have been transferred from Wiltshire: (1) Poulton; (2) Kemble and the two Keynes, Poole and Somerford together with Shorncote; (3) Ashley and Long Newnton. In the south-west, the newer county of Bristol, discussed here with Gloucestershire, includes some territory surveyed in the Somerset folios—the vills of Bedminster and Knowle. Along the north-west, a few places mentioned in the Herefordshire text belong to modern Gloucestershire—Forthampton (described partly under Gloucestershire and partly under Herefordshire), Staunton,

[1] See p. 219 below.

the mysterious *Wiboldingtune*,[1] and Ruardean.[2] Alvington, although on the shores of the Severn estuary, was likewise surveyed in the Hereford-shire folios.[3] Finally, we cannot be certain about the western boundary of the county along the Forest of Dean, unless it was the Wye itself.

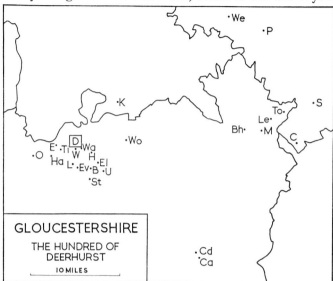

Fig. 3. Gloucestershire: The Deerhurst estates.

The county boundaries are those of *circa* 1960. The hundred is described on fo. 166 of the Domesday Book, and comprised: (1) Holdings at the following places, belonging to the abbey of Westminster, which formed part of the manor of Deerhurst (D): B, Boddington; Bh, Bourton on the Hill; E, Ellings; El, Elmstone; Ev, Evington; H, Hardwicke; Ha, Hasfield; K, Kemerton (in Worcs.); Le, Lemington; M, Moreton in Marsh; O, Oridge (*Tereige*); S, Sutton under Brailes (in Warwick); Ti, Tirley; To, Todenham; W, Wightfield. (2) Holdings at the following places belonging to the abbey of St Denis of Paris; Ca, Calcot; Cd, Coln St Dennis; C, Little Compton (in Warwick); K, Kemerton; L, Leigh; P, Preston on Stour (in Warwick); St, Staverton; U, Uckington; Wa, Walton; We, Welford (in Warwick); Wo, Woolstone. Note that Kemerton appears in both groups and that Boddington, Kemerton and Lemington appear also under Tewkesbury; see p. 5.

[1] *V.C.H. Herefordshire* (London, 1908), I, p. 322 identifies *Wiboldingtune* with Whittington 'opposite Doward on the Wye'. It has been so marked on our maps, but Whittington does not appear on the O.S. map.

[2] The Herefordshire folios (181) also record that 'in *Niware* (Newerne) are $2\frac{1}{2}$ hides which used to come and do service, but Roger de Pistes in the time of Earl William made them part of Gloucestershire'. The Gloucestershire folios make no reference to them, but see C. S. Taylor, *op. cit.* p. 208, and *V.C.H. Herefordshire*, I, p. 319.

[3] For the identification see V. H. Galbraith and J. Tait, *Herefordshire Domesday*, *circa 1160–1170* (Pipe Roll Society, London, 1950), p. 109.

Within the Domesday county there were some forty hundreds varying in size from *Tviferde* with, apparently, about 5 hides, to *Salemanesberie* with as many as 178. The history of the Gloucestershire hundreds seems to be obscure. Some of the Domesday hundreds were not compact territorial units but aggregations of the widely scattered estates of great landowners. Thus Deerhurst hundred included those estates which had

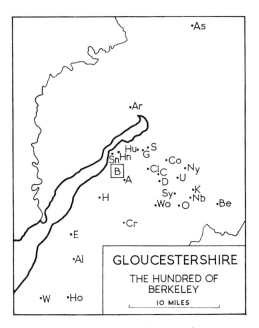

Fig. 4. Gloucestershire: The Berkeley estates.

On fo. 163 of the Domesday Book, the following places are described as members of the manor of Berkeley (B): A, Alkington; Al, Almondsbury; Ar, Arlingham; As, Ashleworth; Be, Beverstone; C, Cam; Cl, Clingre; Co. Coaley; Cr, Cromhall; D, Dursley; E, Elberton; G, Gossington; H, Hill; Hn, Hinton; Ho, Horfield; Hu, Hurst; K, Kingscote; Nb, Newington Bagpath; Ny, Nympsfield; O, Ozleworth; S, Slimbridge; Sn, Sharpness; Sy, Symondshall; U, Uley; W, Weston; Wo, Wotton under Edge.

formerly belonged to the priory of Deerhurst, and which by 1086 were divided between the abbeys of Westminster and St Denis of Paris (Fig. 3). The hundred of Berkeley was likewise scattered in character and only contained portions of the ancient estates of the Crown (Fig. 4). The hundred of Tewkesbury included only lands once held by Brictric son of Algar (Fig. 5). The hundred of *Bernintrev* (later Henbury) was another dispersed hundred that contained a portion of the lands of the bishop of Worcester.[1] Such complications make it difficult to draw a straightforward map of the hundreds of the county.

Another difficulty in handling the Gloucestershire material from a geographical point of view is the fact that the information about two or more places is sometimes combined in one statement. This is particularly

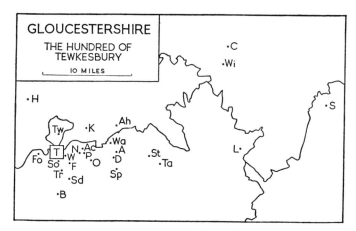

Fig. 5. Gloucestershire: The Tewkesbury estates.

The county boundaries are those of *circa* 1960. The hundred of Tewkesbury (T) is described on fos. 163 and 163 b of the Domesday Book, and included holdings at the following places: A, Alderton; Ah, Ashton under Hill (in Worcs.); Ac, Aston on Carrant; B, Boddington; C, Clifford Chambers (in Warwick); D, Dixton; F, Fiddington; Fo, Forthampton; H, Hanley (in Worcs.); *Hundewuic* (unidentified); K, Kemerton (in Worcs.); L, Lemington; N, Natton; O, Oxenton; P, Pamington; S, Shenington (in Oxford); So, Southwick; Sp, Stanley Pontlarge; St, Stanway; Sd, Stoke Orchard; Ta, Taddington; Tr, Tredington; Tw, Twining; W, Walton Cardiff; Wa, Great Washbourne; Wi, Wincot (in Warwick). Note that Boddington, Kemerton and Lemington appear also under Deerhurst; see p. 3.

true of the non-territorial hundreds of Deerhurst, Berkeley and Tewkesbury, although it is not confined to them. Thus in Deerhurst hundred, we are told that the abbey of St Denis held seven villages (*villas*) which

---

[1] See C. S. Taylor, *op. cit.* p. 33: 'In these cases it is clear that the hundred depended not on any idea of number at all, but merely on the extent of the possessions of the owner, and that it might very naturally increase by the addition of fresh estates, which the lord of the hundred might acquire in the neighbourhood.' Thus on fo. 166 we are told of 8 hides in Kemerton and 3 hides in Boddington which had 'always paid geld and rendered other services in Deerhurst hundred' until Gerard the Chamberlain acquired them; both holdings are described in detail under Tewkesbury on fo. 163 b.

are named (166). Their respective hidages are given separately but their resources are described collectively, so that we have no means of apportioning the details of population, plough-teams, mills, meadow and wood among the seven. For the purpose of constructing distribution maps the details of this and other combined entries have been divided exactly among all the villages concerned. But the item relating to wood in this particular entry, as in some others, cannot be so divided; the wood measured two and a half leagues by one league and two furlongs, and it has been allocated to Uckington, the first-named of the seven villages.[1] Fisheries and mills have been treated likewise where relevant.[2]

SETTLEMENTS AND THEIR DISTRIBUTION

The total number of separate places mentioned in the Domesday Book for the area now included in the modern county of Gloucester seems to be 367, including the four places for which burgesses are recorded—Bristol, Gloucester, Tewkesbury and Winchcomb. This figure, however, may not be quite accurate; for when two or more adjoining villages bear the same surname today, it is not always clear whether more than one unit existed in the eleventh century. There is no indication that, say, the Upper and Lower Slaughter of today existed as separate villages; the Domesday information about them is entered under only one name (*Sclostre*), though there may well have been separate settlements in the eleventh century. The same applies, for example, to the two Swells, the three Rissingtons and four Ampneys.[3] The distinction between the respective units of each of these groups appears later in time. Thus the names Eastleach Martin and Eastleach Turville, so far as the evidence goes, date from the thirteenth century, the former from the dedication of the church, the latter from the Turville family. For some counties the Domesday text occasionally differentiates between the related units of such groups by designating one unit as *alia* or *parva*, but none of the Gloucestershire groups is distinguished in this way.

---

[1] For this entry see p. 24 below.
[2] See pp. 36 and 39 below.
[3] Upper Swell, Lower Swell; Great Rissington, Little Rissington, Wyck Rissington; Ampney Crucis, Ampney St Mary, Ampney St Peter, Down Ampney. The fact that the name Ampney may have covered more than one settlement is suggested by the presence of three priests (and presumably three churches) on the Domesday holdings. Bisley, Guiting and Siddington also had two priests each, see p. 42 below.

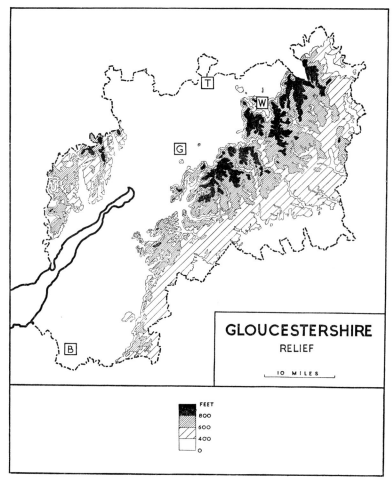

Fig. 6. Gloucestershire: Relief.
Domesday boroughs are indicated by initials: B, Brisol;
G, Gloucester; T, Tewkesbury; W, Winchcomb.

The total of 367 includes about a dozen places for which very little information is given; the record may be incomplete, or the details may have been included with those of a neighbouring village. Thus Harnhill (168), adjoining the Ampneys, answered for 5 hides and it had included three manors in 1066, but we are told nothing of its population or resources either for 1066 or 1086. The entry for Trewsbury (in Coates), on the other hand, tells us that it answered for half a hide, that there was

one plough on the demesne and that its value had increased from 10*s*. to
15*s*., but we are told nothing about its other resources or about its
population (168b). No plough-teams are likewise entered for Newnham
(167), or population for the mysterious *Aldeberie* (168b). Stow on the
Wold is not separately surveyed but the neighbouring parish of Maugersbury is entered in the text as *Malgeresberiae ad Eduuardestow* (165b), and
it seems certain that the eleventh-century resources of the forty-five
acres that comprise the present-day Stow were included with those of
its larger neighbour. An interesting example of incomplete information
is provided by Woodchester, to the south of Stroud. The entry relating
to one holding sets out its resources in the usual manner (170b); but
that relating to the other confesses its ignorance (164), and says that no
one came to give an account of it to the king's commissioners and that
none of its people was present at the making of the description (*De quo
manerio nemo Legatis regis reddidit rationem, nec aliquis eorum venit ad hanc
descriptionem*).

   Not all the Domesday names appear on the present-day map of
Gloucestershire villages. Some are represented by hamlets, by individual
houses and farms, or even by the names of topographical features. Thus
Norcote or *Nortcote* is now the hamlet of Norcott in Preston near Cirencester, and *Hildeslei* is that of Hillsley in Hawkesbury. *Gosintune* is the
hamlet of Gossington in Slimbridge, and *Hirslege* is now Hurst Farm in
the same parish. *Culcortorne* and *Hasedene* are, respectively, the hamlets
of Culkerton and Hazleton Manor in Rodmarton. *Claenhangare* has
become Clingre House and Clingre Farm in the parish of Cam. *Peclesurde*
is represented by Pegglesworth Hill in Dowdeswell, and *Aicote* by Eycote
Wood and Eycotfield in Colesborne. *Frantone* appears on the modern
map as the hamlet of Frampton Mansell and Frampton Common in
Sapperton. *Wigheiete* survives only in the names Wyegate Green and
Wyegate Hill in St Briavels in the Forest of Dean. These are but some of
the changes in the Gloucestershire villages. To them must be added a
number of unidentified names.[1] Whether these will yet be located or
whether the places they represent have completely disappeared, we cannot say.

   On the other hand, some villages on the modern map are not mentioned

---

   [1] The complete list of names that have not been located in the present analysis is as
follows: *Aldeberie* (168b), *Chingestune* (163), *Chire* (163), *Cliftone* (163), *Hundeuuic*
(163b), *Lega* (165) and *Uletone* (167b).

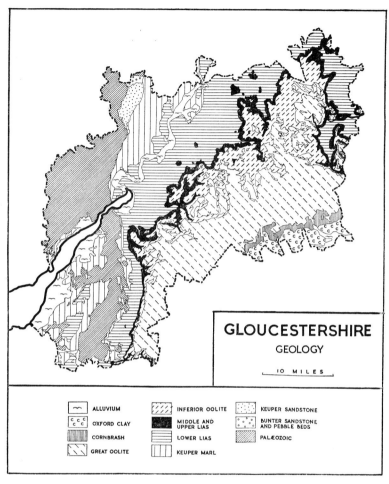

Fig. 7. Gloucestershire: Geology.
Based on Geological Survey Quarter-Inch Sheets, 14, 15, 18 and 19.

in the Domesday Book. Their names do not appear until the twelfth and thirteenth centuries, and, presumably, if they existed in 1086, they are accounted for under the statistics of neighbouring settlements. Thus, as far as record goes, Randwick was first mentioned in 1121, Didbrook in 1248 and Thrupp in 1261. Another example of a unit that came into being after 1086 is that of Cranham, first mentioned in 1148. It is set between Painswick and Brimpsfield, and we know that it was formed partly out of land given from Brimpsfield to Gloucester Abbey, and partly out of a portion of Painswick given to the Church of Cirencester. Some of the places not mentioned in the Domesday Book must have existed, or at any rate been named, in Domesday times because they appear in pre-Domesday documents. The following names, for example, are found in

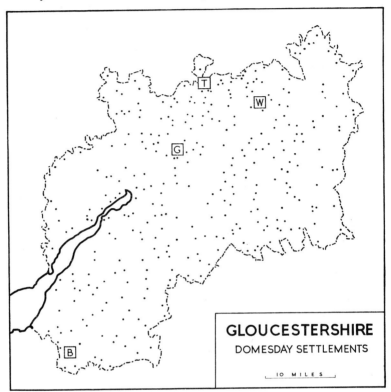

Fig. 8. Gloucestershire: Domesday place-names.
Domesday boroughs are indicated by initials: B, Bristol;
G, Gloucester; T, Tewkesbury; W, Winchcomb.

the charters of the ninth and tenth centuries and again in manuscripts of the twelfth and thirteenth centuries, and yet they do not appear in the Domesday text: Lancaut, North Nibley and Rodborough. Sometimes, although a modern parish is not named in the Domesday Book, it contains hamlets that are mentioned. An interesting example of this is Coates near Cirencester. The earliest record of it, apparently, is from the year 1175, but the area covered by the present parish includes the Domesday settle-

ments of Hullasey, Tarlton, Trewsbury and parts of Oakley.[1] Another
missing parish is Eastington near Stroud, and it contains a Domesday
name represented by the modern hamlet of Alkerton. From this account
it is clear that there have been many changes in the village geography of
the county, and that the list of Domesday names differs considerably from
a list of present-day parishes.

Except for one area, the distribution of Domesday names over the
county is surprisingly even (Fig. 8). Villages were as frequent in the
Cotswold valleys as on the clay plains of the central Vale. The exception
is the Forest of Dean which stands out as an empty area in a striking
fashion.

THE DISTRIBUTION OF PROSPERITY AND POPULATION

Some idea of the nature of the information in the Domesday folios for
Gloucestershire, and of the form in which it is presented, may be obtained
from the account of the village of Cowley situated in the Cotswolds to
the south of Cheltenham and in the valley of the Churn (166). The village
was held entirely by the abbey of Pershore, and so it is described in
a single entry:

> The Church of St Mary of Pershore holds Cowley. There 5 hides pay geld.
> In demesne are 2 plough-teams, and 14 villeins and one bordar with 7 plough-
> teams. There 5 serfs, and a mill worth (de) 50d., and 6 acres of meadow, and
> wood 3 furlongs long and one broad. It is worth 100s.

This entry does not include all the kinds of information that appear
elsewhere in the folios for the county. It does not mention, for example,
the categories of population known as radmen, coliberts and *ancillae*.
There is also no mention of a church or a fishery. A value is stated for
only one date, 1086; presumably the estate was worth the same in 1066,
but there is space before the statement of value and it is just possible that
the value of the estate in 1066 was intended to be inserted there. But
although not comprehensive, it is a fairly representative and straightforward entry, and it does set out the recurring standard items that are
found for most villages. These are four in number: (1) hides; (2) plough-
teams; (3) population; (4) values. The bearing of these four items of

---

[1] The dates in this paragraph are from A. H. Smith, *The Place-Names of Gloucestershire*, 4 vols. (Cambridge, 1964–5).

information upon regional variations in the prosperity of the county must now be considered. It is interesting to note that the Gloucestershire folios, like those for Herefordshire and Worcestershire, hardly ever mention plough-lands; the few stray references are discussed in the section dealing with plough-teams.

(1) *Hides*

The Gloucestershire assessment is stated in terms of hides and virgates, and, very occasionally, of acres: thus on one holding at Culkerton there were 3 virgates and 5 acres (167).[1] In some entries the phrase 'of land' appears; at Cirencester (162b), for example, the king had 5 hides of land (*v hidae terrae*). Sometimes the word *geldantes* (168) or a variant is added, but more usually the number of hides alone is just plainly stated. A rare variant is *pro ii hidis se defendebat haec terra* (162b). Fractions, of both hides and virgates, are common. At Chedworth (164) there were 15 hides of 'wood, field and meadow' (*inter silvam et planum et pratum*), and yet no wood or meadow is specifically mentioned in the entry. There were also 5 hides of 'wood and field' at Churcham and Morton (165b), but here, at any rate, the wood is measured.[2] In the entry for Hambrook (165) there is a curious statement which says that on a holding of one hide there were only 64 acres of land when it was ploughed (*In ista hida quando aratur non sunt nisi lxiiii acrae terrae*).

A glance through the Gloucestershire folios reveals many examples of the five-hide unit: thus Winstone, to the east of Painswick, was held for 5 hides (169b), and Buckland, to the north of Winchcomb for 10 hides (165b). When a village was divided amongst a number of owners, the same feature can sometimes be demonstrated. Dodington, near Chipping Sodbury, was held by two lords whose holdings together amounted to 5 hides:

> Bishop of Coutances (165): 1 hide and a half and a third part of half a hide.
> Roger of Berkeley (168): 3 hides and 2 parts of half a hide.

The number of villages assessed in 1066 at either 5 or 10 hides amounts to 86 out of a total of 367; there are also a number of other assessments of 15, 20 and even 50 hides, although the larger figures are totals covering

---

[1] In the description of the land beyond the Wye, however, the carucate is used see p. 54 below.

[2] See p. 28 below.

composite entries. We cannot be far wrong in saying that the five-hide principle is readily apparent in over one-quarter of the Gloucestershire villages. It is possible that some villages were grouped in blocks for the purpose of assessment, as in Leicestershire.[1] In any case, the absence of information about these groups makes it difficult to be definite about the full extent of the five-hide unit in Gloucestershire.

Whether the five-hide unit can be demonstrated or not, it is clear that the assessment was largely artificial in character, and bore no constant relation to the agricultural resources of a vill. A striking instance of low hidation is that of Painswick. It answered for only one hide, yet it had 53 teams and was valued at £24 in 1086 (167b). The variation among a representative selection of ten-hide vills speaks for itself:

	Teams	Population
Alveston (163)	25	35
Avening (163b)	24	59
Bourton on the Water (165b)	13½	27
Buckland (165b)	15	36
Hawling (170)	12	31
Horton (168)	11	26

A number of exemptions or reductions in assessment are recorded. Thus Olveston and Cold Ashton, held by the church of Bath, were five-hide vills, but in each case we are told that two of the hides had been exempt from tax by a grant of King Edward and King William (165):

Olveston: *Ibi v hidae. Tres geldant ex his et ii non geldant concessu E. et W. regum.*

Cold Ashton: *Ibi v hidae. Harum duae sunt a geldo quietae concessu E. et W. regum. Tres vero geldant.*

The 2 hides and 2 virgates at *Dene* (167b) had been granted free from geld by King Edward in return for keeping the forest (*Has terras concessit rex E. quietas a geldo pro foresta custodienda*). The succeeding entry on fo. 167b describes a holding at Taynton which is not measured in hides and which is stated to be 'free land' (*Haec terra libera est*). At Newent (166) not far from Taynton and *Dene*, there were 6 hides that had never paid geld (*non geldaverunt*); and the ten-hide manor of Roel (166b),

---

[1] See p. 325 below. For the Cambridgeshire blocks, see H. C. Darby, *The Domesday Geography of Eastern England* (Cambridge, 3rd ed., 1971), p. 276.

south-east of Winchcomb, was in the same category (*Hoc manerium nunquam geldavit*). At Lechlade (169) the liability had been reduced from 15 hides to 9 by the king himself (*Ibi xv hidae T.R.E. geldantes. Sed ipse rex concessit vi hidas quietas a geldo*). At Tewkesbury, exemption was on a large scale (163). There were 95 hides; 'of these' we are told '45 were in demesne and were quit from all royal service and tax except service to the lord of the manor'. All these examples are typical of a much larger number, and they serve to show the widespread character of exemption on the Gloucestershire estates.

The assessment (including non-gelding hides) amounted to 2,421 hides, 1¾ virgates, 5 acres, but it must be remembered that this refers to the area included in the modern county. Maitland estimated the number of hides in the Domesday county at 2,388.[1] C. S. Taylor's total for the Domesday county, in his summary tables, came to as much as 2,611 hides 5 acres;[2] but his second estimate, made in a different manner, yielded only 2,595 hides 3 virgates 1 acre, a discrepancy of about one-half per cent. 'Considering the mass of figures to be dealt with,' wrote Taylor, 'this is, perhaps, a very fair approach to exactness.'[3] The fact is that the nature of some of the composite entries makes exact calculation exceedingly difficult. As Maitland himself said, 'two men not unskilled in Domesday might add up the number of hides in a county and arrive at very different results, because they would hold different opinions as to the meaning of certain formulas which are not uncommon'.[4] All the figures can do is to indicate the order of magnitude involved.

(2) *Plough-teams*

The Gloucestershire entries, like those of other counties, draw a distinction between the teams on the demesne and those held by the peasantry. Occasional entries seem to be defective. That for Newnham, for example, mentions 3 villeins and 3 bordars who paid 20s., but says nothing about their ploughs (167); maybe they had none. Occasionally we are specifically told that there was nothing on the demesne, e.g. at Baunton (168b)

[1] F. W. Maitland, *Domesday Book and Beyond* (Cambridge, 1897), p. 400. On another page Maitland declared: 'My 2,388 is I think a trifle too low; but I believe the number lies very close to 2,400 on one side or the other' (p. 457).

[2] C. S. Taylor, *op. cit.* p. 332.

[3] *Ibid.* p. 233. For a discussion of Taylor's figures see F. W. Maitland, *op. cit.* pp. 412 and 457n.

[4] F. W. Maitland, *op. cit.* p. 407.

and Woodchester (170b); at Shipton (167b), on the other hand, it was the peasantry who were without a plough (*sine caruca*). But the record of teams is usually complete and fairly straightforward. Only rarely are the bare facts enlightened by some additional detail. One of these rarities is the statement that, in 1066, on some of the dependencies of Tewkesbury (163) the radmen ploughed and harrowed on the lord's manor (*arabant et herciabant ad curiam domini*). On the Deerhurst estates (166), also in 1066, the radmen ploughed, harrowed, mowed and reaped for the lord (*arabant, herciabant, falcabant, et metebant*).

The Gloucestershire folios make no reference to plough-lands or to potential teams. There are, however, some stray phrases that seem to have crept into the text, and that may indicate the presence of this information in the original returns for the county. At Hambrook (165) there were 4 plough-teams, 2 on the demesne and 2 with the villeins; but there is another statement written in small letters over one of the lines of the entry, saying that there was land for 5 ploughs (*Terra est v carucis*). An entry for Alderton (165b) records 3 plough-teams, and then goes on to say that there could be 3 more (*et adhuc iii possunt esse*). The very next item on fo. 165b relates to a holding with 4 ploughs at Naunton, and we are told that there might be 6 more (*et adhuc vi ibi possent esse*). It seems as if, on each of these three occasions, the compiler had for a moment varied the form of his summary and given us a little extra information. The only other places for which we are given such details are those within the modern county but outside the Domesday county, and so not conforming to the normal type of entry for Gloucestershire. Thus details of plough-lands as well as of plough-teams are given for Ashley (71b), Kemble (67), Long Newnton (67), Poole Keynes (69b), Poulton (68b), Shorncote (73) and Somerford Keynes (66), which were in Domesday Wiltshire; and for Bedminster (86b) and Knowle (98) in Domesday Somerset. Along the northern boundary at Ruardean (185b), in Domesday Herefordshire, the peasantry had 3 ploughs, but there could also be another 3 on the demesne (*In dominio possent esse iii carucae*). Worcestershire was another county for which the Domesday Book frequently recorded the possibility of additional teams, but it so happens that none of the entries relating to the Worcestershire villages now in Gloucestershire includes this formula.

The total number of plough-teams amounted to 3,822 $\frac{3}{28}$, but it must be remembered that this refers to the area included in the modern county.

Maitland estimated the number for the Domesday county at 3,768.[1]
Taylor's figure came to 3,909;[2] but, as we have seen, a definitive total is
hardly possible.[3]

(3) *Population*

The bulk of the population was comprised in the three main categories
of villeins, serfs and bordars. In addition to these were the burgesses
together with a large miscellaneous group that included radmen, coliberts,
priests and others. The details of the groups are summarised on p. 19.
There are two other estimates of population, by Ellis[4] and by Taylor[5]
respectively, but they are comparable neither with one another nor with
the present estimate, which has been made in terms of the modern county.[6]
Definitive accuracy rarely belongs to a count of Domesday population,
and all that can be claimed for the present figures is that they indicate
the order of magnitude involved. These figures are those of recorded
population, and must be multiplied by some factor, say 4 or 5, in order
to obtain the actual population; but this does not affect the relative
density as between one area and another.[7] That is all that a map, such as
Fig. 11, can roughly indicate.

It is impossible for us to say how complete were these Domesday
statistics. But it does seem as if some people had been left uncounted, for
we hear of ploughs worked by unspecified groups of men. We are told
that the radmen at Berkeley (163), the knights at Cerney (168b), the
freemen at Sevenhampton (165) and a Frenchman at Duntisbourne (166b)
cultivated their lands *cum suis hominibus*; and there was also a priest at
Bibury (164b) who held his ploughs *cum suis*. Moreover, as we have seen,
a few entries seem to be defective and contain no reference to population,
but we cannot be certain about the significance of these omissions.[8]

Villeins constituted the most important element in the population,
and amounted to 46% of the total. Half-villeins are entered for Hazleton

---

[1] F. W. Maitland, *op. cit.* p. 401.   [2] C. S. Taylor, *op. cit.* p. 332.

[3] See p. 14 above.

[4] Sir Henry Ellis, *A General Introduction to Domesday Book* (London, 1833), II,
pp. 444–5.   [5] C. S. Taylor, *op. cit.* p. 335.

[6] Taylor's estimate came to 8,239; that of Sir Henry Ellis came to 8,366, but this
cannot be compared with Taylor's estimate for two reasons: (i) it includes tenants-
in-chief, under-tenants, all the urban population and *ancillae*; (ii) it seems to cover the
land beyond the Wye, dealt with in a prefatory annexe to the Gloucestershire folios;
see p. 53, below.

[7] But see p. 430 below for the complication of serfs.

[8] See p. 7 above.

(168) and Hillsley (169b), but in neither case are we given any clue about the other half; they were presumably men who held only half a villein tenement, or those whose land and services were divided between two lords. At Hidcote Bartrim (166) the wives of 4 villeins not long dead had one plough-team (*uxores iiii villanorum nuper defunctorum habent i carucam*). The statistics relating to bordars present no special feature for our purpose. Those relating to serfs are interesting because the percentage of serfs was higher in Gloucestershire than in any other county.[1] They amounted to about one-quarter of the total population (26%); but it is interesting to note that in the district between the Severn and the Wye there were very few serfs and the percentage fell to about five. In many entries a combined number of male and female serfs is given (*servi* and *ancillae*), and such figures have been divided equally between the sexes for purposes of computation; *ancillae* have not been included in the total population. At Hailes (167b) there had been 12 serfs whom the lord had freed (*Ibi erant xii servi quos Willelmus [Leuric] liberos fecit*).

The miscellaneous category was a very varied one. One of the most interesting groups was that of radmen (*radchenistres*) who were found mainly in the western counties of Gloucester, Worcester, Hereford and Shropshire. They did riding service for their lords as well as carried on the agricultural work of their holdings with the help of labourers.[2] They seem to have been freemen, and in the descriptions of the manors of Berkeley (163) and Deerhurst (166) they were definitely described as such.[3] Only a few freemen were specifically mentioned. Another interesting group was that of coliberts who were in some ways intermediate between serfs and villeins.[4] Among the smaller groups there were 5 potters (*figuli*) who paid 44d. at Haresfield (168b), and a smith (*faber*) at Pinbury (166b) and at Quenington (167b). The infrequent mention of craftsmen, such as smiths and carpenters, in the Domesday Book is surprising. The smith at Quenington rendered 2s., but the one at Pinbury was grouped with 8 villeins as owning 3 plough-teams. It may well be that the craftsmen in general were usually reckoned among the villein tenants. At

---

[1] For the problems raised by the entries for serfs, see p. 430 below.

[2] See P. Vinogradoff, *English Society in the Eleventh Century* (Oxford, 1908), pp. 69–71. See p. 128 below.

[3] Berkeley (163): *Ibi xix liberi homines Radchenistres habentes xlviii carucas cum suis hominibus*; Deerhurst (166): *Radchenistres id est liberi homines*.

[4] See (i) F. W. Maitland, *op. cit.* pp. 36–7; (ii) P. Vinogradoff, *op. cit.* pp. 468–9.

Ashley (71 b) there were 2 cottars and at Long Newnton (67) there were 5; there were also 15 coscets at Kemble (67), 7 at Poulton (68b) and 2 at Long Newnton (67), but these were places surveyed in the Wiltshire folios; no mention of either cottars or coscets is made in the Gloucestershire folios. The Herefordshire folios record 4 swineherds at Forthampton (180b), and today the parish is divided between Gloucestershire and Worcestershire. Ruardean, with its one Welshman, is also surveyed in the Herefordshire folios (185 b). The other miscellaneous categories were those of *francigenae* (Frenchmen or foreigners), knights, reeves (*prepositi*), together with priests and burgesses discussed on p. 42 and p. 43 below. Finally, there were simply 'men' (*homines*) who for the most part seem to have paid for their holdings in money or kind rather than by service. The 23 men at Bisley (166b) rendered 44*s.* as well as some honey; the 17 in the market at Berkeley (163) paid a rent; the 6 at Pucklechurch (165) rendered iron (*c massas ferri x minus*); the 2 at Doynton (165) rendered 5*s.*; and the solitary man at Batsford (169b) rendered 6 plough-shares (*sochos*). On the other hand, no render is indicated for the 2 men at Meysey Hampton (166b), the 3 at Brewerne (162b) and the 4 at Upton St Leonards (162b).

(4) *Values*

The value of an estate is normally given in a round number of pounds for two dates, 1066 and 1086. The amount had sometimes risen, sometimes fallen, or had occasionally remained the same. There are a few entries in which only one value is given, e.g. for Avening (163b) and Cowley (166). No one returned any value for two estates in Longtree hundred (166b) and the jury therefore rated them at £8 (*Non fuit qui responderet de his terris sed per homines comitatus appreciantur viii libras*).[1] No one likewise answered for a holding in Swinhead hundred (170) which was rated at 10*s.* (*Non fuit qui de hac terra responderet*). Two holdings at Ampney and Upleadon were said to be scarcely (*vix*) worth 20*s.* and 30*s.* respectively (165 b). At Chedworth and Arlington (164) we are told that the reeves had paid what they would in 1066 (*prepositi quod volebant reddebant T.R.E.*), but that the sum yielded in 1086 was £40. At Dymock, likewise, an uncertain yield had been changed to a fixed sum (164). Some tenants in Slaughter had also been accustomed to render what they would

---

[1] There are other references to *homines de comitatu*, e.g. in entries for Dymock (164) and Westbury on Severn (163). See also an entry relating to Guiting for *comitatus testatur* (167).

and so did not know how to rate their estate (*De hoc manerio reddebant quod volebant vicecomiti T.R.E. Ideo nesciunt appreciari*); their liability, too, was fixed under the new order (163).

Recorded Population of Gloucestershire in 1086

A. *Rural Population*

Villeins	.	.	.	.	.	3,730
Serfs .	.	.	.	.	.	2,052
Bordars	.	.	.	.	.	1,871
Miscellaneous	.	.	.	.	.	401
Total	.	.	.	.	.	8,054

There were also 193 bondwomen (*ancillae*) and 4 wives (*uxores*) who have not been included in the above total.

Details of Miscellaneous Rural Population

Radmen	124	Cottars	7
Coliberts	85	Knights	6
Men	58	Potters	5
Priests	53	Reeves	2
Coscets	24	Smiths	2
Freemen	20	Swineherds	2
Francigenae	12	Welshman	1
		Total	401

B. *Urban Population*

Serfs are also included in the table above.

GLOUCESTER 73 burgesses; 22 *mansiones*; 2 *domus*; 14 *domus vastatae*; 16 *domus desunt*.

WINCHCOMB 32 burgesses.

TEWKESBURY 13 burgesses; 22 serfs and *ancillae* (presumably 11 serfs).

BRISTOL unspecified number of burgesses; 12 *domus*.

The 'values' are usually entered as plain statements of money, but there are entries which indicate considerable differences in the method of reckoning. We are specifically told that the values of some estates were reckoned by the tale or number, when the coins were accepted at their nominal value. But in view of the circulation of debased coins, some values were reckoned by weight, and were also tested by assay; the latter was known as 'white' money. Some of the variations are indicated below:

Slaughter (163): £27 by tale (*xxvii libras ad numerum*).

Alveston (163): £12 by weight (*xii libras ad pensum*).

Berkeley (163): £170 burnt and weighed, i.e. assayed (*clxx libras arsas et pensatas*).

Tidenham (164): £25 of 20*d.* to the ounce and white (*xxv libras de xx in ora et albas*).

Chedworth and Arlington (164): £40 of white money of 20*d.* to the ounce (*xl libras alborum nummorum de xx in ora*).

There was another complication. On some royal estates, the valuation was made partly or entirely in kind, either in produce or entertainment. Thus in 1086, the manors of Cheltenham (162b), Barton by Gloucester (162b) and Cirencester (162b) each paid about £20 and also rendered cows and pigs; moreover, on each manor a charge of providing 3,000 loaves for the hounds had been commuted into a money payment of 16*s.*[1] Westbury on Severn (163), on the other hand, rendered the expense of one night's entertainment (*una nox de firma*); the manor of Bitton with its dependencies of Wapley and Winterbourne did likewise (162b); that of Awre rendered only half a night's entertainment (163), but we are not told about the other half. All these complications make it difficult to total up the liabilities involved and so compare the values of some manors one with another.

Generally speaking the greater the number of plough-teams and men on an estate, the higher its value, but it is impossible to discern any constant relationships as the following figures for four holdings, each yielding £5 in 1086, show:

	Teams	Population	Other resources
Adlestrop (165b)	7	17	Meadow
Bromsberrow (168)	15	20	Wood
Frampton on Severn (168b)	9	27	Mill, wood, meadow
Whaddon (168b)	10	8	Nil

[1] £20 each from Barton and Cheltenham; £20. 5*s.* from Cirencester. Cirencester had also rendered wheat, barley and honey in 1066. There were 23 men at Bisley who rendered 44*s.* and some honey (166b); and there were 6 men at Pucklechurch who rendered iron (165); honey and iron also appear in the description of Gloucester itself (162); a man at Batsford (169b), and a burgess at Quenington (167b), rendered plough-shares. Frederic Seebohm thought that the 1066 render in kind at Cirencester was 'very much like a survival of the Welsh food-rents at one of the cities conquered by the Saxons in 577' (*The English Village Community* (Cambridge, 1926), p. 211). See p. 55 below for account of food-rents in the land beyond the Wye. Deerhurst had also rendered honey in 1066 (166).

It is true that the variations in the arable, as between one manor and another, did not necessarily reflect variations in total resources, but even taking the other resources into account the figures are not easy to explain. Moreover, in the absence of information about conditions in 1066, we are without a clue to the vicissitudes that lay behind the changes over twenty years.

Conclusion

The varied size of the Domesday hundreds and the dispersed nature of some of them do not make them convenient units for the purpose of calculating densities.[1] Some twenty-nine more or less artificial units have therefore been adopted. In forming them, however, local considerations have been borne in mind and, as far as possible, they have been devised so as to keep those of the Cotswolds separate from those of the plain to the west.

Of the four standard formulae, those relating to plough-teams and population are most likely to reflect something of the distribution of wealth and prosperity throughout the county in the eleventh century. Taken together, they supplement one another to provide a general picture (Figs. 9 and 11). The density of plough-teams over the Cotswolds ranges from about three to about five; that of population from about six to eleven. The densities over the plain tend, on the whole, to be lower except in the area to the west of Stroud and around Gloucester itself where the figures equal those of the Cotswolds. But the interesting feature is the lack of any marked contrast between the Cotswolds and the plain. There is no direct Domesday evidence about the Cotswold plateau itself, but it is clear that there was considerable agricultural activity in the valleys that broke its surface, and the densities of both population and plough-teams were, roughly speaking, of the same general order as those of much of midland England. Standing in marked contrast to the rest of the county was the Forest of Dean. Its plough-team density was not much more than one per square mile, and the density of its population not much more than two. It was a relatively empty place and the law of the forest lay heavily upon it.

Figs. 10 and 12 are supplementary to the density maps, but it is necessary to make one reservation about them. As we have seen on p. 6, it is possible that some Domesday names may have covered two or more

[1] See p. 4 above.

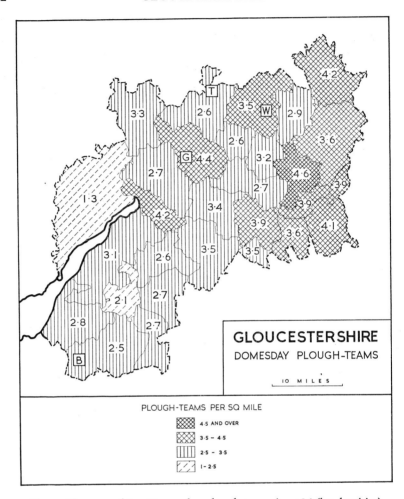

Fig. 9. Gloucestershire: Domesday plough-teams in 1086 (by densities).
Domesday boroughs are indicated by initials: B, Bristol;
G, Gloucester; T, Tewkesbury; W, Winchcomb.

settlements, e.g. the present-day villages of Upper and Lower Slaughter
are represented in the Domesday Book by only one name. A few of the
symbols should therefore appear as two or more smaller symbols, but
this limitation does not affect the main pattern of the maps. Generally
speaking, they confirm and amplify the information of the density maps.

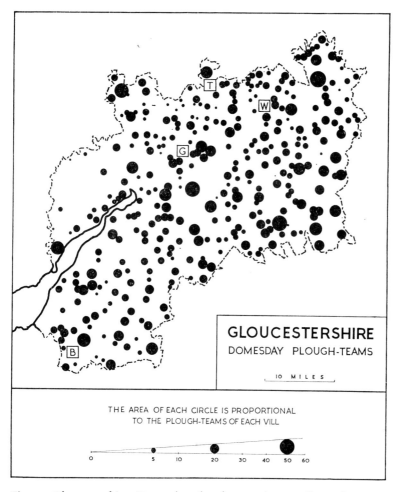

Fig. 10. Gloucestershire: Domesday plough-teams in 1086 (by settlements).
Domesday boroughs are indicated by initials: B, Bristol;
G, Gloucester; T, Tewkesbury; W, Winchcomb.

WOODLAND AND FOREST

Types of entries

The amount of woodland on a holding in Gloucestershire was normally
recorded by giving its length and breadth in terms of leagues and furlongs.
The following examples illustrate the kind of entry that is encountered:

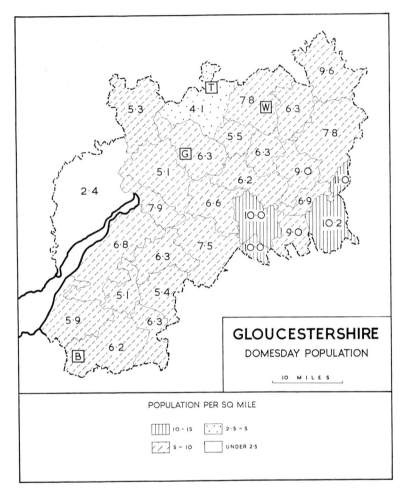

Fig. 11. Gloucestershire: Domesday population in 1086 (by densities).
Domesday boroughs are indicated by initials: B, Bristol;
G, Gloucester; T, Tewkesbury; W, Winchcomb.

Tibberton (167): *Silva iii leuuis longa et una lata.*
Allaston (166b): *Silva dimidia leuua longa et dimidia lata.*
Newnham (167): *Silva ibi ii quarentenis longa et una lata.*
Hawkesbury (166): *Silva de ii leuuis longa et una lata.*
Uckington, Staverton, Coln, Calcot, Little Compton, Preston and Welford
    (166): *Silva ii leuuis et dimidia longa et una leuua et ii quarentenis lata.*

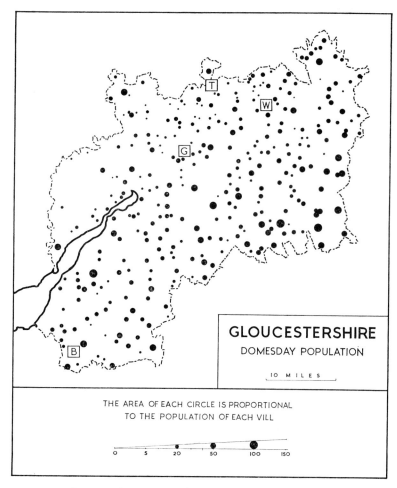

Fig. 12. Gloucestershire: Domesday population in 1086 (by settlements).
Domesday boroughs are indicated by initials: B, Bristol;
G, Gloucester; T, Tewkesbury; W, Winchcomb.

The wood at Tortworth was valued as well as measured: *Silva i leuua longa et dimidia lata reddens v solidos* (169b); and that at Avening (163b) had a hawk's nest (*aira accipitris*). In an entry for Earthcott in Alveston (165) the word *silva* is followed by a blank space where measurements were apparently to be inserted.

The exact significance of all these entries is far from clear, and we

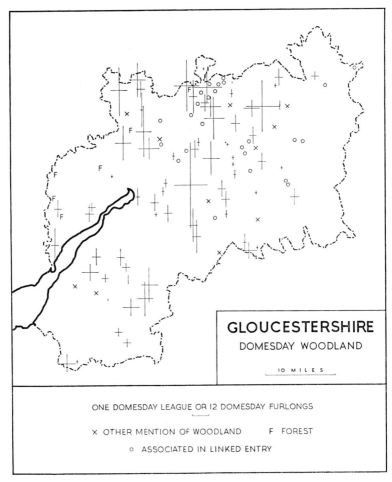

GLOUCESTERSHIRE
DOMESDAY WOODLAND

10 MILES

ONE DOMESDAY LEAGUE OR 12 DOMESDAY FURLONGS

× OTHER MENTION OF WOODLAND   F FOREST

○ ASSOCIATED IN LINKED ENTRY

Fig. 13. Gloucestershire: Domesday woodland in 1086.

Where the wood of a village is entered partly in linear dimensions and partly
in some other way, only the dimensions are shown.

cannot hope to convert them into modern acreages.[1] All we can do is to
plot them diagrammatically as on Fig. 13. Occasionally one set of measurements is found in a combined entry covering a number of villages. This
seems to imply, although not necessarily, some process of addition
whereby the dimensions of separate tracts of woodland were consolidated
into one sum. On Fig. 13 these dimensions have been plotted for the caput

[1] See p. 437 below.

of the manor or for the first-named village in a composite entry, e.g. for Uckington in the example given above. The associated villages in all such entries have also been indicated.

There are four entries that differ from the usual pattern:

Charfield (170): *Silva dimidia leuua longa et lata.*
Tytherington (165): *Silva dimidia leuua in longitudine et latitudine.*
Poole Keynes (69b): *Silva i leuua in longitudine et latitudine.*
Forthampton (180b): *Silva habet iiii leuuas inter longitudinem et latitudinem.*

The first two places were in the hundred of *Bachestane*, and it so happens that there were also two holdings in the same hundred, for each of which only one dimension is given. At Iron Acton (165) there was one furlong of wood (*una quarentena silvae*), and at Wickwar (170) there were six furlongs (*vi quarentae de silva*). C. S. Taylor believed that these entries implied 'areal leagues' or 'areal furlongs', and that an areal league comprised 120 acres.[1] The evidence of the *Liber Exoniensis* for the South-west counties seems to indicate, however, that all these variant entries imply the same facts as the more usual formula, and that they occur occasionally when length and breadth were the same.[2] The wood for Charfield, Tytherington, Poole Keynes and Forthampton has been plotted on this assumption on Fig. 13, but the two single dimensions have been shown as single lines. The record for *Bachestane*, as a whole, seems to indicate some idiosyncrasy. There were only five villages in the hundred, and the wood entries for four of them, as we have seen, are curious; and at the fifth village—Tortworth—a wood one league by a half is entered in the usual way (169b), but a money render is also stated.

While leagues and furlongs form the usual units of measurement, there are some exceptional entries that refer to wood in other ways:

Hawling (170): A wood is there (*Silva est ibi*).
Forthampton (163b): A wood there (*Ibi silva*).
Newent (166): From the wood 30d. (*De silva xxx denarii*).
Bisley (166b): A wood worth 20s. (*Silva de xx solidis*).
Bishop's Cleeve (165): There a small wood (*Ibi silva parvula*).
Sodbury (166b): A small amount of wood (*Silvae aliquantulum*).

[1] C. S. Taylor, *op. cit.* p. 59. See also R. W. Eyton, *A Key to Domesday; the Dorset Survey* (London, 1878), pp. 31–5.
[2] H. C. Darby and R. Welldon Finn (eds), *The Domesday Geography of South-west England* (Cambridge, 1967), pp. 374–5 and 382–3. See p. 437 below.

Cirencester (162b): Two woods worth 50s. (*ii silvas de l solidis*).

Highnam (165b): As much wood as suffices for the manor (*Silva quanta manerio sufficit*).

The Wiltshire folios also refer to 5 acres of wood at Ashley (71b). Some entries link wood with either pasture or meadow. At Chedworth (164) we are told that there were 15 hides of 'wood, field and meadow', and yet the detailed enumeration of its resources makes no reference to wood. There were also 5 hides of 'wood and field' at Churcham and Morton (165b), but here we are also told of wood measuring one league by one league. At Olveston (165) there were meadows and wood for the support of the manor (*prata et silva ad manerium sustinendum*). Another entry (165), covering a group of vills (Foxcote, Colesborne, Hilcote, Dowdeswell, Pegglesworth, Notgrove and Aston Blank) says that there was some meadow and wood but not much (*In quibusdam locis pratum et silva sed non multa*). At Guiting (167b) the wood and pasture rendered 40 hens (*De silva et pastura xl gallinas*). For Bedminster, part of which is now in Gloucestershire, the Somerset folios record meadow and wood separately but go on to say that of this manor the bishop of Coutances held '112 acres of meadow and wood' (86b).

No mention of underwood (*silva minuta*) appears in the Gloucestershire folios, but there is a reference to brushwood in a combined entry (166) for Elmstone Hardwicke, Bourton on the Hill, Todenham and Sutton under Brailes (*Silva i leuua longa et dimidia lata et Brocae iii quarentenis longae et i latae*). The list of variations is completed by the reference to a park (*unus parcus*) at Sodbury (163b) where some wood was also measured.

Distribution of woodland

The main fact about the distribution of wood in Domesday Gloucestershire was the existence of a broad belt along the edge of the Cotswolds from near Winchcomb in the north to Minchinhampton in the south (Fig. 13). As far as we can say, this woodland seems to have been spread over the Inferior Oolite. South of Minchinhampton there was a gap, and the belt was resumed between Wotton under Edge and Chipping Sodbury. It is interesting to notice that on the modern Land Utilisation map this area is still the most wooded in the county, apart, of course, from the Forest of Dean. The rest of the Cotswolds carried only scattered wood, and there also seems to have been hardly any wood on the Oxford Clay in the south-east of the county.

In the main Vale of Gloucester there was a number of scattered wooded areas—around Thornbury in the south; to the west of Stroud in the middle; and in the neighbourhood of Tewkesbury in the north. The Vale as a whole carried but little wood.

The greatest stretch of woodland in eleventh-century Gloucestershire, as today, must have been in the Forest of Dean, and yet Fig. 13 shows very little here. What wood there was appears on the margins, e.g. near Tidenham and Lydney in the south, and around Huntley and Tibberton in the north. Of the interior we are told nothing except that a few places here were 'in the forest'.[1] The word 'forest' in the medieval sense was far from being synonymous with 'woodland', yet there can be no doubt that this royal forest at any rate was largely wooded. The official Forest of Dean in the eleventh century probably covered the area between the Severn, the Wye and the Gloucester–Newent–Ross road.[2] It was practically uninhabited and, as the hunting preserve of the king, it stood outside the Inquest. Only along the margins did the local population have rights distinct from those of the crown. An extension of the wooded area of the Forest stretched northward beyond Newent, and this is reflected in the Domesday entries for Dymock, Upleadon and Bromsberrow.

Fig. 13 may be compared with a map of Domesday woodland prepared by G. B. Grundy in 1936.[3] Grundy attempted to convert the Domesday measurements into modern acres and so obtain a quantitative estimate of the wood-cover in 1086.[4] He then interpreted this partly in terms of geological formations and partly in terms of the distribution of place-names indicating the progress of the Saxon settlement. As an early attempt at plotting Domesday data his map is a most interesting experiment, and the outline of his wooded areas is shown on Fig. 14. In general, Fig. 13 agrees with it, bearing in mind that Grundy made allowance for the unrecorded wood of the Forest of Dean. The differences lie mainly in the absence of wood in the southern part of the Vale, but the Domesday entries for Thornbury (163b), Pucklechurch (165) and a number of other villages certainly mention wood.

[1] See p. 31 below.
[2] G. B. Grundy, 'The Ancient Woodland of Gloucestershire', *Trans. Bristol and Gloucs. Arch. Soc.* (Gloucester, 1936), LVIII, p. 70.   [3] *Ibid.* pp. 65–155.
[4] Grundy assumed that the Domesday league comprised 12 furlongs and that it was 2,420 yards in length, but see p. 435 below.

Forests

The royal forests of Gloucestershire for the most part go unmentioned
in the Domesday Book. There is no reference, for example, to the forests
of Kingswood and Horwood in the Vale to the north of Bristol. There is
one reference to the *silva regis* in the Herefordshire entry relating to

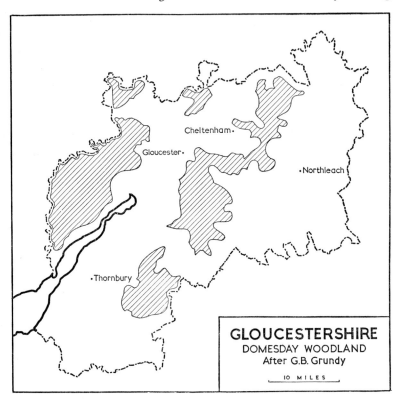

Fig. 14. Gloucestershire: Domesday woodland in 1086 (after G. B. Grundy).
This map is re-drawn from G. B. Grundy, 'The Ancient Woodland of Gloucestershire', *Trans. Bristol and Gloucs. Arch. Soc.* (Gloucester, 1936), LVIII, pp. 65–155.

Forthampton (180b). This entry is interesting because it mentions
swineherds, and a hawk's eyry: 'There 4 swineherds with one plough
render 35 pigs. The wood is 3 leagues in length and breadth. It is in the
enclosure of the king's wood and there [is] a hawk's nest.'[1] The Hereford-

[1] '*Ibi iiii porcarii cum i caruca reddunt xxxv porcos. Silva habet iii leuuas inter longitudinem et latitudinem. In defenso silvae regis est et ibi airea Accipitris.*'

shire folios also speak of waste land at *Stantun* (181) which was in the king's wood (*in silva regis*), and this has been identified with Staunton now near the border in Gloucestershire.[1]

The few references to forests in the Gloucestershire folios are, as might be expected, to the Forest of Dean. We hear of tenants at *Dene* who held their lands free from geld in return for looking after the Forest; Taylor was inclined to equate this place with Mitcheldean, Little Dean and Abenhall, and not with the large parishes of East Dean and West Dean that now occupy the central portion of the Forest.[2] We are also told that part of Taynton was in the Forest, and that by command of King William the forest had been extended to include Wyegate and Hewelsfield near the Wye. Finally, the section dealing with the lands beyond the Wye tells of ships 'going to the wood' who paid toll at the new castle of Chepstow (*Estrighoiel*). The details are as follows:

> *Dene* (167b): *Has terras concessit rex E[dwardus] quietas a geldo pro foresta custodienda.* Plotted at Mitcheldean.
> Taynton (167b): *Ibidem una virgata terrae jacet ad forestam.*
> Wyegate (166b): *Nunc est jussu regis in foresta sua.*
> Hewelsfield (167): *Haec terra jussu regis est in foresta.*
> Chepstow (162): *Castellum de Estrighoiel fecit Willelmus comes, et ejus tempore reddebat xl solidos tantum de navibus in silvam euntibus.*

The Domesday folios for some counties mention *haiae*, or enclosures, and we are sometimes told that they were for catching roe-deer. There are only two references to such hunting activity in the Gloucestershire entries. At Newent (166) there were two enclosures which the king had seized (*Ibi ii haiae quas habet saisitas rex*); and at Churcham and Morton (165b), the abbey of Gloucester had the right of hunting in three enclosures both in 1066 and 1086 (*Silva i leuua longa et una lata. Ibi habuit ecclesia venationem suam per iii haias T.R.E. et tempore Willelmi*). All this does far too little justice to the Norman love of the chase, and to the activity that must have gone on in the royal forests of Gloucester as in those of the realm at large.

---

[1] *V.C.H. Herefordshire*, I, p. 319.
[2] C. S. Taylor, *op. cit.* p. 204, but G. B. Grundy seems to assume that it referred to Little Dean, East Dean and West Dean (*art. cit.* pp. 68–9).

MEADOW

Types of entries

The entries for meadow in the Gloucestershire folios are for the most part comparatively straightforward. For holding after holding the same phrase is repeated monotonously—'n acres of meadow' (*n acrae prati*). The amount of meadow in each vill varied from 2 acres, e.g. at Edgeworth (167b), to over 100 acres at South Cerney where there were exactly 100 acres on one holding (169) and another 30 acres shared with Ampney (168). Large amounts appear in the combined entries relating to those large manors with dependencies; thus 120 acres are entered both for Barton by Gloucester (165b) and for Tewkesbury (163). Generally speaking, amounts above 20 for individual vills are rare. As in the case of other counties no attempt has been made to translate these figures into modern acreages. The Domesday acres have been treated merely as conventional units of measurement, and Fig. 15 has been plotted on that assumption.[1]

While measurement in acres is normal, there are a number of entries that refer to meadow in other ways. Thus in two succeeding entries on fo. 162b we are simply told that there was sufficient meadow for the plough-teams;[2] another two entries elsewhere speak of a certain amount (*aliquantum*); yet another two entries refer merely to 'a small amount'. There are also some entries that give not the measurements but the value of meadow; thus on each of two holdings, at Slaughter and at Longborough and Meon, the meadow yielded 10s., while that at Thornbury yielded as much as 40s. These sums were greatly exceeded at Kempsford and Lechlade where £7. 7s. and £9 respectively were returned, as well as hay or pasture for the oxen; this money must represent a considerable quantity of meadow along the Thames. Finally, there are two entries that mention meadow without specifying either its size or value. These miscellaneous entries are set out below.

Haresfield, Down Hatherley, Sandhurst (162b): *Ibi pratum sufficiens carucis.* Harescombe (162b): *prata carrucis [sic].*

[1] When acres of meadow appear in a composite entry they have been divided amongst the vills concerned, and so indicated. Otherwise, the symbol 'other mention' appears for each vill of a composite entry.

[2] The meadow of Bedfordshire, Buckinghamshire, Cambridgeshire, Hertfordshire and Middlesex was almost entirely measured in this way.

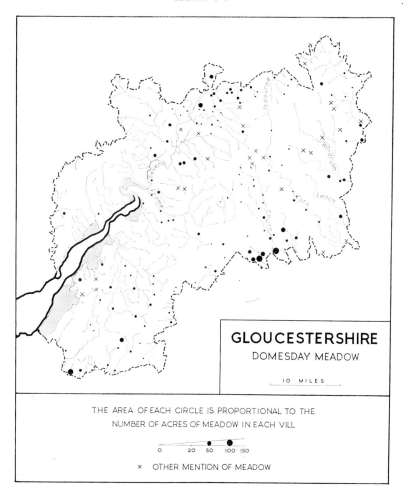

Fig. 15. Gloucestershire: Domesday meadow in 1086.

The area of alluvium along the Severn is indicated only below Gloucester—see Fig. 7. Where the meadow of a village is entered partly in acres and partly in some other way, only the acres are shown.

Maugersbury (165 b): *prati aliquantum.*[1]
Southam, Saberton and Gotherington (165): *aliquantum prati.*
Adlestrop (165 b): *Ibi parvum prati.*
Willersey (166): *parvum prati.*
Slaughter (162 b): *Prata de x solidis.*

[1] The Domesday contraction is *aliqtū*; it might read *aliquantulum.*

Longborough and Meon (163): *Pratum de x solidis.*
Thornbury (163 b): *unum pratum de xl solidis.*
Lechlade (169): *De pratis vii librae et vii solidi praeter[1] fenum boum.*
Kempsford (169): *de pratis ix librae praeter[2] pasturam boum.*
Cirencester (162 b): *et Prata.*
Sezincote (170): *prata.*

In addition to these, there are, as we have seen, a number of entries that link meadow with wood. For Chedworth (164) 15 hides 'of wood, field and meadow' are recorded, but the detailed enumeration of its resources makes no reference to meadow. At Olveston (165) there were meadows and wood for the support of the manor (*prata et silva ad manerium sustinendum*). Another entry (165) covering a group of vills (Foxcote, Colesborne, Hilcote, Dowdeswell, Pegglesworth, Notgrove and Aston Blank) says that there was some meadow and wood but not much (*In quibusdam locis pratum et silva sed non multa*). And for Bedminster, part of which is now in Gloucestershire, the Somerset folios record meadow and wood separately but go on to say that of this manor the bishop of Coutances held '112 acres of meadow and wood' (86 b).

Distribution of meadowland

The river valleys cut into the dip-slope of the Cotswolds were marked by villages with moderate amounts of meadow, each for the most part below 20 acres. Along the Churn, South Cerney was outstanding with 100 acres; and along Ampney Brook the Ampneys stood out with over 60 acres, although this amount was probably distributed among the several villages of that name. In the Coln valley, Winson had 15 acres and Bibury and Quenington 10 each; but, surprisingly enough, no meadow is recorded for the large settlement of Fairford where there were 39 plough-teams. Along the Leach, Eastleach had 38 acres. Along the Windrush, Sherborne had 30 acres and Barrington and Windrush itself 26 and 18 acres respectively. In the extreme south-east of the county, Kempsford and Lechlade, which reached down to the Thames, must have had very considerable amounts of meadow judging from the sums of money it yielded. There was but little meadow in the northern part of the Cotswolds, and there was probably even less than might seem from Fig. 15. The meadow that is recorded here appears in composite entries, e.g. that

---

[1] The Domesday contraction is p̄ter.
[2] The Domesday contraction is p̄ł.

for Southam and two other vills where there was only 'a little', or that for Foxcote and six other vills where there was some meadow 'but not much'.

In the Vale of Gloucester, in the vicinity of Tewkesbury, a number of small streams drain into the Severn (e.g. Carrant Brook and the Swilgate), and here a number of Domesday villages had small amounts of meadow; Deerhurst on the banks of the Severn itself had as much as 60 acres. There was also a substantial amount around Gloucester, but below the city there seems to have been little, and C. S. Taylor suggested that this might have been because the tide-water overflowed the low-lying lands.[1] Farther south in the Vale, in the valley of the Frome that flows through Stroud to the Severn, there was another group of vills mostly with 10 acres apiece. Farther south still, in the plain between the Cotswolds and the Severn, there is a network of small streams draining either westwards to the estuary or southwards to the Bristol Avon. Some of the villages in the area had as much as 20 acres of meadow each; Littleton upon Severn had 30; Pucklechurch had as much as 60.[2]

The Forest of Dean, as might be expected, was a meadowless area, but along the estuary 20 acres are entered for *Ledenei* (now St Briavels) and 10 for Allaston.

PASTURE

Pasture is mentioned in connection with five holdings at four places in the Gloucestershire folios:

Tetbury (168): *pastura de x solidis.*
Shipton Moyne (170): *De pastura ii solidi.*
Shipton Moyne (170): *De pastura ii solidi.*
Guiting (167b): *De silva et pastura xl gallinas.*
Kempsford (169): *de pratis ix librae praeter pasturam boum.*[3]

This meagre list stands in great contrast to the regular entries for pasture that occur in the folios relating to the adjoining counties of Oxford, Wiltshire and Somerset. It can only mean that the record of pasture was deliberately excluded for Gloucestershire, and that these few references

---

[1] C. S. Taylor, *op. cit.* p. 68.

[2] Bedminster, surveyed under the Somerset folios, had 34 acres of meadow and '112 acres of meadow and wood', see p. 34 above.

[3] See p. 89 below for a similar formula at Monkland. It may imply not pasture as distinct from meadow but the fact that grazing after the hay harvest was not included in the valuation. See p. 34 above for *praeter fenum boum* at Lechlade (169).

have crept in either by chance or because at these places the grazing was
not subject to free common right but had to be paid for. The only other
references that might imply the presence of pasture are those relating to
sheep at Cirencester (162b) and to a sheepfold at Kempsford (169).[1]

In order to complete the picture for the modern county we should
mention that the Somerset folios enter 20 acres of pasture for Knowle
(98), and that the Wiltshire folios enter pasture of 80 acres for Long
Newnton (67), of 3 furlongs by 2 for Poole Keynes (69b), of 3 furlongs
by 1 for Poulton (68b) and of 2 furlongs by 1 for Shorncote (73). This
makes a total of nine places with a record of pasture in the area covered
by present-day Gloucestershire.

FISHERIES

Fisheries (*piscariae*) are recorded in connection with, at least, sixteen
places in Gloucestershire. It is difficult to be sure of the exact number of
places because of two composite entries: one fishery was recorded for
Purton, Etloe and Blidisloe (163), and another for Poulton in Awre and
Purton (164). Most entries state only the number of fisheries, but the
renders of some fisheries are occasionally given, e.g. at Allaston there
was one of 12d. (166b); at Gloucester, one of 58d. (163b); at Woolaston,
one of 5s. (166b); at Wyegate, one of 10s. (166b); and four at Madgett
yielded as much as £4 (164). Half-fisheries are mentioned at some places
but, for the most part it seems impossible to combine the fractions in any
intelligent way.[2] The kind of fish is mentioned only twice: at Gloucester,
16 salmon were rendered to the church of St Peter (165b), and at Lechlade
on the Thames there was a fishery yielding 175 eels (169). One of the
Tewkesbury entries seems to refer to a fishery in 1066 (163) and this has
not been counted as additional to that entered for 1086 (163b).

As Fig. 16 shows, most of these fisheries were along the Severn
estuary, and more particularly along that part of the estuary that borders
the Forest of Dean. The outstanding centre was Tidenham with 65
fisheries: 53 of these were in the Severn itself (*in Sauerna*), 5½ in the Wye
(*in Waia*), and it is not clear in which river the remaining ones were
situated. The 5 fisheries at Madgett and the half at *Ledenei* (now St
Briavels) were said to be in the Wye; that at Wyegate must also have been

---

[1] See p. 49 below.

[2] At Brewerne (169b), Hempsted (164), Standish (164b), St Briavels (167), Tidenham (164, 167b), and Westbury hundred (167).

there, although we are not specifically told so. Upstream, the line of the Severn is indicated by fisheries at Standish, Longney, Hempsted, Gloucester, Brewerne and Tewkesbury. Elsewhere, the only other fishery

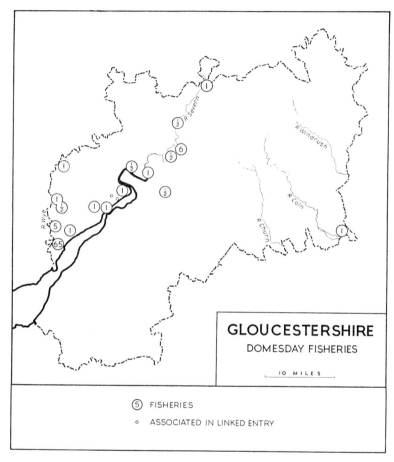

Fig. 16. Gloucestershire: Domesday fisheries in 1086.
The figure in each circle indicates the number of fisheries. This map does not show the half-fishery for an unnamed holding in Westbury hundred, presumably in the Severn (167); nor is it included in the total of 16.

recorded in the county was at Lechlade on the Thames. It is surprising, for example, that none was entered for the eastern bank of the Severn estuary below Longney. There was, it is true, a fishery associated with Thornbury, but we are specifically told that this was at Gloucester (163b).

SALT

The Gloucestershire folios mention salt in connection with ten places. The holdings at six of these places are specifically linked with *Wich* (i.e. Droitwich), and it is obvious that the brine springs of Worcestershire formed an important element in the economy of the medieval Gloucester-shire village. The units used to measure salt are the *summa* (horse-load), the *mitta*, the *sextarium* and the *mensura*. The entries for four places (Awre, Guiting, Toddington and Chedworth) do not mention *Wich*. It is possible that at Awre salt was manufactured on the spot from the waters of the Severn estuary, but at the other three places it seems certain that the salt must have come from the Worcestershire springs. To this information in the Gloucestershire folios one further item must be added. A reference to *Wich*, in the Worcestershire section of the Domesday Book, speaks of holdings there belonging to the King and to the Church of St Peter at Gloucester (174). It is conceivable, therefore, that perhaps we should add the city of Gloucester to the list of those Gloucestershire places with interests in the Worcestershire salt industry, but it has not been so indicated on Fig. 87. All the relevant entries are set out below:

Tewkesbury (163): *una salina apud Wicham pertinens ad manerium (T.R.E.).*
Tewkesbury (163 b): *uns salina apud Wicham.*
Stanway (163 b): *apud Wicham una salina.*
Thornbury (163 b): *ad Wiche xl sextaria salis vel xx denarii.*
Sodbury (163 b): *Ad hoc manerium pertinet una virgata in Wiche quae reddebat xxv sextaria salis, Ursus vicecomes ita vastavit homines quod modo reddere non possunt salem.*
Mickleton (166): *xxiiii mensurae salis de Wich.*
Rockhampton (168 b): *salina ad Wich de iiii summis salis.*
Awre (163): *salina de xxx summis salis.*
Guiting (167): *Ibi v salinae reddentes xx summas salis.*
Guiting (167 b): *salina de xx solidis et xii summas [sic] salis.*
Toddington (169): *De una salina l mittas salis.*
Chedworth (164): *theloneum salis quod veniebat ad aulam.*
Gloucester (174): *Æcclesia St Petri de Glouuecestre tenet dimidiam hidam in Wich et est in eadem consuetudine qua est et dimidia hida regis quae est in Wich pertinens ad Glouuecestre.*[1]

[1] Another Worcestershire entry, under *Terra Regis*, connects Gloucester and Droitwich, but again does not specifically mention salt: *In Wich est dimidia hida quae pertinet ad aulam de Glouuecestre* (172 b).

It is interesting to note the relation of these places to the roads or tracks along which the salt was carried on packhorses from Droitwich. Many of the details of this traffic are lost to us, but it has left its mark on names that appear on the One Inch Ordnance Survey map. In Gloucestershire, names such as Salt Way, Saltway Farm, Salter's Hill, Saltway Plantation, Salterley Grange, and Saltridge Hill speak for themselves.[1] The main salt routes through Gloucestershire seem to have been five in number:

(1) From Weston Subedge towards Mickleton, then to Evenlode.

(2) From Weston Subedge via Donnington to Southrop.

(3) From Dumbleton through Toddington and Stanway, then by the Guitings and Chedworth southward to the Thames.

(4) From Tewkesbury to Cheltenham and so to Leckhampton, Painswick, Stroud and Wotton under Edge where it divided: (a) one route forking to Rockhampton and Thornbury; (b) the other route going to Chipping Sodbury.

(5) From Tewkesbury to Gloucester and so to Awre.

Many of these names appear on the list on p. 38, but the Domesday Book tells us nothing about the traffic and its problems. There is, however, one illuminating entry in the description of the royal manor of Chedworth, part of whose profits, we are told, came from the toll of salt that was brought to the hall (164).

MILLS

Mills are mentioned in connection with at least 126 of the 367 Domesday settlements within the area covered by modern Gloucestershire.[2] It is difficult to be sure of the exact total because mills sometimes appear in composite entries covering a number of places. Thus a mill of 20d. is entered for Henbury, Redwick, Stoke Bishop and Yate, all to the north of Bristol (164b); three mills of 13s. 4d. are likewise entered for Foxcote and six other villages (165). For each composite entry, the mill or mills have been plotted for the first-named place.[3] The associated villages in all such entries have also been indicated on Fig. 17.

[1] See the discussion in: (i) A. Mawer and F. M. Stenton, *The Place-Names of Worcestershire* (Cambridge, 1927), pp. 4–9; (ii) A. H. Smith, *The Place-Names of Gloucestershire*, part I, pp. 19–20. See pp. 257–8 below.

[2] They also appear for two anonymous holdings; two mills in Bradley hundred rendered 20s. (163), and the mill in Blidisloe hundred has no value stated for it (169).

[3] The figure of 126 results from counting as one group the mills of a composite entry. When more than one mill is recorded in such an entry, and it is assumed that

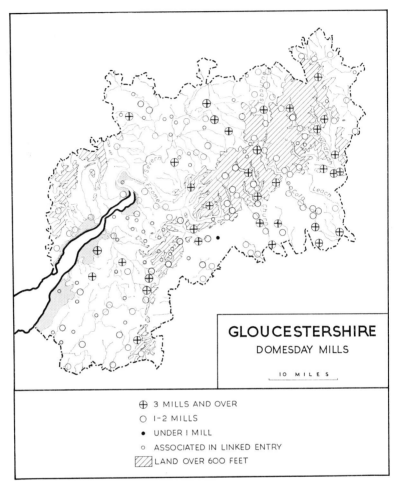

Fig. 17. Gloucestershire: Domesday mills in 1086.
The area of alluvium along the Severn is indicated only below
Gloucester—see Fig. 7.

In each entry the number of mills is given, and also their annual value
ranging from one mill worth only *6d.* at Saintbury (169b) to others worth
much larger sums; the mill at Hatherop rendered 15*s.* (169); two mills at
Barton by Bristol rendered 27*s.* (163); and another two at King's Stanley

each was at a separate place, the total of places with mills is 142. If all the places named
in the relevant composite entries are counted, the total becomes 177, a very artificial
figure.

35*s*. (169b). The mill at Rudford (170b) rendered as much grain as it could (*molendinum reddit annonam quantum potest lucrari*). A number of half-mills are recorded. On one holding at Iron Acton, for example, half a mill rendered 16*d*. (165), and on another holding 1½ mills rendered 64*d*. (170). It seems as if there were two mills there rendering 2*s*. 8*d*. and 4*s*. respectively. It is not always possible, however, to assemble the fractions in such a comprehensible manner. There was half a mill of 30*d*. at Hasleton (168), but we are given no clue to the other half.

Domesday Mills in Gloucestershire in 1086

Under 1 mill	1 settlement	5 mills	5 settlements
1 mill	63 settlements	6 mills	1 settlement
2 mills	28 settlements	8 mills	2 settlements
3 mills	15 settlements	12 mills	1 settlement
4 mills	10 settlements		

The group of 6 mills was entered for Sudeley Manor to the south of Winchcomb (169); one of 8 for Minchinhampton (166b), and another of 8 for the 21 berewicks of Berkeley (163). The group of 12 entered for Blockley (173), on a tributary of the Stour, is surprising, but, judging from the form of the entry, it is possible that this number was shared with Ditchford and Church Icomb. The boroughs were but poorly represented. There was one mill at Gloucester, and there had been 2 at Tewkesbury which do not appear for 1086. No mills were recorded for Bristol unless we can assume that the 2 at Barton were in Bristol itself; neither were any recorded for Winchcomb, yet the closely related Evesham Abbey Survey enters no fewer than 3.[1]

Fig. 17 shows the alignment of the mill sites along the eastern valleys of the Cotswolds, those of the Churn, the Coln, the Leach, the Windrush and their associated streams; there were many villages with 3 or more mills. The western valleys also had a good number. Sudeley Manor, with 6 mills, is on the Isbourne above Winchcomb. Farther south, Cheltenham had 5 mills. But the most important of these western streams were those of the Stroud valleys through which the Frome and its tributaries run. Thus there were 8 mills at Minchinhampton, and 5 each at Bisley and Avening, and there were also other centres with fewer but very profitable mills; as we have seen, the two mills at King's Stanley were the wealthiest in the

[1] Sir Henry Ellis, *op. cit.* II, pp. 446–7.

county. The mills of the Vale are frequently recorded in composite entries, as Fig. 17 shows. Finally, in the poorly cultivated district west of the Severn, mills were few in number and their renders were small in amount.

CHURCHES

Churches (*ecclesiae*) are mentioned in connection with only ten places in Gloucestershire, including that at Bristol (163) and the *monasterium* at Stanway (163b). The geld liability of some of these churches is sometimes stated, e.g. the church at Cheltenham was rated at $1\frac{1}{2}$ hides (162b), and that at Awre at only 1 virgate (163). Priests are mentioned in association with only four of these churches—at Ampney (169b), Littleton upon Severn (165), Cheltenham (162b) and Stow on the Wold (165b).[1] There were, however, forty-three other places for which priests are recorded, so we may guess that the Domesday Book gives an indication of churches at a total of fifty-three places in the area covered by the modern county. Four of these places had more than one priest; Ampney had three, and it is interesting to note that the name covers a group of parishes today (164, 166b, 169b). Guiting likewise had two priests and there are two parishes of that name today (167, 167b). Siddington also had two priests (168, 169b); it is only one parish today but there were formerly two parishes each with a church—St Mary and St Peter. Bisley had two but it was a very populous place and, moreover, it may have included Stroud (166b). A priest sometimes held a substantial amount of land—3 hides and 4 teams at Bibury (164b), 1 hide and 2 teams at Bishop's Cleeve (165)—but usually the priest of a village was grouped with the villeins and bordars: thus at Barrington there were '14 villeins and a priest and 2 bordars with 9 ploughs' (169).

Even the total figure of fifty-three can represent only a fraction of the parish churches of eleventh-century Gloucestershire. Neither churches nor priests are specifically recorded at Gloucester itself or at Tewkesbury and Winchcomb. The church at Bristol is entered only incidentally. We can only conclude that the greater number throughout the county went unmentioned.

---

[1] The church of St Edward at Stow on the Wold was entered under Maugersbury, see p. 8 above.

URBAN LIFE

Four places in Gloucestershire seem to have been regarded as boroughs. Gloucester itself is described as a city (*civitas*); Winchcomb is called a borough (*burgus*); and we also hear of burgesses at Tewkesbury and Bristol. The information for all four places is very unsatisfactory, and it provides us with hardly any indication of their life and activities. A market, for example, is entered only for Tewkesbury. This evidence, slender as it is, is set out below.

Gloucester

The city of Gloucester must always be of especial interest to the Domesday student, for it was here, in the midwinter of 1085, that King William had that 'deep speech' with his wise men which resulted in the making of the great Survey. If, however, we expect to find a satisfactory account of the city in the text of the Domesday Book, we shall be disappointed. The description of Gloucestershire starts with an account of the city (*civitas*) on fo. 162, and this main account is supplemented by two entries on fols. 163b and 165b. We are told at once that in 1066 the city had rendered £36 by tale together with honey and also iron and rods of iron drawn out for making nails for the king's ships (*et xxxvi dicras ferri et c virgas ferreas ductiles ad clavos navium regis*); these latter renders presumably reflect the proximity of the Forest of Dean. By 1086 this diversity of render had been consolidated into one money payment of £60 at the rate of 20d. to the ounce, and there was also a payment of £20 for the mint. We are not told how many burgesses or houses were in the city, and so can form no estimate of its population. The greater part of the main entry consists of an account of the houses (*mansiones* or *domus*) detached from the king's demesne in the city; 23 of these are specified but there were also others. Moreover, 16 houses (*domus*) had been destroyed (*quae modo desunt*) to make room for a castle, and there were also 14 houses waste (*wastatae*). There is no account of the houses yet remaining to the king nor of any others.

Some Gloucester burgesses are recorded as belonging to rural manors and also (in the past tense) to Tewkesbury; these are set out not in the entries for Gloucester but in those of their respective manors which are very widely scattered through the county (Fig. 18). These 'contributory burgesses' for the most part rendered money, but one returned iron,

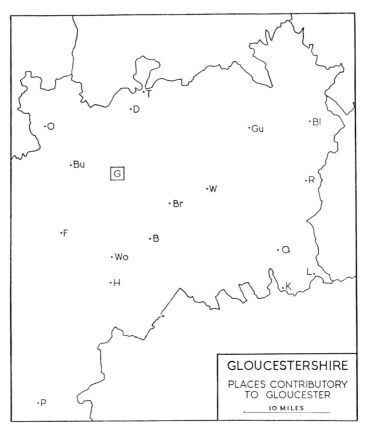

Fig. 18. Gloucestershire: Places contributory to Gloucester (G).

The county boundaries are those of *circa* 1960. B, Bisley; Br, Brimpsfield; Bl, Broadwell; Bu, Bulley; D, Deerhurst; F, Frampton; Gu, Guiting; H, Horsley; K, Kempsford; L, Lechlade; O, Oxenhall; P, Pucklechurch; Q, Quenington; R, Rissington; T, Tewkesbury (T.R.E.); W, Withington; Wo, Woodchester.

another returned plough-shares and a third nothing at all. At Horsley, a house, not a burgess, was entered. The full list of seventy-three burgesses and one house for 1086 is as follows:

Bisley (166b): 11 burgesses rendering 66*d*.
Brimpsfield (168b): 5 burgesses of (*de*) 2*s*.
Broadwell (166): 4 burgesses in Gloucester and 1 in Winchcomb render 27*d*.
Bulley (169): 1 burgess renders 18*d*.
Deerhurst (166): 30 burgesses render 15*s*. 8*d*.

Frampton on Severn (168b): 1 burgess of 6*d*.
Guiting (167b): 2 burgesses of 10*d*.
Horsley (166b): 1 house of 6*d*.
Kempsford (169): 7 burgesses render 2*s*.
Lechlade (169): 1 burgess without render (*sine censu*).
Oxenhall (167b): 3 burgesses of 15*d*.
Pucklechurch (165): 1 burgess renders 5*d*.
Quenington (167b): 1 burgess renders 4 plough-shares (*soccos*).
Rissington (168): 1 burgess of 3*d*.
Withington (165): 4 burgesses render 7½*d*.
Woodchester (170b): 1 burgess renders 20 [pieces of] iron (*xx ferra*).

The entry for Tewkesbury says: *In Gloucestre erant viii burgenses* (163). Could it refer not to Tewkesbury itself, but to a component of the manor?

As we are told nothing of the commercial life of the city beyond the existence of a mint, so are we told little or nothing about its other resources. In 1066, the king's demesne had been inhabited and cultivated (*hospitatum et vestitum*), and it was likewise cultivated afterwards when Count William had it, but with what teams and peasantry we are not told. The influence of the Severn makes itself felt even through this dearth of information. There was one fishery the render of which was not specified (162), and another belonging to Thornbury that yielded 58*d*. (163b); and we are also told that 'T.R.E. St Peter had, from his own burgesses in Gloucester, 19*s*. 5*d*. and 16 salmon; he has now as many salmon and 50*s*. There is a mill of 12*s*. and 4 fisheries for the table of the monks' (165b). None of the entries relating to Gloucester specifically mentions a church or churches, but the Abbey of St Peter appears as owning considerable territory in the county. All this forms a most unsatisfactory body of evidence. The incompleteness of the Domesday information is shown by the fact that the Evesham Abbey Survey, closely related to the Domesday account, enters no fewer than 613 burgesses and ten churches for Gloucester.[1]

Winchcomb

The Domesday account of the borough (*burgus*) of Winchcomb consists of one entry on fo. 162b. It is very brief and uninformative, and merely states the money renders of the borough in 1066 and 1086. No reference is made to the burgesses or to any other category of

---

[1] Sir Henry Ellis, *op. cit.* II, p. 446.

population. Like Gloucester it had contributory burgesses recorded in connection with various rural manors around (Fig. 19). Two of these are no longer within the boundaries of the county: Childs Wickham

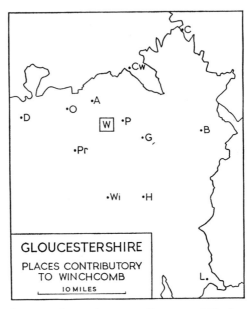

Fig. 19. Gloucestershire: Places contributory to Winchcomb (W).

The county boundaries are those of *circa* 1960. A, Alderton; B, Broadwell; D, Deerhurst; G, Guiting; H, Hampnett; L, Lechlade; O, Oxenton; P, Pinnock; Pr, Prestbury; Wi, Withington. To these must be added, Cw, Childs Wickham, now in Worcestershire, and C, Clopton, now in Warwickshire.

has been transferred to Worcestershire and Clopton to Warwickshire. The full list of thirty-two burgesses is as follows:

Alderton, etc. (163 b): 1 burgess.[1]
Broadwell (166): 4 burgesses in Gloucester and 1 in Winchcomb render 27*d*.
Clopton (167): 1 burgess.
Deerhurst (166): 2 burgesses render 10*d*.
Guiting (167): 2 burgesses render 11*s*. 4*d*.
Guiting (167b): 3 burgesses of (*de*) 32*d*.
Hampnett (168): 10 burgesses render 65*d*.
Lechlade (169): 2 burgesses render 16*d*.

[1] This appears in a composite entry relating to Alderton, Dixton and *Hundeuuic*; Dixton is now in the parish of Alderton, *Hundeuuic* is unidentified.

Oxenton (163 b): 3 burgesses render 40*d*.
Pinnock (170 b): 1 burgess renders 8*d*.
Prestbury (165): 1 burgess renders 18*d*.
Childs Wickham (168): 1 burgess of 16*d*.
Withington (165): 1 burgess renders 3*s*.[1]

We are given no clue to the activities—agricultural or commercial—of the inhabitants of the borough. No specific mention is made of the abbey, and yet it appears as a prominent landowner in the county. Furthermore, the Evesham Abbey Survey, closely related to the Domesday account, enters no fewer than 116 burgesses and 3 mills for Winchcomb.[2]

Tewkesbury

The account of Tewkesbury, on fols. 163 and 163 b, is a difficult one because the place was the head of a large and dispersed manor,[3] and many details are entered collectively for all the constituents of the manor as a whole. The two entries that seem to refer to Tewkesbury itself read as follows:

Fo. 163. In the head of the manor (*in capite manerii*) there were 12 ploughs on the demesne, and 50 serfs and bondwomen (*ancillae*); and 16 bordars lived around the hall (*circa aulam manebant*); and 2 mills of 20*s.*; and one fishery; and one salt-pan at Droitwich belonging to the manor.

Fo. 163 b. At Tewkesbury there are now 13 burgesses rendering 20*s.* a year.[4] A market which the queen established there (*constituit ibi*) renders 11*s.* 8*d.* There is one plough more (*plus*) and 22 serfs and bondwomen. One fishery; and one salt-pan at Droitwich.

The first statement refers to 1066 but its 12 teams were apparently still there in 1086 as well as the additional team entered for that date. The details of population at either date cannot have been complete. It is impossible to estimate the size of the borough. The market indicates commerce; the plough-teams, and the fishery tell of other activities. There may have been wood and meadow; at any rate some was entered for the manor as a whole. There must also have been many other things.

[1] This appears under the general heading of the manor of Withington, but following an account of Foxcote, Colesborne, Hilcote, Dowdeswell, Pegglesworth, Notgrove and Aston Blank.

[2] Sir Henry Ellis, *op. cit.* II, pp. 446–7.

[3] In the hundred of Tewkesbury, see p. 5 above.

[4] For the relation of the 13 burgesses to the 8 contributory to Gloucester in 1066 (p. 45 above), see A. Ballard, *The Domesday Boroughs* (Oxford, 1904), pp. 97–8.

Bristol

The Domesday account of Bristol is as slender as that of the other boroughs of Gloucestershire. The main entry on fo. 163 reads as follows:

This manor [of Barton] and Bristol renders to the king 110 marks of silver. The burgesses say that Bishop G. has 35 marks of silver and one mark of gold besides the king's rent.

As the manor of *Bertune apud Bristou,* and its member, Mangotsfield, seem to have been rural settlements, Bristol itself must have been responsible for the greater part of the render. A church is also mentioned. It was obviously a place of some substance yet we are told nothing more about it beyond two references to contributory properties in the manors around. An entry for Westbury on Trym states that '2 houses in Bristol rendered 16*d.*' (164b); and another for Bishopsworth in the Somerset folios also speaks of 10 houses at Bristol (88). It is a meagre body of evidence.

MISCELLANEOUS INFORMATION

Markets

The entries relating to markets in the Gloucestershire folios are four in number:

(1) Berkeley (163): There is a market in which 17 men dwell (*Ibi unum forum in quo manent xvii homines*).
(2) Cirencester (162b): From a new market 20*s.* (*De novo foro xx solidi*).
(3) Tewkesbury (163b): A market which the queen established there renders 11*s.* 8*d.* (*Mercatum quod regina constituit ibi reddit xi solidos et viii denarios*).
(4) Thornbury (163b): There, a market of 20*s.* (*Ibi forum de xx solidis*).

It is a short list, and burgesses are recorded for only one of these places— Tewkesbury. No markets are entered for the other boroughs—Bristol, Gloucester and Winchcomb.

Other references

Waste is recorded for only two places in 1086. At Whittington (182)[1] 3 hides were waste: *Wastae sunt et wastae fuerunt.* Part of the five hides of Awre (163) were also waste: *De eodem manerio jacet wastata dimidia*

---

[1] For the identification of Whittington, see p. 3 above.

hida. There had also been one waste hide at Staunton (181) but it was 'in the king's wood' in 1086.

There is only one reference to the growing of the vine. At Stonehouse (166b), not far from Stroud, there were 2 arpents of vineyard (*Ibi ii arpenʒ vineae*). C. S. Taylor noted that 'there could be few warmer spots in the county than the south-western slope of the hill above Stonehouse'.[1]

Among other miscellaneous items must be mentioned the sheepfold at Kempsford (169) which rendered 120 weys of cheese, presumably made from ewes' milk (*de ovili cxx pensas caseorum*). There is also a mention of sheep in the entry relating to Cirencester (162b): here, the queen had the wool of the sheep (*Lanam ovium regina habebat*). Bedminster (86b) and Knowle (98), being in Domesday Somerset, are described also in the Exon Domesday (90b, 447) where they have livestock entered for them.

At Sharpness (163), Earl William had built a small castle (*unum castellulum*); there was also a castle at Gloucester (162), and another at Chepstow beyond the Wye (162). Renders of iron and of rods of iron drawn out for making nails for the king's ships are mentioned for the city of Gloucester (162), and they presumably reflect the proximity of the Forest of Dean (*xxxvi dicras ferri et c virgas ferreas ductiles ad clavos navium regis*). So do the twenty blooms of iron (*xx blomas ferri*) entered for Alvington on the shores of the Severn (185b).[2] There were also 6 men at Pucklechurch (165) to the north-east of Bristol, who rendered ninety 'masses' of iron (*vi homines reddunt c massas ferri x minus*).

Finally, the entries that set out renders in kind mention a number of commodities. At Cirencester (162b) we hear of wheat, malt, honey, cows and pigs; and cows and pigs were also rendered at Cheltenham (162b) and Barton by Gloucester (162b).[3] At Bisley there were 23 men who rendered honey (166b); and renders of honey and iron appear in the description of Gloucester itself (162).[4] Eight sesters of honey were also rendered at Alvington (185b). At Guiting, the wood and pasture rendered 40 hens (167b). No sheep are mentioned amongst these renders in kind.

---

[1] C. S. Taylor, *op. cit.* p. 70.

[2] The entry for Alvington occurs in the Herefordshire folios (185b), see p. 3 above.

[3] See p. 20 above.   [4] See p. 45 above.

Gloucestershire is traditionally divided into three areas in accordance with its natural features—the Cotswolds, the Vale and the Forest (Fig. 20). Within each of these areas there are variations, but the broad division into three will suffice for the purpose of summarising the geography of the county in the eleventh century.

(1) *The Cotswolds*

The Cotswolds form the outstanding feature of the relief of the county, and occupy the greater part of its eastern half. They are marked on the west by a steep escarpment which, for the most part, is about 600–800 ft. above sea-level, but which rises to just over 1,000 ft. From this western edge a wide dip-slope, falling to below 300 ft., extends south-eastwards towards the Thames. This sloping plateau surface is broken by a series of streams that flow to join the Thames in valleys for the most part some 200 ft. below the level of the surrounding countryside; the main valleys are those of the Churn, the Coln, the Leach, the Windrush (with its tributary the Dikler) and the Evenlode. The light soils of the plateau reflect its Oolitic Limestone character, and are generally thin. There are local variations, and heavier soils are found in places, more particularly on the Forest Marble and Cornbrash formations of the south-east. Here, towards the border of the county, these formations dip beneath the Oxford Clay through which the Thames flows.

Before enclosure the Cotswold plateau was a land of sheep-walk, but it is clear that there was also considerable agricultural activity throughout the area as a whole in the eleventh century. Its general prosperity, as indicated by the densities of plough-teams and population, show it to be of a rather similar grade to much of midland England. The valleys were marked by numerous villages each with a small amount of meadow, generally of the order of from ten to 20 acres. Along the Thames itself, Lechlade and Kempsford seem to have had considerable quantities of meadow, but these were not measured in acres and so comparison with other villages is impossible. Many of the valley villages also had mills; some had three or more mills. The region as a whole carried but little wood, even on the Oxford Clay outcrop of the south-east. In the north, however, there was a belt of scattered woodland stretching from near

Winchcomb in the north to Minchinhampton in the south, and spread over the Inferior Oolite. South of Minchinhampton there was a gap, and the belt was then resumed between Wotton under Edge and Chipping Sodbury. It is interesting to note that on the modern Land Utilisation map this area is still the most wooded in the county, apart of course, from the Forest of Dean.

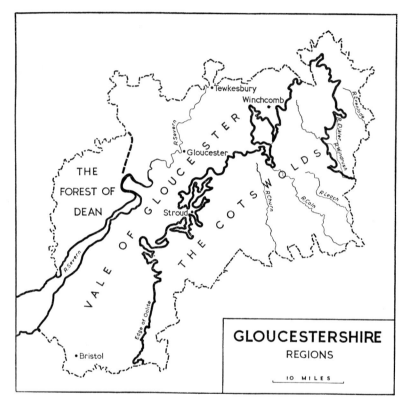

Fig. 20. Gloucestershire: Regional subdivisions.

(2) *The Vale of Gloucester*

The Vale of Gloucester, stretching below the Cotswold escarpment, lies almost entirely below 400 ft. apart from occasional outliers of Oolitic Limestone. It has been excavated mainly in Lower Lias clays and Keuper Marl, and its soils are for the most part heavy and deep. Local variations, however, include loams and sands, and in the south of the area inliers of

Palaeozoic rocks diversify the soil and surface. The even distribution of
Domesday names over the Vale tells its own story of fairly uniform
agricultural utilisation. The densities of Domesday plough-teams and
population were rather similar to those of the Cotswolds but were
inclined in places to be lower. The close network of streams that covers
the Vale suggests that there might have been a fair amount of meadowland; that suggestion is borne out by the meadow map, but the amounts
assigned to each village were not large—usually between 10 and 20 acres
and only rarely above. As might be expected, mills were not as frequent
as in the Cotswold valleys. The original vegetation of the region was
presumably damp oak wood, but by the eleventh century most of this
must have been cleared. All that remained, or at any rate all the Domesday
Book recorded, were some scattered areas of woodland—around Thornbury in the south; to the west of Stroud in the middle; and in the neighbourhood of Tewkesbury in the north. The Vale as a whole carried but
little wood. The line of the Severn above Longney is indicated by fisheries,
but it is surprising that none was entered for the eastern bank (as for the
western bank) of the estuary below Longney. Nor is there any reference
to the marsh that bordered the estuary at many places.

(3) *The Forest of Dean*

The Forest of Dean is an upland area for the most part over 400 ft. above
sea-level, rising in places to over 800 ft., and in a few summits to 950 ft.
The general character of most of the area is that of alternating ridges and
deep valleys that trend N.N.W.–S.S.E., and that are frequently too steep
for cultivation. The rocks are Carboniferous or older in age; and the soils
vary considerably from coarse sands to clays and tend to be shallow and
infertile. Agriculturally, it is therefore an unrewarding region, and these
natural conditions were emphasised by its forest status. It is not surprising
that the densities of plough-teams and population were only about one-
quarter of those for the rest of the county. Even these low figures were
due largely to the villages on the margins of the area; the interior stands
out on the map of Domesday names as an empty tract. No mills were to
be found in the area and hardly any meadow. A distinctive element in the
economy of some villages was provided by the fisheries of the Severn and
the Wye. It must have been a heavily wooded region in the eleventh
century, but the Domesday Book records only small amounts of wood

for the marginal villages. As the hunting preserve of the king, the main part of the Forest stood outside the Inquest. The Domesday Book likewise does not mention the iron of the region, although renders of iron at the city of Gloucester may reflect iron-working activity in the Forest.

APPENDIX: THE WELSH BORDERLAND

One of the interesting features of the Gloucestershire folios is the account of the country beyond the Wye which occurs after that of the city of Gloucester and before that of Winchcomb (162). This territory was part of the Welsh district of Gwent, and it had remained in Welsh hands until it was conquered in 1065 by Harold who annexed it to his earldom of Hereford. After the coming of the Normans it passed into the hands of William Fitz Osbern who was created the new earl of Hereford by King William, and who entered upon his inheritance with vigour. It was he who built the castle of *Estrighoiel* on the west bank of the Wye at Chepstow. He was succeeded by his son Roger in 1071, but Roger rebelled in 1075 and so forfeited his lands to the king. It was thus as a royal possession that the area beyond the Wye was described in what can best be called a preliminary annexe to the main description of Gloucestershire.

The somewhat anomalous account seems to be an untidy one. None of the entries appears to give a full Domesday description of the holdings to which it refers. Some omit the assessment; others omit to mention plough-teams. Furthermore, entry after entry omits any mention of population; what references there are, however, specify villeins, half-villeins, bordars, serfs, bondwomen, men (*homines*), one knight and two Welshmen. Woodland and meadow are never entered, but there is a solitary reference to pannage for swine. There are also sporadic references to mills and fisheries. The six mills were quite substantial in size, ranging from two of (*de*) 10s. to one rendering 15s. The entries relating to fisheries are three in number; there were seven fisheries in the Wye and in what is apparently the Usk (*in Waie et Huscha*); there were another three in the Wye; and there was half a fishery as part of a holding somewhere beyond the Usk (*ultra Huscham*). The unsatisfactory character of these entries is increased by the fact that many do not refer to specific localities but are couched in general terms, e.g. 'in Wales' or 'between the Usk and the Wye' or, as we have just seen, 'beyond the Usk'. In view of this, it is clear that the nature of the evidence does not permit an ordered account of the land beyond the Wye.

In spite of its unsatisfactory nature, the evidence does show that here, beyond the Wye, was a district with a social order quite different from that of England. One fact that immediately strikes us is the absence of any reference to hidation. Where assessments are entered, they are not in terms of hides, as in the entries of the main body of the Gloucestershire text, but in carucates. Nor is this difficult to understand in view of the history of the district. It was only a recent accession to the English realm, and it had never been hidated. When the Normans divided it out, they used carucates. We even have a glimpse of the process by which the Norman lords divided their spoil when we are told that William Fitz Osbern 'gave to Ralph de Limesi 50 carucates of land as it occurs in Normandy', which possibly means as it was measured in Normandy (*l carucatas terrae sicut fit in Normannia*).

Some entries, however, make no reference even to carucates, and they appear to refer to territory on which the Welsh land system had remained unaltered. Instead of being divided into carucates, the territory was still divided on the basis of vills. About seventy or eighty of these vills are indicated; no more precise estimate is possible, and there may well have been more. Three hardwicks (*harduices*) are also mentioned—*Lamecare, Poteschiuet* and *Dinan*; but the vills themselves are rarely mentioned by name. They are recorded in groups, each group under a reeve or *prepositus*:

Under Waswic, the *prepositus*, are 13 vills; under Elmvi are 14 vills; under Blei 13 vills; and under Idhel 14 vills. These render 47 sesters of honey, 40 pigs, 41 cows and 28s. for the hawks. Under the same *prepositi* are 4 vills wasted by King Caraduech.[1]

The last entry in the section likewise refers to another group:

The same Alured has in Wales 7 vills which were in the demesne of Earl William and Roger his son. These render 6 sesters of honey and 6 pigs and 10s.

Ten single vills held by different lords are also mentioned, and Ballard has pointed out that if we add the four wasted vills to these, another group of fourteen vills is formed.[2] These arrangements have been discussed at

---

[1] These may well have been wasted as the result of an incident mentioned in the *Anglo-Saxon Chronicle* under the year 1065. This was a raid by Caradoc to destroy a building erected by Harold at Portskewett.

[2] A. Ballard, *The Domesday Inquest* (London, 1906), p. 199.

length by Seebohm in the light of the evidence of the ancient laws of Wales. The laws show a grouping of 'vills' or 'trevs', each group being under a 'prepositus' or 'maer'; they also show how each group rendered certain specified food rents. In Wales, food rents replaced the manorial services to be found in England, and there was no demesne land to be cultivated by the common plough. It is clear, wrote Seebohm, that 'the river Wye separated by a sharp line the Saxon land, on which the manorial land system prevailed, from the Welsh land, on which the Welsh tribal land system prevailed'.[1]

BIBLIOGRAPHICAL NOTE

(1) It is interesting to note that Samuel Rudder's *A new history of Gloucestershire* (Cirencester, 1779) contains a facsimile copy of the Domesday folios for the county.

Early in the next century came a translation of the text: William Bawdwen, *Dom Boc. A Translation of the Record called Domesday, so far as relates to the counties of Middlesex, Hertford, Buckingham, Oxford and Gloucester* (Doncaster, 1812).

(2) Some use of the Domesday statistics for Gloucestershire was made by Sir Matthew Hale in *The primitive origination of mankind, considered and examined according to the light of nature* (London, 1677). He compared the evidence of the Domesday Book with that of subsequent times in order to demonstrate an increase of population. He gave Domesday figures for the city of Gloucester and for several other places in the county. The treatment is not full, but it must be remembered that the Domesday text was much less accessible than it was after the great edition of 1783. It is an interesting item in the history of the use of Domesday evidence.

(3) A pioneer study is C. S. Taylor's 'An Analysis of the Domesday Survey of Gloucestershire', *Bristol and Gloucestershire Archaeological Society* (Bristol, 1887–9). Although out of date this is still valuable, and will always be of interest as one of the early attempts to set out the information of the Domesday Book in tabular form.

(4) Various aspects of Domesday Gloucestershire are discussed in the following papers:

A. S. ELLIS, 'On the Landholders of Gloucestershire named in Domesday Book', *Trans. Bristol and Gloucs. Arch. Soc.* (Bristol, 1880), IV, pp. 86–198.

C. S. TAYLOR, 'The Pre-Domesday hide of Gloucestershire', *Trans. Bristol and Gloucs. Arch. Soc.* (Bristol, 1894), XVIII, pp. 288–319.

[1] F. Seebohm, *The English Village Community* (Cambridge, 1926; first edition, London, 1883), p. 208. For Welsh law at Caerleon, see p. 112 below.

C. S. Taylor, 'Note on the entry in Domesday Book relating to Westbury-on-Severn', *Trans. Bristol and Gloucs. Arch. Soc.* (Bristol, 1913), XXXVI, pp. 182–90.

C. S. Taylor, 'The Norman Settlement of Gloucestershire', *Trans. Bristol and Gloucs. Arch. Soc.* (Bristol, 1917), XL, pp. 57–88.

G. B. Grundy, 'The Ancient Woodland of Gloucestershire', *Trans. Bristol and Gloucs. Arch. Soc.* (Gloucester, 1936), LVIII, pp. 65–155.

(5) Other works of interest to the Domesday study of the county are as follows:

C. S. Taylor, 'The Northern Boundary of Gloucestershire', *Trans. Bristol and Gloucs. Arch. Soc.* (Bristol, 1909), XXXII, pp. 109–39.

Charles Oman, 'Concerning some Gloucestershire Boundaries', in *Essays in History presented to Reginald Lane Poole*, ed. by H. W. C. Davis (Oxford, 1927).

L. E. W. O. Fullbrook-Leggatt, 'Saxon Gloucestershire', *Trans. Bristol and Gloucs. Arch. Soc.* (Gloucester, 1935), LVII, pp. 110–35.

G. B. Grundy, *Saxon Charters and Field-Names of Gloucestershire*, published in two parts by the Bristol and Gloucestershire Archaeological Society (Gloucester, 1935–6).

C. E. Hart, 'The origin and the geographical extent of the Hundred of St Briavels in Gloucestershire', *Trans. Bristol and Gloucs. Arch. Soc.* (Gloucester, 1945), LXVI, pp. 138–65.

C. E. Hart, 'The Metes and Bounds of the Forest of Dean', *Trans. Bristol and Gloucs. Arch. Soc.* (Gloucester, 1945), LXVI, pp. 166–207.

(6) A valuable aid to the study of the Domesday county is A. H. Smith, *The Place-Names of Gloucestershire*, 4 vols. (Cambridge, 1964–5).

CHAPTER II

HEREFORDSHIRE

BY C. W. ATKIN, M.A.

The Domesday folios for Herefordshire are of more than usual interest because of the frontier position of the county. There were no fixed western boundaries; there was only 'a border', wrote Round, 'which was ever shifting with the ebb and flow of conquest.... What Harold had recovered with his light infantry, what William Fitz Osbern and his mailed horsemen could hold at the lance's point, that, at the moment of the great Survey, was all part of Herefordshire—no more and no less.'[1] In general, the newly-won districts were reckoned in carucates, while the older English territory was hidated. A number of places described, both in carucates and hides, now lie in Wales, beyond the present limits of the county,[2] but it is important to note that two parts of the present county were then predominantly Welsh in nature. Of these, Archenfield lay in the south of the county and was a more or less autonomous Welsh district under the overlordship of the English king, while Ewias was a recently conquered district in the west, between the Golden Valley and Wales itself; the Domesday statistics for both these two areas are incomplete, and this must be remembered when looking at the maps.

A transcript of the Herefordshire folios was made during the reign of Henry II, most probably between 1160 and 1170. Written in the royal scriptorium, says Professor Galbraith, 'it attempts to trace the contemporary holders of the properties mentioned in Domesday Book. The new names are added in the margin and a good many hands have been at work on the manuscript. There can be no doubt that the book was in use at the Exchequer for most of the reign, perhaps even longer.'[3] By some chance it has survived, as Balliol MS. 350, and nothing like it is known for any other county. It 'adds nothing to the text of Domesday, and even omits things not considered necessary, such as the numbered rubrics giving

[1] J. H. Round in *V.C.H. Herefordshire* (London, 1908), I, p. 263.
[2] See p. 111 below for an account of the Welsh borderland.
[3] V. H. Galbraith and J. Tait, *Herefordshire Domesday, circa 1160–1170* (Pipe Roll Society, London, 1950), p. xiv.

the names of the tenants',[1] but it is of great value for the light which it throws upon the identification of many Domesday place-names.

Apart from the vagueness of the western boundary, the Domesday county of Hereford corresponds more or less with the modern county,

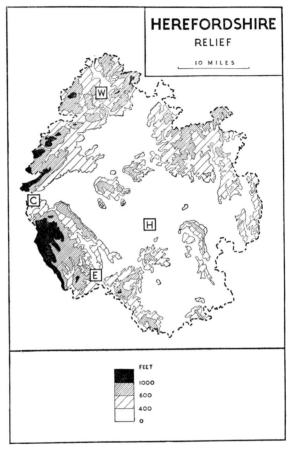

Fig. 21. Herefordshire: Relief.
Domesday boroughs are indicated by initials: C, Clifford; E, Ewias Harold;
H, Hereford; W, Wigmore.

but there are some important differences. In the north, twelve places were entered in the Shropshire folios; but Ludford, now in Shropshire, was then part of Herefordshire. On the Worcestershire boundary, Rochford was a detached portion of Herefordshire in 1086, but became

[1] V. H. Galbraith and J. Tait, *op. cit.* p. xvii.

part of Worcestershire in 1837, and Stoke Bliss likewise in 1897. Mathon, divided with Worcestershire, became wholly part of Herefordshire in the same year. Also transferred from Worcestershire to Herefordshire were Edvin Loach in 1893, and Acton Beauchamp in 1897. Fo. 180b contains under Herefordshire a list of places in Worcestershire, with one in Gloucestershire, whose revenues had been annexed to Hereford by William Fitz Osbern when he was Earl Palatine.[1] In the south, Alvington, Ruardean, Staunton and the mysterious *Wiboldingtune*[2] were surveyed under Herefordshire though they are now in Gloucestershire. Finally, the Domesday folios for Herefordshire include references to places in the modern counties of Monmouth and Radnor.

One of the difficulties presented by the Domesday account of Hereford-shire is that, although there seem to have been sixteen hundreds in 1086,[3] the names of only five of them have survived as those of modern hundreds, of which there are eleven. Moreover, it has not yet been possible to complete the reconstruction of the Domesday hundred boundaries, and these hundreds could not therefore be adopted for the plotting of densities. Another complication is the fact that the information about two or more places is sometimes combined into one statement. The outstanding example is that of the manor of Leominster with at least sixteen members (Fig. 22). Other composite entries cover merely two or three places. For the purpose of constructing distribution maps the details of each of these entries have been divided exactly among the villages concerned. But wood (measured in linear dimensions), fisheries and mills have been plotted for the first-named place in each composite entry.[4]

SETTLEMENTS AND THEIR DISTRIBUTION

The total number of separate places mentioned in Domesday Book for the area included in the modern county of Hereford appears to be approxi-mately 313, including the four places which had, or seem to have had,

[1] Bushley, Eldersfield, Feckenham, Hanley Castle, Hollow Court, Martley, Pull Court, Queenhill and Suckley in Worcestershire, and Forthampton in Gloucestershire. See *V.C.H. Herefordshire*, I, p. 272. See also pp. 220–1 below.

[2] For the identification with Whittington, see p. 3 above.

[3] Including the single reference (179b) to the hundred of *Lene*. This is a district name of Welsh origin—see E. Ekwall, *The Concise Oxford Dictionary of English Place-Names*, 3rd ed. (Oxford, 1947), p. 282.

[4] See pp. 82, 91 and 98 below.

burgesses—Hereford, Clifford, Ewias Harold and Wigmore.[1] This figure, however, can hardly be accurate because there are many instances of two adjoining villages bearing the same surname today although it is never clear whether more than one separate unit existed in the eleventh century. For example, the Domesday *Tedesthorne*, or *Tetistorp* has become the modern parishes of Tedstone Wafer and Tedstone Delamere. Or, again, the Domesday *Lentehale* (with variants) is today represented by the parishes of Earls Leinthall and Starkes Leinthall. The distinction between the respective units of these groups appears in the thirteenth century. For some counties the Domesday text occasionally differentiates between the related units of such groups by designating one unit as *alia* or *parva*; none of the Herefordshire groups is distinguished in this way.

The total of 313 includes about fifteen places for which very little information is given; the record may be incomplete, or the details may have been included with those of another village. Thus, the entry for Kingstone (179b) relates that 'to this manor belonged T.R.E. a part of the land of Cusop', although we are told nothing of what was there, and there is no other entry for Cusop. Some places, like Woonton, in Laysters (184), and Pipe (182b), each had plough-teams but no recorded population, and we are left to conjecture who worked the teams at these places. Sometimes the reverse is true, and men are recorded, but no plough-teams. Street, in Kingsland, had 3 villeins and was worth 15s. but, it would seem, had no teams (184b). Yatton was held for one hide and returned 30s., but we are given no details of its resources or population (181). Several of these problematical entries refer to places which must have been very small, and which never developed into substantial settlements.

Of the 313 Domesday places, 45% do not appear on the modern parish map of Herefordshire. Some of them are represented by hamlets,

---

[1] Including four places not named in Domesday Book, but which can be identified:
Fo. 179. Unnamed manor of Ilbert: Hungerstone (V. H. Galbraith and J. Tait, *op. cit.* pp. 6 and 83).
Fo. 181. Unnamed manor of Gilbert son of Torold in Archenfield: Ballingham (V. H. Galbraith and J. Tait, *op. cit.* pp. 19 and 87).
Fo. 181. Unnamed vill of Werestan in Archenfield: Harewood (V. H. Galbraith and J. Tait, *op. cit.* pp. 19 and 87).
Fo. 186. Unnamed manor of Alvred de Merleberge: Pencombe (V. H. Galbraith and J. Tait, *op. cit.* pp. 59 and 113).
Two other unnamed Domesday holdings can similarly be identified: Dormington and Ocle. But these also had named holdings. See V. H. Galbraith and J. Tait, *op. cit.* pp. 32 and 92, 69 and 120.

by individual houses and farms, or by the names of topographical features. For instance, the Domesday *Baissan* (181) is now the hamlet of Baysham, in Sellack. *Wilmestune* (187) has survived as Wilmastone Farm in Peterchurch, while *Gadredehope* (180) is similarly preserved as Gattertop in

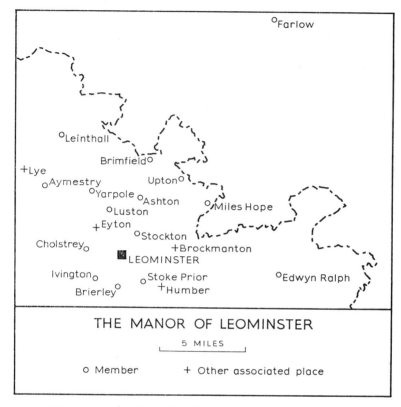

Fig. 22. Herefordshire: The Domesday manor of Leominster.

The manor of Leominster comprised itself and 16 named members for which combined statements are given for 1066 and for 1086. Forming part of the entry is another combined statement for 1086 which covers 7 places, 4 of which do not appear among the 16 named members. The total number of places associated with Leominster would therefore seem to be 20: one of the 16 members (*Mersetone*) is unidentified; another (Farlow) lies in Shropshire.

Hope under Dinmore. *Sargeberie* (185) or *Salberga* (181b) is now Sawbury Hill in Bredenbury, and *Bageberge* (182) is represented by Backbury Hill in Mordiford.

Finally there are 28 Domesday names which remain unidentified or

which cannot be identified with certainty.[1] Fortunately most of the places
were quite small; only three had values exceeding 20s. apiece in 1086, and
another six were waste; one other (*Brocote*) had been absorbed into the
forest (181). The maps of plough-teams and population are affected
slightly by the impossibility of plotting these places, but they had between
them only 53½ teams and a recorded population of only 94.

On the other hand, some villages on the modern map are not mentioned
in the Domesday Book. Their names did not appear until later in the
Middle Ages, and, presumably, if they existed in 1086, they are accounted
for under the statistics of neighbouring settlements. The names of most
of them first appear, as far as record goes, in the twelfth or thirteenth
centuries:[2] Abbey Dore in 1147, Llangarren and Kentchurch in 1130,
and Pencoyd in 1291, are typical of the Welsh part of the modern county.
The earliest record of a number of villages in the Bromyard region occurs
about the same time; thus Wacton cannot be traced before 1242, nor
Brockhampton near Bromyard before 1251. Other examples are scattered
about the county. Sometimes, although a modern parish is not named in
the Domesday Book, it contains hamlets that are mentioned. The present
parish of Allensmore contains the three Domesday names of Cobhall,
Mawfield and Winnall, but the earliest record of the parish name itself
does not occur until 1249: it is named after Alan de Plokenet, who
reclaimed the moor.[3]

In many counties, despite the complications of extinct villages and
additional parishes, the distribution of Domesday names is remarkably
similar to that of the present-day villages. In Herefordshire this is not
so. The Domesday list of place-names is far from complete in Archenfield
and the western parts of the county; and, also, as we have seen, very many
Domesday names now appear as subsidiary settlements, hamlets or farms
in parishes that bear another name and not necessarily a Domesday one.

---

[1] The complete list is: *Ach* (181b), *Alac* (179b, 180b), *Alcamestune* (187),
Almundestune (187), *Beltrov* (187), *Bernoldune* (187), *Brocote* (181), *Burcestanestune*
(186b), *Chetestor* (187), *Chipelai* (185), *Curdeslege* (187b), *Edwardestune* (184), *Etone*
(187), *Hanlie* (186b), *Lacre* (187b), *Lege* (187), *Mateurdin* (181, 187b), *Mersetone* (180),
Merstone (187), *Merstune* (187b), *Penebecdoc* (181), *Pletune* (184b), *Ruuenore* (186b),
Stane (181b), *Wadetune* (184), *Wapleford* (179b), *Westelet* (182b), *Winetune* (182b).

[2] Balliol MS. 350 (*circa* 1160–70) is the earliest record in which many of these places
are separately distinguished.

[3] For the place-names in this paragraph, see (i) A. T. Bannister, *The Place-Names
of Herefordshire* (Hereford, 1916); (ii) E. Ekwall, *The Concise Oxford Dictionary of
English Place-Names*, 3rd ed. (Oxford, 1947).

Herefordshire is thus a county to which one can hardly apply Maitland's generalisation that a place 'mentioned in Domesday Book will probably be recognised as a vill in the thirteenth, a civil parish in the nineteenth century'.[1] It may be that so many exceptions to Maitland's statement are found in Herefordshire because its rural settlement is largely dispersed in hamlets and scattered farms. Where a parish is centred upon a nucleated village, that village is more likely to survive the vicissitudes of the centuries and so maintain its identity. On a general view, the distribution of Domesday names is fairly uniform over the central part of the county except where they thin out on the Wormsley and Dinmore Hills (Fig. 23). Names are likewise less frequent on the Bromyard Upland to the north-east, and on the uplands along the western border of the county.

THE DISTRIBUTION OF PROSPERITY AND POPULATION

Some idea of the nature of the information in the Herefordshire folios, and of the form in which it is presented, may be obtained from the following entries. In nearly two-thirds of the places in the county, the manor was co-extensive with the vill. Most of these places were rather small, and Putley, west of Ledbury, is typical of them:

Fo. 184. The same Roger [de Laci] holds Putley and William of him. Thostin held [it]. There 1 hide pays geld (*ibi i hida geldat*). On the demesne are two ploughs and two villeins and 1 bordar with 2 ploughs. There 2 serfs. It is, and was, worth 20*s.*

On the other hand, Ashperton, in the Frome valley, was divided between four tenants-in-chief, and provides an example of a large vill:

Fo. 183 b. The same Ralph [de Todeni] holds one virgate of land in Radelau Hundred, and Bristoald the priest of him. Ulwi held [it] as a manor and could betake himself whither he would. This land pays geld. It is called Ashperton. There is half a plough and 1 villein and 1 bordar. It is worth 4*s.*

Fo. 185 b. The same William [Fitz Baderon] holds Ashperton. Wlwi held [it] of Earl Harold and could betake himself whither he would. There 5½ hides pay geld. On the demesne are 4 ploughs and 6 villeins and 2 bordars with 3 ploughs and 13 serfs, and 20 acres of meadow. The wood 1 league between (*inter*) length and breadth. T.R.E. it was worth 11*s.* Now as much.

---

[1] F. W. Maitland, *Domesday Book and Beyond* (Cambridge, 1897), p. 12.

Fo. 186b. Durrand of Gloucester holds Ashperton and Ralph of him. Ernui, a thegn of Earl Harold held [it] and could betake himself whither he would. There 3 virgates pay geld. On the demesne are 2 ploughs, and 2 bordars and 4 serfs and 2 bondwomen. It was, and is, worth 20s.

Fo. 187b. Madoc holds of the King Ashperton. Godric held [it]. There 1 hide pays geld, and 1 plough on the demesne, and 2 serfs and 1 villein and 1 bordar with 1 plough. It was, and is, worth 15s.

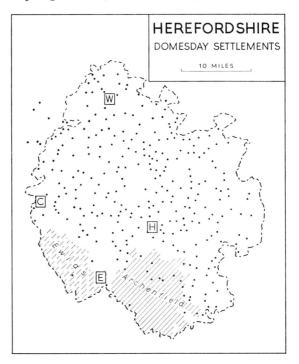

Fig. 23. Herefordshire: Domesday place-names.

Domesday boroughs are indicated by initials: C, Clifford; E, Ewias Harold; H, Hereford; W, Wigmore. The place-names to the west of the modern county appear in the folios for the Domesday county. They have not been included in our total of 313. The map shows the Welsh districts of Ewias and Archenfield, for which the statistics are incomplete.

These entries do not include all the items that are given for some places: there is no mention, for example, of Welshmen, of mills or of forest. But, although not comprehensive, the entries are representative enough, and they do contain four recurring standard items that are found for most villages: (1) hides; (2) plough-teams; (3) population; and (4) values.

The bearing of these four items of information upon regional variations in the prosperity of the county must now be considered. First, however, the almost complete omission of any reference to plough-lands must be noted. In this, Herefordshire resembles the adjoining counties of Gloucester and Worcester. Many entries, however, do tell us something about the potential, as opposed to the existing, arable, and these are discussed under the heading of plough-lands below.

(1) *Hides*

There is a marked contrast in Herefordshire between the greater part of the county, which was assessed in terms of hides, and those parts which were not. Generally, the former may be regarded as long-established areas of English settlement; the latter seem to be the areas more recently won from the Welsh.

The hidated district. The assessment was stated in terms of hides and virgates, except at two places where the hide was divided into 'parts' or thirds. At Chadnor, in Dilwyn (183), there was 'a third part of a hide' (*tercia pars unius hidae*), and at the lost *Pletune* (184b) there were 'two-thirds of a hide' (*ibi ii partes unius hidae*). In three entries, 'Welsh hides' are mentioned. At Westwood, in Llanwarne, there were '6 hides. One of these has Welsh custom and the others English' (181). This was one of the manors within the boundaries of (*in fine*) Archenfield. Just beyond the eastern border of Archenfield was Brockhampton (181 b) with '5 English hides paying geld, and 3 Welsh hides rendering 6s. to the canons yearly' (*v hidae anglicae geldantes et iii hidae Waliscae reddentes vi solidos canonicis per annum*). There was also a Welsh hide (*hida Walesea*) on an anonymous holding in *Stradel* hundred (182b), to the west, which was waste *T.R.E.*

J. H. Round declared that 'the ancient system of assessment, based on the 5-hide unit...is fairly well illustrated' in Herefordshire, and he pointed to some good examples, espeially among the royal and ecclesiastical estates.[1] Of places having 5 hides, we might take as examples Wilmastone in Peterchurch (187), and Brinsop (186); while Much Cowarne (186), Eardisland (179b), Kingsland (179b) and Pencombe (186) each had 15 hides. The largest was the royal manor of Leominster with 80 hides (180). Sometimes the five-hide unit was composed of two or

---

[1] *V.C.H. Herefordshire*, I, p. 270.

more holdings within the same vill: e.g. Shelwick, in Holmer (182), comprised one holding of 2 hides and another of 3 hides. Leintwardine was divided between a holding of ¾ hide (258) and one of 4¼ hides (260). Nevertheless, the five-hide unit is not readily apparent in the majority of entries. It is possible, however, that some villages were grouped in blocks for the purpose of assessment, as in Leicestershire.[1] In any case, the absence of information about these groups makes it difficult to be definite about the full extent of the five-hide unit in Herefordshire.

Whatever may have been the incidence of the five-hide unit, it is clear that the assessment was largely artificial in character and bore no constant relation to the agricultural resources of a vill. The variation among a representative selection of five-hide vills speaks for itself:

	Teams	Population
Bredwardine (186)	4	17
Castle Frome (184)	10	19
Ledbury (182)	31	18
Linton (179b)	16	23
Monkland (183)	9	22

In assessing the potential capacity of some holdings, the Survey seems to assume an arbitrary ratio of 2 plough-lands per hide. Thus the eleven waste manors 'on the marches of Wales' (186b) had a total assessment of 18 hides and land for 36 ploughs. Similarly, the summary for the *Stradelei* Valley speaks of 56 hides which 112 teams could plough (187). Other examples include Almeley (182b) and Grendon Bishop (185) each with 4 hides and 8 plough-lands.

A number of exemptions from the payment of geld are recorded; and the following examples illustrate some of the circumstances in which this occurred. At Bartestree (183), there were 2 hides of which one paid geld 'as the county witnesses' (*Ibi ii hidae. Una earum geldat teste comitatu*); and exactly the same phrase is applied to an unnamed berewick of the same manor.[2] The eleven waste manors 'on the marches of Wales' (186b), with 18 hides, are specifically described as having 'never paid geld' (*Nunquam geldavit*). At Bromyard (182b), there were 30 hides of which

---

[1] See p. 325 below. For the Cambridgeshire blocks, see H. C. Darby, *The Domesday Geography of Eastern England* (Cambridge, 3rd ed., 1971), p. 276.

[2] Identified in Balliol MS. 350 as part of Dormington (see V. H. Galbraith and J. Tait, *op. cit.* pp. 32 and 92).

3 were waste, but the others paid geld (*In Bromgerbe sunt xxx hidae. Ex his iii hidae sunt et fuerunt wastae. Aliae geldant*). 'Free hides' are also recorded: one of the hides of Brampton Abbotts (182b) was 'free from geld and all customary dues' (*Ibi ii hidae. Una geldat, alia est libera a geldo et ab omni consuetudine*); while at Bowley, in Bodenham (183), there was one hide free from geld and the king's service (*Ibi i hida libera a geldo et regis servitio*). There is an example of an actual reduction of the assessment at Much Cowarne (186), where King William exempted 6 of the 15 hides from geld (*Ibi xv hidae geldant sed rex W. condonavit vi hidas quietas a geldo*). But most of the non-gelding hides in Herefordshire were waste; free hides and reductions of assessment were, by comparison, unusual.

The unhidated district. The unhidated lands lay to the west of Offa's Dyke, in those parts of the county where the English advance had been comparatively recent.[1] But it must be pointed out that not all land on the 'Welsh' side of the Dyke was unhidated. There were even traces of the five-hide unit among the villages of the Upper Wye and Upper Dore area, e.g. at Bredwardine (186), Monnington on Wye (183) and Wilmastone (187).

To the west lay the castellany of Clifford, outside the hundredal system, not liable to customary dues and unhidated. Its land was described by formulae unusual in the Herefordshire folios: *terra ad n carucas* (183) or *n carucatae terrae* (184). To the south lay the castellany of Ewias Harold, also unhidated and reckoned in carucates of which there was a total of 9 with 32 acres (184, 185, 186). Its Welsh nature was such that it remained within a Welsh diocese until 1852. Further south still was Archenfield (179, 181). It was under Welsh law in 1086 and remained in a Welsh diocese until 1131. It, too, was unhidated; but the king's men rendered their own peculiar dues to the king on the basis of Welsh custom, and they were liable for military service in Wales. At Garway (*Lagademar*) there were 4 carucates. Some estates in Archenfield had passed into Norman hands but like those of the Welsh, these were unhidated.[2]

One of the characteristics of the unhidated lands was that they almost

[1] See (1) Sir Cyril Fox, *Offa's Dyke* (British Academy, London, 1955); (2) *V.C.H. Herefordshire*, I, p. 264.

[2] Carucates are mentioned only once to the east of the Wye. At Ocle (184) there was a seven-hide holding from which Walter de Laci gave 2 carucates of land (*ii car' terrae*).

all rendered money or honey rents, or both. Thus in Archenfield, on the Norman holding of Baysham, 14 men with 7 ploughs rendered 5s. as custom (181). In Ewias, at Longtown (184), Roger de Laci had '15 sesters on honey and 15 swine when the men are there, and pleas over them' (*xv sextarios melis et xv porcos quando homines sunt ibi et placita super eos*).[1]

Total. The assessment (including waste and non-gelding hides) amounted to 1,191 hides, 3½ virgates and there were also 24 carucates, 32 acres;[2] but it must be remembered that this refers to the area included in the modern county. F. W. Maitland counted 1,324 hides, but this was for the Domesday county.[3] In any case, the nature of some of the entries makes it impossible to arrive at an exact figure.

(2) *Plough-lands*

Apart from a few exceptions, plough-lands are not recorded in the Herefordshire folios. We have already considered the recording of plough-lands in connection with waste vills, and the occasional mention of them in the unhidated areas,[4] but there are just a few other entries where they appear. Thus, at Lincumbe, in Westhide (187), there were 3 virgates with land for one plough, valued at 7s., but no plough-team is recorded (*Ibi iii virgatae. Terra est i caruca. Valuit ix solidos. Modo vii solidos*). Sutton (183), on the other hand, has a peculiar formula: *Terra est ii carucis. In dominio sunt.* Or again at Bowley (183), 2 ploughs were said to be on the demesne, but there was land for 4 ploughs (*In dominio sunt ii carucae.... Terra est iiii carucis*).

Another formula occurs in connection with about 15% of the entries that record teams, and it implies that the potential arable was greater than that being cultivated in 1086. Thus, at Byford (185), there were 3 ploughs at work, but we are also told that '4 more ploughs could be there' (*iiii carucae plus possent ibi esse*); and again at Stoke Lacy, described on the same folio, there were 9 ploughs at work, but another 6 could also be employed (*et aliae vi adhuc possent esse*). The precise wording of the formula varies a little from entry to entry: e.g. at Bunshill (187b), 'there are 3 villeins and 1 bordar with 1 plough, and there could be 3 ploughs there' (*et iii carucae possent ibi esse*); as it stands, this implies land for

---

[1] For Celtic food rents, see p. 55 above, and p. 106 below.
[2] This excludes the 2 carucates (part of 7 hides) at Ocle (184).
[3] F. W. Maitland, *op. cit.* p. 400.
[4] See pp. 66–7 above.

a total of 3 plough-teams, but it may well be that *plus* or *adhuc* is understood, and that land for a total of 4 teams is meant. Such entries as these do not occur for vills where plough-lands are recorded, and they seem to represent an alternative way of providing the same information. Unfortunately, we are still left without any information about the plough-lands in the majority of Herefordshire places. It may be that the arable in these places was being tilled to capacity by the existing teams, but we cannot be sure.

In the entries for those Herefordshire manors surveyed in the Shropshire folios, however, plough-lands were systematically recorded. Thus, Picot de Sai held 3 virgates at Leintwardine (258), and there was 'land for 2 ploughs' (*Terra est ii carucis*). The same Picot held 3 hides and 3 virgates in Adley, and there was 'land for 8 ploughs' (258b). Altogether, there were twelve of these Herefordshire vills recorded under Shropshire.

(3) *Plough-teams*

The Herefordshire entries, like those of other counties, draw a distinction between the plough-teams held in demesne and those held by the peasantry. On many of the small manors in this county, however, the lord retained the whole manor in demesne, or else let it to the peasantry, as the following entries show:

Fo. 187. The same Hugh [Lasne] holds one manor of 1 hide not geldable (*non geldante*). Leflet held [it]. On the demesne are 2 ploughs and 3 bordars and 4 serfs. T.R.E. it was worth 30s., now 20s.[1]

Fo. 184. The same Roger [de Laci] holds Woonton in Laysters and Gerald of him. Ernui held [it]. There 3 virgates pay geld, and there is 1 plough there. It was worth 6s. Now 4s.

Fo. 260. The same Ralph [de Mortemer] holds Adforton. Edric held [it]. There 3 hides pay geld. There is land for 6 ploughs. 5 villeins and 6 bordars and 1 radman with 4 ploughs among them all. It was worth 8s. Now 5s.

Another small departure from the normal formula is found in three entries where the number of oxen is given instead of the number of plough-teams; in each case it is apparently to avoid the use of fractions:

(1) Yarsop (187b): There 2 serfs and 4 beasts (*animalia*).

---

[1] Identified as Livers Ocle in Balliol MS. 350 (see V. H. Galbraith and J. Tait, *op. cit.* pp. 69 and 120).

(2) Garway (*Lagademar*)[1] in Archenfield (181): There 3 bordars have 3 oxen (*boves*).

(3) Newton, near Bredwardine (186b): On the demesne are 3 oxen (*boves*) and 3 villeins and 1 bordar with 1 plough.

The entry for Cobhall (184) tells us that it paid geld for one hide, and then later adds that 'there are 4 bordars with 1 hide', obviously an error for 1 plough-team which has been included in our count.

There are also three entries which record the unusual circumstance of surplus ploughs, and these contrast with the many entries where there could be more ploughs:

(1) Mansell[2] (182b): 'On the demesne is 1 plough and another idle (*In dominio est una caruca et alia ociosa*) and 9 villeins and a reeve and 4 bordars with 6½ ploughs.'

(2) Preston on Wye (181b): 'The villeins have more ploughs than the land needs' (*Villani plus habent carucas quam arabilem terram*) although 'there could be one more plough in demesne' (*una caruca plus posset in dominio esse*).

(3) Bartestree (183): 'On the demesne are 3 ploughs and 3 serfs and a reeve with 1 plough.... There is land for 3 ploughs' (*Terra est iii carucis*).

Three other entries make a comparison between the number of plough-teams in 1086 and at some earlier date:

(1) Leominster (180): There 'were 80 hides and on the demesne 30 ploughs. There were there 8 reeves and 8 beadles (*bedelli*) and 8 radmen and 238 villeins and 75 bordars and 82 between (*inter*) the serfs and bond-women. All these together had 230 ploughs. The villeins ploughed 140 acres of the lord's land and sowed them with their own wheat-seed.... Now the King has in this manor on the demesne 60 hides and 29 ploughs; and 6 priests and 6 radmen and 7 reeves and 7 beadles and 224 villeins and 81 bordars and 25 between the serfs and the bond-women. Among them all they have 201 ploughs'. (These figures refer to the total for Leominster and its sixteen members).

(2) Wellington (187): 'On the demesne are 2 ploughs. Formerly (*Antea*) there were 5 ploughs, and 9 villeins and 8 bordars and a priest and a reeve and a smith and 4 radmen. Among them all they have 8 ploughs.

[1] For the identification, see V. H. Galbraith and J. Tait, *op. cit.* pp. 19 and 87.
[2] This part of Mansell was distinguished as Bishopstone in Balliol MS. 350 (*ibid.* pp. 28 and 90–1).

...T.R.E. it was worth £8. Now £7. There were more ploughs there than there are now' (*Plus erant ibi carucae quam nunc sunt*).

(3) Rotherwas (186b): 'On the demesne are 2 ploughs and 2 villeins and 3 bordars with 2 ploughs. T.R.E. it was worth £6. Now £3. There were there 10 villeins with 13 ploughs' (*Ibi fuerunt x villani cum xiii carucis*).

Such statistics for an earlier date are rare, and we are usually given only a static picture of a local agriculture which was in fact fluctuating widely in fortune. There are just a few other entries where the usual bare facts are enlightened by a little additional detail. At the unnamed berewick of Bartestree (183), which has been identified as part of Dormington,[1] it is noted, apparently as an unusual circumstance, that there was 'a Radman with land without a plough' (*unus Radman cum terra sine caruca*). At Much Marcle (179b), there were '36 villeins and 10 bordars with 40 ploughs. These villeins plough and sow with their own seed 80 acres of wheat and as many with oats, save 9 acres' (*hi villani arant et seminant de proprio semine quater xx acras frumenti et totidem ad avenas praeter ix acras*).

The present count has yielded a total of 2,462¼ plough-teams in 1086, but it must be remembered that this refers to the area included in the modern county. Maitland estimated the total for the Domesday county at 2,479 plough-teams.[2]

(4) *Population*

The main bulk of the population was comprised in the six principal categories of villeins, bordars, serfs, oxmen (*bovarii*), men (*homines*), and king's men (of Archenfield). The remaining categories amounted to less than a tenth of the recorded population and included radmen, Welshmen, priests, reeves and others. The details of these groups are summarised on p. 72. Sir Henry Ellis's grand total for the county came to 5,368 but his estimate was made in terms of the Domesday county and included tenants-in-chief, under-tenants and urban population.[3] No strict comparison with the present total of 4,450 for the rural population is therefore possible. Moreover, varying interpretations of many entries are inevitable; all that can be claimed for the present figures is that they indicate the order of magnitude involved. One other point must always

---

[1] See footnote 1, p. 60 above.   [2] F. W. Maitland, *op. cit.* p. 401.

[3] Sir Henry Ellis, *General Introduction to Domesday Book* (London, 1833), II, pp. 452–5.

Recorded Population of Herefordshire in 1086

A. Rural Population

Villeins	1,728
Bordars	1,285
Serfs	730
Oxmen (*bovarii*)	142
Men (*homines*)	136
King's men (Archenfield)	96
Miscellaneous	333
Total	**4,450**

There were also 112 bondwomen (*ancillae*) who have not been included in the above total.

Details of Miscellaneous Rural Population

Radmen	56	Free oxmen	11
Priests	45	Beadles	9
Welshmen	38	*Servientes regis*	9
Reeves	34	*Hospites*	7
Smiths	25	*Clerici*	3
Francigenae	23	Carpenter	1
Buri	19	Cowman	1
Cottars	19	Knight (*Miles*)	1
Coliberts	16	Swineherd	1
Freemen	15		
		Total	**333**

B. Urban Population

All categories of population apart from burgesses and priests are included in the table above.

HEREFORD — The statistics for Hereford are particularly fragmentary, and a total of only 29 is recorded for 1086. See p. 103 below.

CLIFFORD — 16 burgesses; 13 bordars; 4 oxmen; 5 Welshmen; 6 serfs; 4 bondwomen.

EWIAS HAROLD — 2 burgesses; 31 bordars; 1 priest; 4 oxmen; 1 man; 13 Welshmen; 3 serfs.

WIGMORE — 4 serfs.

be remembered. The figures, apart from those relating to bondwomen, presumably represent heads of households. Whatever factor should be used to obtain *actual* population from *recorded* population, the proportions between the different categories and between different areas remain unaffected.[1]

It is impossible for us to say how complete these statistics were, but it is apparent from the Herefordshire folios that numbers of people had not been counted. The 96 King's men in Archenfield (181) held 73 ploughs 'with their men' (*cum suis hominibus*). There were also unspecified numbers of men (*homines*) at Bromyard (182b) and at Ledbury (182), and an unspecified number of bordars at Cradley (182), while at Eastnor (182) there were '2 bordars with certain others' (*cum aliquibus aliis*).[2] Sometimes an entry appears to be defective, and contains no reference to population.[3] At Hampton Wafer (180), there was 'half a hide. On the demesne is one plough'. But the entry mentions no people. On the other hand, the Survey says of Almeley (182b) that 'the men of another vill work in this vill' (*Alterius villae homines laborant in hac villa*), and Balliol MS. 350 has a marginal note against this item. *Upcote i hida.*[4] Yet Domesday Book itself contains no record at all of Upcott which is now a hamlet in Almeley parish.

Villeins constituted the most important element in the population, and amounted to 39% of the total; bordars came next in number, with about 29% of the population. The entries for both these categories are fairly straightforward. The serfs amounted to about one-sixth of the total population. In a few entries a combined number of male and female serfs is given (*servi* and *ancillae*), and such figures have been divided equally between the sexes for purposes of computation; *ancillae* have not been included in the total population. The status of the radmen has already been discussed in connection with Gloucestershire;[5] there were 56 of them in Herefordshire, and their numbers may be compared with the total of only 15 *liberi homines*. There were 45 priests, a relatively high number as compared with some counties. Those radmen, freemen, priests, and knights and others who were apparently sub-tenants have not been included in our count, but the decision whether or not to include has

---

[1] But see p. 430 below for the complication of serfs.

[2] Note also the indefinite number of men at Hopley's Green and Lyonshall; see p. 74 below.

[3] See p. 60 above.

[4] V. H. Galbraith and J. Tait, *op. cit.* pp. 32 and 92.

[5] See p. 17 above.

often been a difficult one. Another relatively large group was that of the 34 reeves (*prepositi*). There were 25 smiths, and one carpenter; the mason (*cementarius*) at Eastnor (182) has been excluded as an apparent sub-tenant. J. H. Round has dealt at some length with the problem of the status of oxmen (*bovarii*), of whom there were 142 in the county, together with 11 *bovarii liberi*.[1]

Welshmen were recorded at the following places, (Fig. 24):

Bach, in Dorstone (187) .	.	.	8 Welshmen
Bacton (184) .	.	.	3 Welshmen
Bredwardine (186) .	.	.	1 Welshman
Kings Caple (181) .	.	.	5 Welshmen
Clifford (183)	.	.	5 Welshmen
Eardisley (184b) .	.	.	1 Welshman
Ewias Harold (184, 186) .	.	.	13 Welshmen
Mainaure, in Birch (181)	.	.	1 Welshman
Westwood, in Llanwarne (181)	.	.	1 Welshman
Willersley and Winforton (183)	.		1 Welshman[2]
Cleeve, in Ross (179b) .	.	.	'so many Welshmen as have 8 ploughs'

Since Archenfield still had Welsh law and custom, the King's men there presumably were also Welsh, together 'with their men' (181), and it is quite possible that some other 'men' elsewhere were also Welsh.

Certain entries seem to suggest the holding of land apart from the communal arrangements of the townships, and prominent amongst these, are three references to settlers (*hospites*) paying rent.[3]

Hopley's Green in *Elsedune* hd (184b): There, are men rendering 10s. and 8d. for the lands on which they have settled. There is nothing else there (*Ibi sunt homines reddentes x solidos et viii denarios pro suis hospitiis. Nil aliud ibi est*).

Lyonshall (184b): from certain men settled there 100d. are received as long as they themselves wish (*de quibusdam hominibus ibi hospitatis habentur c denarii quamdiu ipsi voluerint*).

[1] *V.C.H. Herefordshire*, I, p. 288. See p. 128 below.

[2] Not included in our count because apparently a sub-tenant.

[3] Maitland wrote that these *hospites* were in fact or theory 'colonists whom the lord has invited on to his land', *Domesday Book and Beyond*, p. 60. See also P. Vinogradoff, *English Society in the Eleventh Century*, pp. 231–2. For *hospites* in Shropshire, see pp. 128–9 below.

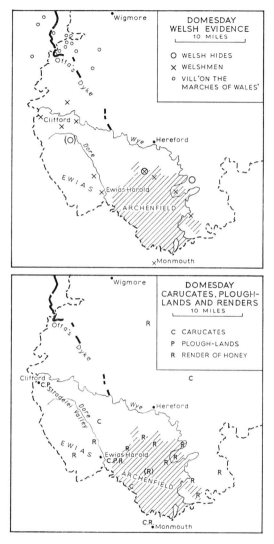

Fig. 24. Herefordshire: Domesday Welsh evidence.
For discussion of the evidence, see pp. 67–8, 74 and 106. The boundary
is that of the modern county.

Letton (184b): On the demesne is one plough, and a priest and 7 settlers with 1 plough render 5s. (*In dominio est una caruca et presbiter et vii hospites cum i caruca reddunt v solidos*).

(5) *Values*

The value of an estate is normally given for two dates; thus the entry for Preston Wynne records that 'T.R.E. it was worth 65s. now 5s. less' (181b). Quite often, the value was recorded in a phrase similar to that used at Ashperton, 'It was, and is, worth 15s.' (187b); but the formula for many of the smaller holdings makes no reference to any year other than 1086, and this may be exemplified by the entry for Pipe, which just says 'It is worth 5s.' (182b). On the other hand, the differences recorded in value between 1066 and 1086 occasionally suggest that the facts have been carefully considered. At Wellington, there was a decrease of 3 ploughs on the demesne, and the value was reduced from £8 to £7 (187); and the entry for Bishop's Frome runs: 'T.R.E. it was worth 70s. Now 3s. more' (184b). An intermediate value is sometimes encountered in runs of consecutive entries, particularly on fo. 182b.

The 'values' are usually entered as plain statements of money; but the city of Hereford and seven of the royal manors were valued in assayed money to which was added, at Lugwardine (in a marginal note), one ounce of gold, and at Kingstone, a hawk. The variations in the formulae relating to these royal manors are indicated below:

Lugwardine (179b): *x libras modo de albis denariis et unam unciam auri.*
Kingstone (179b): *l solidos de candidis denariis et unum accipitrem.*
Eardisland (179b): *xii libras de candidis denariis.*
Stanford (180): *c solidos de albis denariis.*

The valuation of Linton (179b) is unusual: 'there were 5 hides there and they rendered the fourth part of one night's entertainment. Now it is greatly decreased' (*Ibi erant v hidae et reddebant quartam partem firmae unius noctis. Modo est valde imminutum*). Where the manors are small, it is sometimes difficult to distinguish between 'valuations' made partly in kind, and similar 'renders' made by tenants to their lords. Usually such renders are of honey,[1] but the hawk which formed part of the valuation of Kingstone is paralleled at Bach (187) where 8 Welshmen rendered one hawk and 2 dogs (*unun accipitrem et ii canes*).

[1] See p. 106.

Generally speaking, the greater the number of plough-teams and men on a holding, the higher its value; but the anomalies are so numerous as to defy any attempt to find a consistent relation between resources and value. At Moreton Jeffreys, for example, 5 plough-teams and 7 men, on 4 hides, were valued at £5 (181 b); at Moreton on Lugg 5½ plough-teams and 15 men, on 4 hides, were valued at £3 (182); but at Donnington 8 plough-teams and 12 men, on one hide, were valued only at 25s. (181 b). Certain big estates seem to have been valued somewhat arbitrarily at the rate of £1 per hide; Stoke Lacy had 10 hides, with 9 plough-teams and 33 men, and was valued at £10 (185); while 60 hides of Leominster manor (180), were valued at £60, but this was in addition to the maintenance of the nuns of the dissolved Abbey and it was believed that the manor could pay £120 if it were freed from other claims (*si deliberatum esset*). In some cases the value of a berewick was included in that of its parent manor, and Leominster, with its sixteen members, is the most important example. On a smaller scale, we have such items as the berewick of Wilton (179 b) valued with the manor of Cleeve, in Ross, at a total of £9. 10s. Similarly, Walford, Ross and Upton Bishop were valued together at £14 (182).

The Herefordshire entries do not throw any great light on the question of what exactly was implied by 'value', but what evidence there is seems to suggest that it was a statement of annual rent. A Leintwardine entry (258) states that 'it renders 5s.' (*Reddit v solidos*) and many other entries similarly give the value as a 'render'. The entry for the royal manor of Kingsland states that 'it is farmed for £13 and 3s.' (179 b); while the entry for Lugwardine (179 b) records that T.R.E. it was 'not put out to rent and therefore it is not known how much it was then worth' (*T.R.E. non fuit positum ad firmam et ideo ignoratur quantum tunc valuit*).

Conclusion

The Herefordshire hundreds are unsuitable units for which to calculate densities, partly because of the confusion surrounding their Domesday boundaries, and partly because the entries for some forty holdings do not specify in which hundred they lay. The densities have therefore been calculated on the basis of the ten divisions shown in Figs. 25 and 27.

Of the four standard formulae, those relating to plough-teams and population are most likely to reflect something of the distribution of

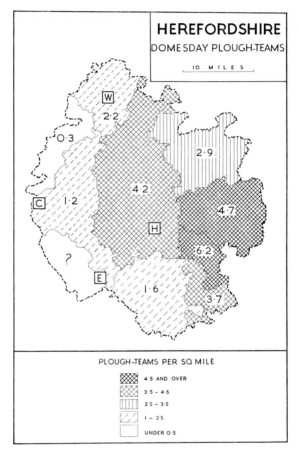

Fig. 25. Herefordshire: Domesday plough-teams in 1086 (by densities).
Domesday boroughs are indicated by initials: C, Clifford; E, Ewias Harold; H, Hereford; W, Wigmore. There is insufficient evidence to calculate a density for Ewias.

wealth and prosperity throughout the county. Taken together, they supplement one another, and, when they are compared, certain common features stand out. The most important of these features is a contrast between the east and the west of the county. In the east, the plough-teams average 5 per square mile, decreasing to about one in the west, and to less than half in the badly ravaged north-west. The density of population shows a like contrast. In the eastern and central parts of the county, the recorded population is about 8 per square mile, falling to 3 in the west,

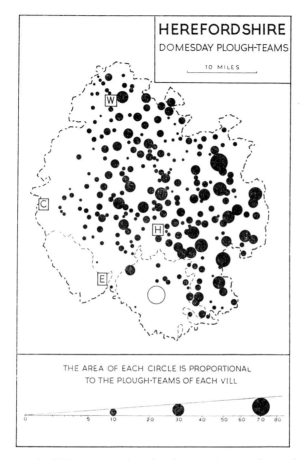

Fig. 26. Herefordshire: Domesday plough-teams in 1086 (by settlements).
Domesday boroughs are indicated by initials: C, Clifford; E, Ewias Harold; H, Hereford; W, Wigmore. The map shows the boundaries of the Welsh districts of Ewias and Archenfield for which the statistics are incomplete. The open symbol in Archenfield shows the combined total of the plough-teams of the 'king's men', the location of which is uncertain.

and to under one in the north-west. This major contrast is due primarily to the unsettled state of life along the Welsh border, and to the effects of Welsh raids. The more detailed differences, on the other hand, reflect the geography of the county. The north of the county had not been wasted by the Welsh, and the lower densities here must be due to the broken topography and poorer soils of the Silurian upland. The

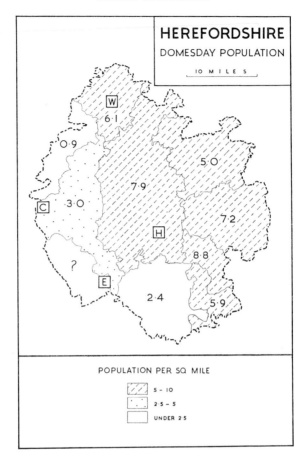

Fig. 27. Herefordshire: Domesday population in 1086 (by densities).
Domesday boroughs are indicated by initials: C, Clifford; E, Ewias Harold; H, Hereford; W, Wigmore. There is insufficient evidence to calculate a density for Ewias.

Bromyard upland, in the north-east, have many steep slopes and narrow valleys, and these difficult conditions appear to be reflected in the lower densities. Towards the south-east, lower densities are apparent in the Woolhope Silurian inlier, and are still lower in the extreme south-east on the sandy 'ryelands' soils around Ross.

Figs. 26 and 28 are supplementary to the density maps, but it is necessary to make one reservation about them. As we have seen, it is possible that some Domesday names may have covered two or more

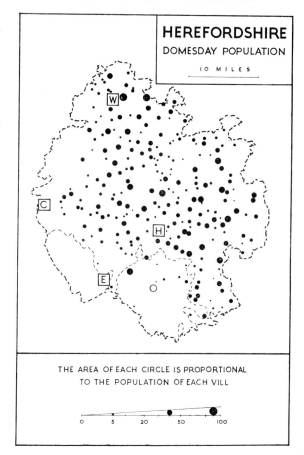

HEREFORDSHIRE

DOMESDAY POPULATION

10 MILES

THE AREA OF EACH CIRCLE IS PROPORTIONAL
TO THE POPULATION OF EACH VILL

0   5   20   50   100

Fig. 28. Herefordshire: Domesday population in 1086 (by settlements).
Domesday boroughs are indicated by initials: C, Clifford; E, Ewias Harold; H, Hereford; W, Wigmore. The map shows the boundaries of the Welsh districts of Ewias
and Archenfield for which the statistics are incomplete. The open symbol in Archenfield shows the combined total of the 'king's men', the location of whom is uncertain.

settlements, e.g. the present-day Sutton St Michael and Sutton St Nicholas
are represented in Domesday Book by only one name. Only where the
evidence specifically indicates the existence of two or more settlements
have they been separately indicated in Figs. 26 and 28. A few of the
symbols should therefore appear as two or more smaller symbols, but
the limitation does not affect the main pattern of the maps. Generally
speaking, they confirm and amplify the information of the density maps.

Types of entries

The most frequent method of measuring the amount of woodland on a holding in Herefordshire is in terms of linear dimensions. The entry for Downton (183b) is characteristic: 'wood half a league long and 5 furlongs broad' (*Silva dimidia leuua longa et v quarentenis lata*). A variant of this formula occurs at Holme Lacy (181b), where there was wood 'half a league long and as much wide' (*Silva dimidia leuua longa et tantundem lata*). The exact significance of these linear measurements is not clear, and we cannot hope to convert them into modern acreages. All we can do is to plot them diagrammatically as on Fig. 29. There is only one composite entry for Herefordshire wood which gives dimensions—that for Leominster and its sixteen members, with 'wood 6 leagues long and 3 leagues wide' (180). Presumably this implies some process of addition whereby the dimensions of separate tracts of wood were consolidated into one sum. It has been plotted for Leominster and, where possible, the members have also been indicated.

There are nine entries that differ somewhat from the usual pattern described above. Four state the amount of wood 'between length and breadth'; thus the entry for Milton in Pembridge (186b) runs: *Silvae iiii quarentenarum inter longitudinem et latitudinem.*[1] In the other five entries, there is no mention of length and breadth, and we find only single dimensions, as in these two examples:

Brampton Bryan (260b): *Silvae dimidia leuua.*
Elton (183b): *ii quarentenae silvae.*

With the exception of Elton, all the single dimensions are from the Shropshire folios where the formula is very common.[2] It may well be that all nine entries imply the same facts as the more usual formula, and that they were sometimes used when length and breadth were the same.[3] They have been plotted on this assumption on Fig. 29, but single dimensions appear as single lines to indicate where they occur.

[1] The other three entries are for Ashperton (185b), *Merstone* (187) and Shobden (183b).
[2] The other three entries are for Adley (258b), Leintwardine (260) and Lingen (260). For the Shropshire entries see p. 136 below.
[3] See p. 437 below.

There are many entries that record wood in other ways, mostly in terms of what they rendered, although there are also references to pannage and to large woods and small woods. The wood at Marden, for example, rendered 20s. (179b), and this may be contrasted with that at Little Marcle which rendered only 17d. (184b). The entry for Bromyard (182b) records 'wood there rendering nothing' (*Silva est ibi nil reddens*), and similar entries occur for other places. There are also places where linear dimensions are linked with renders, thus at Cradley (182) there was a 'wood one league long and half a league broad and it renders nothing' (*Silva una leuua longa et dimidia leuua lata et nil reddit*). At Kingsland (179b) the render of 8s. was 'from wood and pasture', and at Much Cowarne (186) 'meadow and wood' rendered nothing.

Pannage is mentioned twice. Pembridge (186) had 'wood for 160 swine if it had borne mast' (*si fructificasset*).[1] The single entry for Leominster (180) and its sixteen members mentions a 'wood 6 leagues long and 3 leagues wide rendering 22s....Each villein having 10 pigs gives 1 pig for pannage' (*Silva vi leuuis longa et iii leuuis lata reddens xxii solidos.... Quisque villanus habens x porcos dat unum porcum de pasnagio*). This entry, stating so clearly the ratio of render to total for mast swine, appears to be unique in the Domesday Book. But there are some references which give details for grass swine (*de herbagio* or *pro pastura*) in Sussex and Surrey, with ratios of 1 in 6, 1 in 7 and 1 in 10.[2]

The wood on several holdings is recorded in very general terms, or is mentioned only incidentally in other connections. At Burrington (183b) there was 'very little wood' (*paululum silvae*), at Covenhope (183b) there was a 'very small wood' (*parvula silva*), and at Titley (186b) another small wood (*silvula*), while Stretford (186) merely had 'a wood'. There were, on the other hand, 'large woods' at Ailey (187), *Bernoldune* (187) and Rushock (185b). Eardisley (184b) was said to be situated in the middle of a certain wood (*in medio cujusdam silvae*). There are two items which occur only once in the county, and the entry for each is rather vague. Byton (186b) had 'one brushwood' (*una broce*), and at Weobley (184b), in addition to a wood half a league long and 4 furlongs broad there was a park (*ibi est parcus*). Wood is sometimes recorded in

---

[1] Pembridge near Leominster—see V. H. Galbraith and J. Tait, *op. cit.* pp. 59 and 114.

[2] H. C. Darby and Eila M. J. Campbell (eds), *The Domesday Geography of South-east England* (Cambridge, 1962), p. 597.

connection with hays and salt, which are discussed below on pp. 88 and 93 respectively.

Some of the most interesting references to wood in the Herefordshire folios are those which mention assarts. They are four in number, and they are unique in the whole of the Domesday Book:

> Much Marcle (179b): 'In the same manor are 58 acres of land reclaimed from the wood and the reeve and 2 other men hold several acres of this same land' (*In eodem manerio sunt lviii acrae terrae projecte de silva et prepositus et alii ii homines tenent plures acras de ipsa terra*). In this entry *essarʒ* is interlined above *projecte*.
>
> Leominster Manor (180): 'Of the land reclaimed from the wood the profits are 17*s.* and 4*d.*' (*De exsartis silvae exeunt xvii solidi et iiii denarii*).
>
> Weobley (184b): 'The wood is half a league long and 4 furlongs broad. A park is there and assart land for 1 plough renders 11*s.* and 9*d.*' (*et terra ad i carucam de Essarʒ reddit xi solidos et ix denarios*).
>
> Fernhill (184b): 'Wood there half a league long and 4 furlongs broad, and assart land for 1 plough renders 54*d.*' (*Silva ibi dimidia leuua longa et iiii quarentenis lata et terra ad i carucam de Essarʒ reddit liiii denarios*).

In addition to these four entries, there are others which suggest reclamation, if not from woodland, at least from waste land. These refer to holdings where land was held, apparently in severalty, on similar money rents to those of the assarts.[1] The holdings of the *hospites* were of this kind;[2] but the other examples occur, e.g. at the end of a long entry for Leintwardine (260) we are told that 'two men there render 4*s.* for hire of land' (*de locatione terrae*).

The clearing of the Herefordshire woodland was not, however, an uninterrupted process. Villages ravaged by the Welsh quickly demonstrated the saying that 'oaks are the weeds of Herefordshire'.[3] In the devastated north-western part of the county and in the adjacent parts of Radnorshire, Osbern Fitz Richard held eleven waste manors, with land for 36 ploughs (186b), and we are told that 'on these waste lands have grown up woods in which Osbern hunts, and thence he has whatever he can take. Nothing else' (*In his wastis terris excreverunt silvae in quibus isdem Osbernus venationem exercet et inde habet quod capere potest. Nil*

---

[1] F. W. Maitland, *op. cit.* p. 57.

[2] See (i) p. 74 above; (ii) *V.C.H. Herefordshire*, I, p. 293.

[3] Quoted by T. Rowlandson, *Journal of the Royal Agricultural Society* (London, 1853), XIV, p. 451.

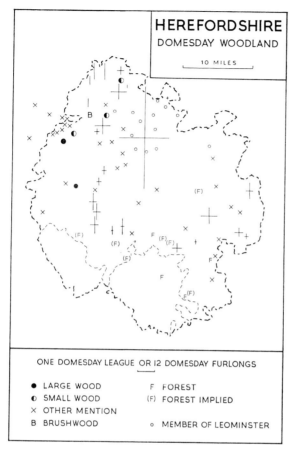

Fig. 29. Herefordshire: Domesday woodland in 1086.
The map shows the boundaries of the Welsh districts of Ewias and
Archenfield for which the statistics are incomplete.

aliud). These lands were lying waste in 1066, and so it is likely that they
had been sacked in 1055, when Griffith ap Llewellyn raided Hereford,
and if this were so the trees would thus be some thirty years old by the
time of the Survey.[1] The entry for Harewood (187), in Clifford, also
suggests conversion to woodland: 'this land has all been converted into
woodland. It was waste and renders nothing' (*Haec terra in silvam est tota
redacta. Wasta fuit et nil reddit*).

[1] *V.C.H. Herefordshire*, I, p. 264.

Distribution of woodland

Fig. 29 shows that the wood entries for Herefordshire are widely distributed, but, in general, they are most frequent for localities in the west. Wood is mentioned in connection with about one-fifth of the settlements of the county, but the amounts recorded are fairly small, with the exception of that in the large composite entry for Leominster. It must be remembered that the wood has been plotted on the sites of place-names, whereas it might well have been located elsewhere within a parish. This limitation does not affect the general pattern of the map, and it is clear that there was far less wood in Herefordshire than in the other border counties of Cheshire, Shropshire and Gloucestershire. Fig. 29, of course, tells us nothing about the woodland—as opposed to the forest—of Ewias and Archenfield, for the statistics relating to these areas are incomplete. The place-name evidence for the county shows frequent wood elements, and we can only suppose that extensive clearing had taken place since Saxon days.[1]

Forests

The Herefordshire folios do not contain any entries devoted primarily to the forests, but there are incidental references in the entries for thirteen places and these convey an interesting, though no doubt imperfect, impression of the extent of the Herefordshire forests in the eleventh century (Fig. 29). The full list is as follows:

Harewood[2] (181): The forest (*foresta*) renders half a sester of honey and 6*d.*
Brocote (181): They [2½ hides] were waste and up to the present are in the king's wood (*et sunt adhuc in silva regis*).
Bullingham (184): The wood is in the king's forest (*in foresta regis*).
Bullingham (186): There is wood but it is placed in the king's forest (*sed posita in foresta regis*).
Burton, in Holme Lacy (181 b): The wood of this manor the king has in his demesne (*in suo dominio*).
Cleeve, in Ross (179b): Of this manor there is in the forest of King William (*in foresta regis Willelmi*) as much land as rendered T.R.E. 6 sesters of honey and 6 sheep with their lambs.

[1] A. T. Bannister, *The Place-Names of Herefordshire* (Hereford, 1916).
[2] In Archenfield. This identification is from V. H. Galbraith and J. Tait, *op. cit.* pp. 19 and 87.

Much Cowarne (186): One hide of this land lies in the king's wood (*in silva regis*).

Didley, in St Devereux, and *Stane* (181 b): Of these 9 hides one part is in the jurisdiction of Alured's castle at Ewias Harold and the other part in the king's enclosure (*in defensione regis*).

Dinedor (183): The wood of this place the king has in demesne (*habet rex in dominio*).

Madley (181 b): Wood there half a league long and one furlong in breadth. This wood is in the king's enclosure (*in defensu regis*).

In Stradel hd (182b): Of this land the greater part is in the king's enclosure (*in defensu regis*).

Ross (182): The wood is in the king's enclosure (*in defensu regis*).

Turlestane[1] (179b): Of this manor [of Marcle] is one hide at *Turlestane* which T.R.E. rendered 50 masses of iron and 6 salmon. Now this land is in the forest (*Modo est haec terra in foresta*).

Eight of these entries relate woodland to the forest, and although *forest* is essentially a legal term, it is quite clear that the Herefordshire forests of 1086 were intimately associated with woodland. Much Cowarne, to the north-east of Hereford, and *Stradel* hundred, on the Welsh border near Clifford, appear to have been remote from any other recorded forest; but none of the remaining places for which forest is mentioned is far from the borders of Archenfield, and there may possibly have been a continuous belt of forest in this part of the county. The yield of the Archenfield Forest itself was small, and there is no evidence as to its location or extent. Ross and Cleeve are rather closer to the Forest of Dean than to the forests recorded along the northern Archenfield boundary. The entry for Cleeve shows that the forest here had been extended since 1066; and at Bullingham, just south of Hereford, the forest had obviously been extended at some unspecified date to include the woodland of both the manors there. The entry for the royal manor of Kingstone (179b) includes the interesting item of a 'wood there called Treville rendering no dues except venison. The villeins living there T.R.E. carried venison to Hereford and they did no other service' (*Ibi Silva nomine Triveline nullam reddens consuetudinem nisi venationem. Villani T.R.E. ibi manentes portabant venationem ad Hereford nec aliud servitium faciebant*). At a later date, Treville was the name of an extensive forest. The lord of Kilpeck, near Treville,

---

[1] The exact site of *Turlestane* is unknown, but it appears to have been within the present parish of Much Marcle, and has been so regarded in this study.

William Fitz Norman, appears to have been the Royal Forester, paying £15 to the king for the forests which he held (181).

Hays

Hays (*haiae*), or enclosures, in the wood, are mentioned in connection with twelve places in Herefordshire:

> *Bernoldune* (187), on the Welsh border but unidentified:[1] The wood there is large but its extent has not been told. There is one hay in which he [Hugh Lasne] takes what he can. The other land is waste (*Silva est ibi magna, sed quantitas non fuit dicta. Ibi est una haia, in qua quod potest capere captat. Alia terra est wasta*).
> Colwall (182): one hay.
> Cradley (182): There, one hay.
> Downton (183b): 2 hays are there.
> Eastnor (182): 2 hays.
> Lingen (260): 3 hays for taking roe-deer (*iii haiae capreolis capiendis*).
> Lye, in Aymestry (260): There, 2 hays.
> *Mateurdin* (187b): It was and is waste, but nevertheless Robert holds there of Grifin one hay (*Wasta fuit et est, sed tamen tenet ibi Robertus de Grifino unam haiam*).
> Rushock, in Kington (185b): One hay is there in a large wood (*Ibi est una haia in una magna silva*).
> Titley (186b): It was and is waste. Nevertheless 1 hay is there in a small wood (*Wasta fuit et est. Est tamen ibi i haia in silvula*).
> Upton Bishop (182): 1 hay.
> Walford, near Ross (182): 3 hays.

Almost all these entries relate to the north and north-west of Herefordshire, where the amount of wood and waste land gave ample opportunities for private hunting, but some of them lay along the line of the Malverns and near Ross. The various entries show that a close connection sometimes existed between hays, woodland and waste; but Colwall, at the foot of the Malverns, was quite a prosperous vill for which no wood was recorded. The entry for Lingen explains that the hays were for taking roe-deer; but, nevertheless, a wood used for hunting did not necessarily contain a hay, for at Ailey, in Eardisley (187), there was 'a large wood for hunting' (*Silva magna ad venandum*) without any record of hays. Neither

[1] But for a possible identification with Barland south of Presteigne in Radnorshire, see Lord Rennell of Rodd, '*Bernoldune* and Barland', *Trans. Radnorshire Soc.* (Llandrindod Wells, 1961), XXXII, pp. 1–2.

are hays mentioned in connection with the forests, although the entry for Hereford (179) records that 'when the king was pursuing the chase, from each house according to custom went one man to his station in the wood' (*ad stabilitionem in silva*), i.e. to prepare the hays for taking deer.[1]

MEADOW

Types of entries

The entries for meadow in Herefordshire are for the most part comparatively straightforward, and four-fifths of them repeat the formula '*n* acres of meadow' (*n acrae prati*). As in the case of other counties, no attempt has been made to translate these figures into modern acreages. The Domesday acres have been treated merely as conventional units of measurement, and Fig. 30 has been plotted on that assumption. The amount of meadow in each vill varied considerably, from one acre at Norton Canon (182b) to 28 acres at Hampton Bishop (182). There were several vills with 2 acres, but more usual amounts are represented by Eastnor with 6 acres (182) and Lyde with 8 acres (182b), while a few vills had over 20 acres. At Maund Bryan (185b), and several other places, there was merely 'meadow for the oxen' (*pratum bobus*). A variant of the formula appears at Bodenham (184), where there was meadow 'sufficient for the oxen' (*Pratum est bobus tantum*). At Monkland (183), there was 'from meadow 5s. besides pasture for the oxen' (*De prato v solidi praeter pasturam bovum*). The only other mention of pasture in the county was at Kingsland (179b), where 'the wood and pasture rendered 8s.' (*De silva et pastura viii solidi*). At Much Cowarne (186) the 'meadow and wood' rendered nothing, and at Marden (179b) there were 48 acres *inter planam terram et pratum*. There are also other variants. The meadow at Stretford (186), for example, rendered 3s. (*pratum reddit iii solidos*); and the entry for Bartestree (183) appears to be incomplete, for the phrase 'meadow there' (*Ibi pratum*) is followed by a blank in the manuscript.

Distribution of meadowland

In view of the great importance of meadow in the economy of the medieval village, it is a little surprising to discover that it was recorded for only about one-sixth of the villages of the county, and this in spite of the generally well-watered nature of the countryside. Almost the whole

[1] See A. Ballard, *The Domesday Inquest* (London, 1906), p. 167.

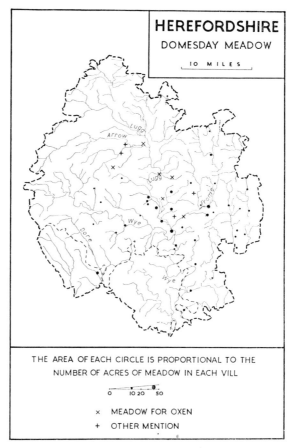

THE AREA OF EACH CIRCLE IS PROPORTIONAL TO THE
NUMBER OF ACRES OF MEADOW IN EACH VILL

×   MEADOW FOR OXEN
+   OTHER MENTION

Fig. 30. Herefordshire: Domesday meadow in 1086.
The map shows the boundaries of the Welsh districts of Ewias and
Archenfield for which the statistics are incomplete.

of the recorded meadow was concentrated in the centre and east of the
county, especially in the Lugg and Frome valleys, towards their confluence
(Fig. 30). It extended up to the Malverns in the east, and along the Wye to
the south, but no meadow was entered for the sixty-hide manor of
Leominster. A certain amount of meadow lay scattered to the west of
Leominster and Hereford, and there was also some along the northern
and western boundary—at Little Hereford (182b), and Ewias Harold
(186). Meadow was thus notably lacking in Archenfield, in the Golden
Valley, in north-western Herefordshire and in the Bromyard district.

The distribution, though not the total amount, of meadow therefore accords reasonably well with what might be expected from the disposition of the river systems and the state of settlement at that time.

FISHERIES

Fisheries (*piscariae*) are specifically recorded for 1086 in connection with at least nine places in Herefordshire. Most of the entries state only the number of fisheries, but for some a render is also given. Only one fraction is mentioned—half a fishery, at Cleeve in Ross. It is difficult to be sure of the exact number of places because of two composite entries:

Ford, Broadfield and Sarnesfield (180): a fishery rendering 600 eels.
Broadward (180): a fishery yielding (*de*) 500 eels.
Fownhope (187): 3 fisheries rendering 300 eels.
Cleeve and Wilton (179b): a fishery rendering nothing (*nil reddens*).
Cleeve (180): half a fishery.
Downton (183b): a fishery.
Ewias Harold (186): 3 fisheries.
Howle, in Walford (181): a fishery.
Marden (179b): a fishery which pays no rent (*piscaria sine censu*).
Dinedor (183): fishery implied by the statement: *In aqua vero nemo piscatur sine licentia*.

There was also an important fishery in the Wye (*in Waia*), but its location was not stated although it was valued at £6 (184).[1] There is one reference to conditions in 1066 in the entry for Much Marcle (179b), which states that in King Edward's time, the hide of this manor at *Turlestane* rendered 6 salmon, but that in 1086 it was in the forest. In addition to places for which fisheries are specifically mentioned, there were at least nine other places with mills rendering eels (counting Leominster and its 16 members as one).[2] These places are likewise indicated on Fig. 31. Altogether there were thus at least 18 places for which fisheries were recorded or implied.

Of the $13\frac{1}{2}$ fisheries recorded in the county, $6\frac{1}{2}$ were located on the Wye below Hereford, and the unnamed fishery was also somewhere on

[1] The margin of Balliol MS. 350 has *Piscaria de Hodenac* (V. H. Galbraith and J. Tait, *op. cit.* p. 41).
[2] Not only were there renders of eels for both 1066 and 1086 from the 8 mills of Leominster, but the men of the manor, at the latter date, made a customary payment of 17s. for fish.

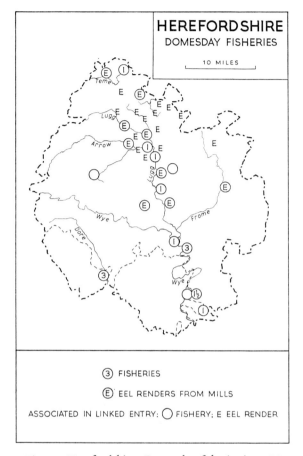

Fig. 31. Herefordshire: Domesday fisheries in 1086.

The important fishery in the Wye (180) cannot be plotted. The map shows the boundaries of the Welsh districts of Ewias and Archenfield for which the statistics are incomplete.

this river. Three fisheries were on the Lugg between Leominster and Hereford; three were on the Dore at Ewias Harold; and the remaining one was on the Teme in the extreme north of the county. The mills rendering eels were mostly on the Lugg and its tributaries in the Leominster basin. It is noteworthy that no fisheries were recorded on the river Frome, that is apart from the eel renders at two places.

SALT

The Herefordshire folios mention salt in connection with eleven places. Each entry specifically mentions the salt-pans at *Wich* (Droitwich) in the neighbouring county of Worcestershire:

Leominster (180): Wood 6 leagues long and 3 leagues broad rendering 22*s*. Out of these 5*s*. are given for buying wood in Droitwich and 30 mitts of salt are had from there (*Ex his dantur v solidi ad ligna emenda in Wich et habentur xxx mittae salis inde*).

Much Marcle (179 b): There wood rendering 5*s*. which are given to Droitwich for 60 mitts of salt (*Ibi Silva reddens v solidos qui dantur ad Wich pro lx mittis salis*).

Cleeve, in Ross (180): and then there belonged there 25 mitts of salt from Droitwich (*et tunc pertinebant ibidem xxv mittae salis de Wich*).

Marden (179 b): From the salt-pans at Droitwich 9 loads of salt or 9*d*. (*De salinis in Wich ix summae salis aut ix denarii*).

Wellington (187): At Droitwich he has 17 mitts of salt for 30*d*. (*Ad Wich habet xvii mittas salis pro xxx denariis*).

Tupsley (182): A salt-pan at Droitwich rendering 16 mitts of salt (*Salina ad Wich reddens xvi mittas salis*).

Moreton Jeffreys (181 b): and one salt-pan in Droitwich (*et una salina in Wich*).

Eastnor (182): and part of a salt-pan in Droitwich (*et pars salinae in Wich*).

Backbury, in Mordiford (182): and part of a salt-pan in Droitwich (*et Pars salinae in Wich*).

Ledbury (182): and part of a salt-pan in Droitwich (*et pars salinae In Wich*).

Ullingswick (181 b): and part of a salt-pan in Droitwich (*et pars salinae in Wich*).

At Leominster there was also a render of 8*s*. for salt (*ad sal*), and J. H. Round suggested that this sum represented the commutation of a former service of carting wood to Droitwich for use as fuel in the saltworks.[1] Round also suggested that the loss of salt privileges implied in the entry for Cleeve was due to the loss of woodland when part of the manor was placed in the forest.[2] Certainly the Leominster and Marcle entries quite clearly relate the salt privileges to the manorial woodland. There is an obvious lack of uniformity about the arrangements, and there seems to be no consistency in the value placed upon a load of salt. The places

[1] *V.C.H. Herefordshire*, I, p. 296 n. [2] *Ibid*. p. 295.

mentioned are scattered through the eastern and central parts of the county, and there does not appear to be any particular significance in their distribution.[1]

WASTE

Many entries in the account of Herefordshire bear witness to Welsh raiding and border warfare. There is specific mention of one such raid in the entry that tells how King Grifin and Blein had devastated Archenfield in the time of King Edward (181); presumably this refers to events in 1055 when the Welsh plundered the city of Hereford itself.[2] By 1086 there was still a substantial number of manors that had not yet recovered.

There is no doubt about the date implied by the phrase *wasta est*; it must refer to 1086. The phrase *T.R.E. wasta fuit*, on the other hand, can mean only 1066. But the phrase *wasta fuit* (without T.R.E.) is not so clear.[3] Does it refer to 1066 or to a date between this year and 1086, or to both dates? We cannot be certain, and on Fig. 32 it has been shown as a separate category. But what of those entries (e.g. for Newton, Chickward and Eardisley) which merely state the fact of waste (apparently for 1086) and make no reference to earlier conditions? Are we to assume that they imply land that had always been waste—in 1066, later and in 1086? The following representative entries indicate the ambiguities involved:

Lulham (181 b): *T.R.E. wasta fuit. Modo valet x libras.*
Tibberton (181 b): *T.R.E. wasta erat. Modo valet iii libras.*
Letton (260): *Wasta fuit et post valuit v solidos. Modo x solidos.*[4]
Kinnersley (183 b): *Wasta fuit. Modo valet xxx solidos.*
Sawbury (185): *Wasta fuit et est.*
Laysters (186 b): *Wasta est et fuit.*
Laysters (187 b): *Tunc valebat xv solidos. Modo est wasta.*
Newton near Leominster (180): *Haec terra est wasta.*

[1] See p. 258 below.
[2] *V.C.H. Herefordshire*, I, pp. 352 *et seq.*
[3] The entry for Thornbury (186) is unusual in form. Early on it says *Wasta fuit haec terra*, and from the context this would seem to refer to 1066. The entry then states the number of men and teams in 1086, and ends with *Valebat xiii solidos et iiii denarios. Modo ii solidos.* Does the *valebat* refer to an intermediate date? The preceding entry also refers to Thornbury and gives values for three dates which is unusual in the Herefordshire folios. In spite of the doubt about *valebat*, the vill has been shown on Fig. 32 as partly waste in '1066 and/or *c.* 1070'.
[4] From the Shropshire folios where three dates frequently appear for waste.

Chickward (181): *Ibi i hida et iii virgatae terrae wastae.*
Eardisley (181): *Ibi sunt ii hidae et dimidia wastae.*

At Grendon Bishop (185) we are simply told: *Nichil ibi habetur.* It is not specifically said to be waste, and so has not been marked as such.

Only four entries specifically ascribe waste to an intermediate date:

Pembridge (186): *T.R.E. valebat xvi libras et post fuit wasta. Modo valet x libras et x solidos.*

Walford near Leintwardine (260): *Valebat xii solidos. Modo viii solidos. Wastam invenit.*[1]

Lye (260): *T.R.E. valebat v solidos. Modo vii solidos. Wastam invenit.*[1]

See of Hereford (181b): *Robertus episcopus quando venit ad episcopatum invenit xl hidas vastatas et ita sunt adhuc.*

In addition, waste at an intermediate date (*quando Willelmus recepit*) is also mentioned for Caerleon outside the county (185b).

A number of waste holdings had been placed in the forest or had become overgrown with wood. Thus of Harewood (187) we read: *Haec terra in silva est tota redacta. Wasta fuit et nil reddit.* Eleven waste manors in the north-west of the county had also been covered with wood in which Osbern Fitz Richard hunted (186b).[2] The waste at *Brocote* (181) had been placed *in silva regis.* Part of the nine waste hides at Didley and *Stane* (181b) had also been placed *in defensione regis*; so had waste land in *Stradel* hundred (182b).[3]

That some vills were only partly waste, we may judge from two types of evidence. In the first place, there were those vills comprising two or more holdings, of which at least one was waste. Thus of two holdings at Newton near Leominster, one was waste in 1086 (180), and the other was tilled (185b). In the second place, single holdings were themselves partly waste, as the following two examples show:

Huntington (182): 10 hides. *Ex his iiii sunt wastae.*

Walford near Ross (182): People and teams, yet *villani reddunt x solidos pro wasta terra.*

The entry for *Curdeslege* (187b) is unusual. It was and had been waste except for 3 acres of land lately ploughed there (*Wasta fuit et est praeter iii acras terrae nuper ibi aratas*), but we are not told by whom or with what oxen; nor is a value given for the holding.

[1] From the Shropshire folios where three dates frequently appear for waste.
[2] See p. 84.   [3] See pp. 86–7.

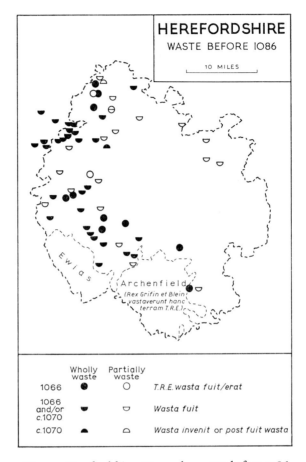

Fig. 32. Herefordshire: Domesday waste before 1086.

Note: (1) Lye is unusual in that one entry mentions *wasta erant* (187) and another *wastam invenit* (260)—hence the unusual symbol in the north of the county. (2) The map does not show those vills said to be waste in 1086 but with absolutely no reference to their earlier condition. There were nine of them in the area to the north of Clifford. It is conceivable that they, too, were waste before 1086.

Even when a holding was waste, it was not necessarily completely devoid of value or resources. Here are four representative entries relating to such holdings:

Brampton Abbotts (182b): *Wasta fuit et est. Tamen reddit v solidos.*
Upleadon (184b): *Wasta est. Valet iiii solidos.*

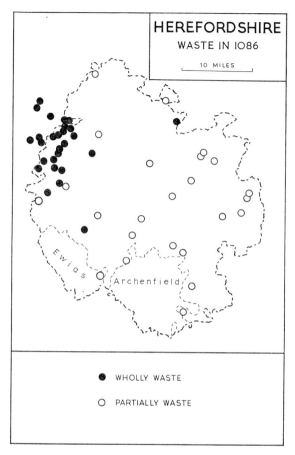

Fig. 33. Herefordshire: Domesday waste in 1086.

Rushock (185 b): *Wasta fuit et est. Ibi est una haia in una magna silva.*
Titley (186b): *Wasta fuit et est. Est tamen ibi i haia in silvula.*

These and similar holdings have been regarded not as partly waste, but as entirely waste in the sense that all their arable seems to have been devastated and that they had no population. They do not necessarily appear as 'wholly waste' on Figs. 32 and 33 because some vills had fully stocked holdings in addition to their devastated land.

Fig. 32 shows the places that were waste before 1086. Some certainly had been waste in 1066; others also may have been waste at that time,

4

but how many depends upon the interpretation given to the formula *wasta fuit*. As might be expected, the greater number lay in the western part of the county. Considerable recovery had taken place by 1086 but, as Fig. 33 shows, many places were still waste, and partly waste places were even more frequent in the eastern half of the county. The table below gives the number of places, including unidentified vills, affected in the area covered by the modern county.

	Wholly waste	Partly waste
1066 (minimum)	10	2
1066 (possible total)	42	21
1086	25	25

These figures can only be uncertain for the reasons already stated. To attempt estimates for an intermediate date would be even more hazardous. The extent of the recovery can be seen more clearly on Fig. 34 which shows the number of teams at work in 1086 on holdings which had been waste at some earlier date.

It may well be that Fig. 33 does not give a complete picture of the empty villages of the county in 1086. A few holdings not specifically described as waste had no population or teams entered for them e.g. Lincumbe (187) with one plough-land but no teams or population. Some places, waste before 1086, had been placed in the forest, and they, too, do not appear on Fig. 33.

MILLS

Mills are recorded for 1086 in connection with at least 72 of the 313 Domesday settlements within the area of the modern county. It is difficult to be sure of the exact total because eight mills are recorded for Leominster and its sixteen members (180), and they have been plotted at Leominster itself. A further two mills are recorded in a composite entry for seven places, and these have been plotted at the first-named of the places (180).[1] The associated villages in both these entries have been indicated on Fig. 35.

The usual type of entry states the number of mills and their monetary value, supplemented in about a fifth of the entries by a render of eels. The

[1] The figure of 72 results from counting the Leominster and the other group each as one. If it is assumed that each mill was at a separate place, the total of places with mills is 80. By counting all the places named in the two composite entries, the total becomes 89, a very artificial figure.

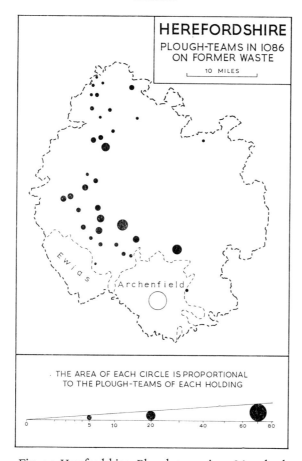

Fig. 34. Herefordshire: Plough-teams in 1086 on land
which had previously been waste.
The open symbol in Archenfield refers to the combined total of the teams of the
'king's men', the location of which is uncertain.

value is often a multiple of the *ora* of 16*d*.;[1] but it varies in other entries
between wide limits, of which the extremes are illustrated by Linton with
a mill worth 8*d*. (179 b) and Leinthall with a mill worth 30*s*. (183 b). None
of the three mills at Aston Ingham (186), Letton (184 b) and Avenbury
(183) rendered anything. Burghill (186) is typical of the entries with eels,
having a 'mill worth 20*s*. and 25 sticks (*stiches*) of eels'. There was a mill
at Richards Castle (185) 'rendering 4 muids of grain (*modios annonae*)

[1] *V.C.H. Herefordshire*, I, p. 293.

4-2

and 15 sticks of eels'. Clifford (183) similarly had a mill 'rendering 3 muids of grain', and at Little Marcle (184b) there was a 'mill rendering grain'. At Much Marcle (179b) there was a mill which rendered nothing 'beyond the sustenance of him who keeps it' (*nisi tantum victum ejus qui eum*

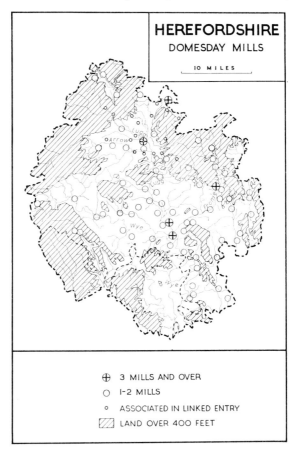

Fig. 35. Herefordshire: Domesday mills in 1086.
The map shows the boundaries of the Welsh districts of Ewias and
Archenfield for which the statistics are incomplete.

custodit). Two and a half mills are recorded for Hampton Bishop (182), but the ownership of the other half does not appear to be entered. Another example of divided ownership is afforded by Bullingham which comprised three holdings; two had each a third of two mills rendering 14s. 8d.

(184, 186); the remaining holding had a third of one mill also rendering 14*s*. 8*d*. (186b). Is the mill in the third entry an error, and was there a round total of 2 mills in the village?

Domesday Mills in Herefordshire in 1086

1 mill	59 settlements	4 mills	2 settlements
2 mills	8 settlements	5 mills	1 settlement
3 mills	1 settlement	8 mills	1 settlement

The groups of four mills were entered for Bishop's Frome (181 b, 184 b, 185), and for Lugwardine (179 b). The group of five was entered for Little Hereford (182 b), but there was apparently some dispute as to the ownership of some of them, for the entry records that there is 'a mill worth 6*s*. 8*d*. There are there 4 mills, a moiety of which belongs by right to the said manor' (*molinum de vi solidis et viii denariis. Ibi sunt iiii molini quorum medietas recte pertinet ad predictum manerium*). The group of 8 was entered for Leominster and its sixteen members (180).

Fig. 35 shows that most of the mill sites were aligned along the principal river valleys in the more populous eastern and central parts of the county. They occurred in the west of the county only at a few places in the valleys of the Teme, Wye and Dore.

CHURCHES

Churches are mentioned in connection with thirteen places in Herefordshire, including Archenfield where the king had three churches. The churches of Hereford itself are not mentioned in the account of the city, but three appear incidentally as landholders. The possessions of the cathedral are variously called 'the lands of the Bishop' (179), or the 'lands belonging to the Canons of Hereford' (181 b). We know that the collegiate Church of St Guthlac was then situated on the site of the Castle Green, and it duly appears as the holder of ten manors (182 b). Thirdly, the Church of St Peter was founded and endowed by Walter de Laci, and the Domesday account confirms this with the statement that 'Saint Peter of Hereford holds Frome [in Mordiford]. Walter de Laci gave it to the Church with the consent of King William' (182 b). Apart from Hereford, priests are mentioned in connection with all the churches. There were twenty-eight other places for which priests alone are recorded, so that Domesday Book appears to give an indication of churches at a total of forty-one places in the area covered by the modern county. Three of these places each had two priests, e.g. Avenbury (183). The precise number of priests in Archenfield was not recorded, but they had the possibly somewhat hazardous privilege of bearing the King's embassies into Wales (179). At Bishop's Frome (181b) Domesday Book refers to 'the priest of the vill' (*presbyter villae*), but on the Cathedral lands it is not always easy to distinguish the village priest from priests sitting as under-tenants of the bishop. At Ledbury, for example, a priest held 2½ hides from the bishop, and he has been excluded from the count (182). It is almost certain that the total figure of forty-one can represent only a proportion of the parish churches of eleventh-century Herefordshire. The greater number throughout the county went unrecorded.

URBAN LIFE

Hereford is the only modern town in the county which is recorded as a Domesday borough. The remaining Domesday boroughs were quite small and were connected with the three Norman castles of Clifford, Ewias Harold and Wigmore. One interesting feature that Herefordshire shares with some other counties is that some urban houses or burgages are recorded as belonging to rural manors. The burgages are entered under their respective manors, and are not referred to in the accounts of the boroughs. Three of these references concern Hereford itself; the other two are as follows:

> Coddington (182): To this manor belong 3 houses (*masurae*) in Worcester rendering 30*d*.
> *Hanlie* (186b): one burgess rendering 4*d*. (*de iv denariis*).

Coddington is in the east of the county, but not actually on the boundary, and its connection with Worcester is obscure; the manor was in the hands of the bishop of Hereford. The *Hanlie* entry is accompanied by a marginal note *In Wimstrui Hund*, which would place it in the same part of the county as Coddington, but Domesday Book does not say to which borough the burgess belonged.

Hereford

The city (*civitas*) of Hereford is described in three entries. The first, on fo. 179, opens the survey of the county by enumerating the holdings of the king in the city; the other two give an account of the holdings of the bishop (181 b, 182 b). Taken together, the three entries provide a very unsatisfactory body of information. We are not given a complete account of the population of the city, either for 1066 or 1086, or for 1079 'when Robert came into the bishopric'. The fragmentary nature of the evidence can be seen from the following table:

	1066	1079	1086
Homines	103	—	—
Burgenses	27	—	—
Masurae	98	60	—
Contributory *burgenses*	—	—	10 + ?
French burgesses	—	—	?
Moneyers	7	—	—
Smiths	6	—	—
Villeins	—	—	3
Bordars	—	—	6
Serfs	—	—	2
Buri	—	—	5
Bondwomen (*ancillae*)	—	—	3

From these fragments, it is clearly impossible to form any estimate of the population in 1086.[1]

The rural manors with burgesses in Hereford were as follows:

Much Marcle (179b): At Hereford are 4 burgesses rendering to this manor 18 plough-shares (*socos carrucis*).
Burghill (186): In Hereford 5 burgesses render to this manor 52*d*.
Prior's Frome (185): one burgess in Hereford renders 12*d*.

J. H. Round[2] used the presence of contributory burgesses at Hereford to assist him in rejecting Ballard's arguments in favour of the 'garrison theory' of the borough.[3] According to this theory, first advanced by Maitland,[4] these houses were occupied by burgesses charged with the

---

[1] Not counted are the incidental references to burgesses in the entry for Marden (179b) and the unnamed entry next to that for Maund (184).
[2] *V.C.H. Herefordshire*, I, p. 296.
[3] A. Ballard, *The Domesday Boroughs* (Oxford, 1904), pp. 11–35, and *The Domesday Inquest* (London, 1906), pp. 174–8.
[4] F. W. Maitland, *Domesday Book and Beyond*, pp. 180 *et seq.*

duty of maintaining the walls on behalf of the rural magnates; but Round rejected the three cases recorded for Hereford as the *reductio ad absurdum* of this theory, in view of the great military importance of the city.

There is little indication of the occupations of the burgesses, but the presence of seven moneyers in 1066 seems to suggest a thriving commercial life. The record of French burgesses in 1086 suggests also that there must have been sufficient trade after the Conquest to make their settlement worth while. None of the three Hereford entries mention a market; but the account of Eaton Bishop says that it had been exchanged for 'land in which the market now is, and for 3 hides at Lydney' (181 b). This market was presumably in Hereford itself.

The proximity of the Forest of Dean may have been reflected in the custom which required each smith to make 120 horseshoes from the king's iron. Another customary payment was that of 10*d.* due from every man whose wife brewed ale 'within or without the city'. Mention has already been made, of the churches of the city.[1] The Domesday account of the agricultural activity connected with the city is equally meagre. The bishop held 10 ploughs in demesne, on his lands 'about the gates', and there were 5½ other ploughs; there is no mention of woodland, meadow, or mills.

Clifford

The main account of the borough (*burgus*) of Clifford appears on fo. 183, but there is a subsidiary entry on fo. 184. The settlement was situated in the Wye valley guarding the way from the Welsh highlands into the plain of Herefordshire, and here Earl William had built a castle on some wasted land. Its remote situation is indicated by the fact that, though it was *de regno Angliae*, it did not belong to any hundred (*non subjacet alicui hundret*). The arable amounted to 26 plough-lands, and there were 13 teams at work there. The population comprised 16 burgesses, 13 bordars, 6 serfs, 4 oxmen, 5 Welshmen and 4 bondwomen. There was also a mill rendering 3 muids of grain. No mention is made of wood, meadow or fisheries, nor are we told of any commerce by which the burgesses lived. All that we can be certain about is that here was a small and very agricultural borough. Four carucates are said to be waste *in castellaria.*

[1] See p. 101 above.

Ewias Harold

Ewias Harold (184, 185, 186), at the southern extremity of the Golden Valley, and astride an easy route from Hereford to the Usk, seems to have been primarily a military stronghold. It is not specifically called a borough, nor are there burgesses recorded there; but we are told of 2 messuages in the castle (*In castello habet ii masuras*) which may have represented the beginnings of a borough. Three tenants-in-chief held there, and seven knights appear to be recorded. The rural population consisted of 31 bordars, 3 serfs, 4 oxmen, 13 Welshmen and 1 'man'. There were 16 plough-teams and 22 acres of meadow, and 3 fisheries. The Welshmen rendered 11 sesters of honey between them. Three churches are recorded, but only one priest. Four carucates are said to be waste *in castellaria*.

Wigmore

Wigmore Castle appears to have been built after the Conquest on 2 hides of waste land, although the relevant entries do not agree about the details:

Fo. 179b. Of the manor of Kingsland, 'Ralph de Mortemer holds one member *Mereston* for 2 hides'.

Fo. 180. 'Ralph de Mortemer holds Wigmore. Elward held [it]. There half a hide. Wigmore Castle lies there' (*sedet in ea*).

Fo. 183b. 'Ralph de Mortemer holds Wigmore Castle. Earl William erected (*fecit*) it on the wasted land which is called *Merestun*, which Gunuert held T.R.E. There 2 hides pay geld. On the demesne Ralf has 2 ploughs and 4 serfs. The borough (*Burgus*) which is there renders £7.'

The render in the third of these entries suggests a prosperous community, but of its size and activity we are told nothing.

MISCELLANEOUS INFORMATION

Iron

In 1086, there were twenty-five smiths distributed between twenty-three settlements in the area of the modern county. There is, however, nothing to suggest that they were connected with the Forest of Dean iron industry, and for the most part they were scattered through the more densely populated areas. Bodenham (184 and 186b) and Broadward, in

Leominster (180 and 185b), each had two smiths; otherwise there was only one in each place. There had been six smiths in Hereford in 1066, but the incomplete record for 1086 does not mention them. A hide at *Turlestane* rendered '50 masses of iron' (*l massas ferri*) T.R.E., but it had since been incorporated in the forest (179b).[1]

Honey

Honey renders are mentioned in connection with at least fifteen places, mainly in Archenfield (Fig. 24), e.g. 15 sesters of honey were rendered at Kilpeck (181). In addition to the separate mention of individual places, we are told that 96 king's men in Archenfield rendered 41 sesters of honey (181), and we hear of penalties for those who tried to conceal their honey (179). At seven of the fifteen places, the renders were made by Welshmen.

Sheep

Although there are only eight references to sheep, they are of particular interest in view of the importance of Herefordshire sheep in the later Middle Ages. In the central plain, 6 coliberts at Eardisland rendered 3 sesters of wheat and barley and $2\frac{1}{2}$ sheep with their lambs (179b), but the other entries all refer to Archenfield and the Ross region. The account of Archenfield says that a Welshman paid 2s. fine for the theft of a sheep (179); and later (181) we are told that the king's men there paid '20s. for the sheep which they were wont to give'. For Linton we hear of customary dues of honey and sheep (179b) and again of a render of 6 sheep with their lambs (179b). Five Welshmen at Kings Caple rendered 5 sheep with their lambs (181). Land which, in 1066, had rendered 6 sheep with their lambs to Cleeve (Ross) had been incorporated in the forest by 1086 (179b). Finally, at Elvastone (181) there was a render of 3 sheep. It will thus be seen that, apart from Eardisland, all these places were in the part of the county later famed for Ryelands sheep. They were also in an area that was still predominantly Welsh, and some of these presentations of sheep may have represented survivals of the old Celtic *gwestwa* or food-rent.[2]

---

[1] For the location of *Turlestane*, see p. 87 n.
[2] See Vinogradoff, *English Society in the Eleventh Century* (Oxford, 1908), p. 386.

Other references

The only reference to a market is the exchange of lands (181 b) to provide a site for a market presumed to have been at Hereford.[1] There were castles at Clifford (183, 184), Ewias Harold (184, 185, 186), Richards Castle (*Auretone*, 185) and Wigmore (179 b, 180, 183 b). There are also two very unusual references to *domus defensabilis* at Ailey (187) and at Eardisley (184 b).

Pasture (*pastura*) is recorded for Kingsland (179 b) and Monkland (183).

REGIONAL SUMMARY

From the surrounding hills and upland, Herefordshire may be seen to consist of a central, undulating lowland crossed by two belts of hills, and rising outwards towards a broken rim through an irregular, more sharply accidented zone. The county thus falls naturally into a central district, bordered by a number of regions which are differentiated from each other by their varying landforms and soils. In Domesday times, the western borders of the county were also profoundly affected by the influence of the Welsh, which added a cultural element to the physical factors that distinguished the various parts of the county. Altogether, nine regions have been recognised, predominantly of a physical nature in the east, but much modified in the south and west by political and cultural considerations (Fig. 36).

(1) *The Central Plain*

The central part of Herefordshire is a gently rolling plain, between 150 and 400 ft. above sea-level, except where it is crossed by the line of Wormsley and Dinmore Hills, which rise to over 700 ft. The northern part is drained by the Lugg, which then flows southwards to its confluence with the Wye, just below Hereford. The soils are mostly heavy loams, derived from the Old Red Sandstone, and varied locally by drift. The region was already well-settled in the eleventh century; the villages were closely aligned along the river valleys, becoming more widely spaced towards the hills. The densities of plough-teams and population indicate the general prosperity of the region, which contrasts sharply with the poorer regions to the west. The extent of the Wye, Lugg and Arrow

---

[1] See p. 104 above.

flood-plains suggests that there might have been a considerable amount of meadowland; and more meadow is, in fact, recorded here than for any of the other regions, although, even so, it is found at only one-quarter of the villages. Mills were fairly widely distributed throughout the area. Although the original vegetation of the region was presumably mostly damp oakwood, comparatively little wood was recorded in 1086. There

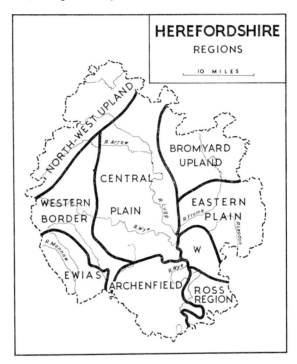

Fig. 36. Herefordshire: Regional subdivisions.
W indicates the Woolhope Region.

was some adjacent to Wormsley and Dinmore Hills, and a fair amount of wood and forest on the hilly ground south of the Wye; otherwise there were only a few scattered places with wood.

(2) The Western Border

The central plain extends westwards up the Wye valley, and in 1086 this extension was a region of small settlements lying across the main route from Hereford into Wales. It is bounded in the north by the Silurian upland, and in the south it reaches into the Golden Valley. The soils are mostly loams developed either on stony drift brought down the Wye valley, or else on the Old Red Sandstone. Some places were wholly wasted in 1086, and the densities of plough-teams and recorded population were low—one plough-team and three persons per square mile, respectively. No meadow or fisheries were recorded, and there was but little wood. Two mills were located on the Wye, and two others in the Golden Valley. The region is thus one which is characterised as much by the way in which it reflects the conditions of eleventh-century border life as by the nature of its physical geography.

(3) The North-west Upland

The Silurian 'edges' of Shropshire continue their N.E.–S.W. trend across northern Herefordshire, and exceed 1,000 ft. in places. The soils are closely related to the rocks on which they are developed, and are mostly stony loams, with some intractable clays. Drainage is by the Teme and the upper Lugg, the Wigmore basin being an area of indeterminate drainage lying between the two rivers. The settlements were sited around the margins of the Wigmore basin, and on the south-facing dip-slopes of the upland, but owing to Welsh raids there was a sharp contrast in Domesday times between (1) the devastated western part, with under one man and less than half a plough-team to the square mile, and (2) the relatively prosperous northern part, centred on the Wigmore basin, with six people and two plough-teams per square mile. The meadow entries seem deficient, for there is only one entry of five acres for the whole region. There were a few mills and some fisheries, mainly along the Lugg and Teme. Small amounts of woodland were fairly general, except in the Wigmore basin.

(4) The Bromyard Upland

In the north-east of the county, around Bromyard, much of the land is over 500 ft., and some reaches 800 ft. This upland of Old Red Sandstone is sharply dissected by narrow valleys mostly draining to the Frome. The soils tend to be light loams, shallow and poor on the higher land, but deeper in the valleys. Some large Domesday vills were situated along the upper Frome, but settlement generally was rather thin; the population averaged only five per square mile, and the plough-teams three. Only twenty-four acres of meadow were recorded for the region, and wood is only mentioned once, but there were a number of mills sited along the rivers.

(5) *The Eastern Plain*

The most prosperous part of the county was the eastward extension of the central plain up the lower Frome valley, and across the low watershed into the upper Leadon basin. The valleys have a heavy loam soil derived from the red marl, while more varied soils are developed on the older rocks of the Malvern foothills. This region was closely settled in the eleventh century, with about seven people per square mile, and five teams. The amount of meadow was second only to that of the Central Plain, although the total amount was still very small. There were a number of mills, and there was also a moderate amount of woodland.

(6) *The Woolhope Region*

South-west of the Eastern Plain is the Woolhope Silurian inlier, which has thin limestone soils on the higher land, and heavy clays in the valleys. Although small, it forms a distinct region, and in 1086 its economy appears to have been dominated by the large manor of Woolhope. The density of both population and plough-teams was high for the county; about nine and six per square mile, respectively. It also had its fair share of meadow, mills, fisheries and woodland.

(7) *The Ross Region*

In the south-east of the county, around Ross, was that part of the Ryelands which lay outside Archenfield. Here was a relatively prosperous district, with an average of six people and four plough-teams per square mile. Almost all the woodland mentioned for this region had been absorbed into the forest, and there was little meadow; but there were several fisheries and mills.

(8) *Archenfield*

Southern Herefordshire, or the Ryelands, forms a distinctive region with light, sandy soils. It extends from the Monnow in the west, to the Forest of Dean in the south and east. It is drained to the south-east by several small tributaries of the Wye, and rises from just under 100 ft. on the Wye to over 600 ft. overlooking the Monnow, and to 1,200 ft. on Garway Hill. The greater part of the area, lying west of the Wye, was in Domesday times occupied by the semi-autonomous Welsh district of Archenfield. A Domesday entry (181) records that 'King Grifin and Blein laid waste this land T.R.E.' (i.e. in 1055). We cannot be certain of its condition in 1086, but it seems to have been a poor area with under three people and about 1½ teams per square mile. Honey renders were characteristic of the region (Fig. 24).

(9) *The Upland of Ewias*

The upland of Ewias lies between the Golden Valley and the Black Mountains, and is deeply incised by the valleys of the upper Monnow, the Escley Brook and Dulas Brook. These valleys are cut down through the hard Upper Old Red Sandstone of the hills to the underlying red marl, and there are also patches of drift. Ewias is an ancient Welsh territorial name, and Domesday Book records so little about it that it must still have been largely in Welsh hands in the eleventh century. The Norman castellany of Ewias was dominated by the castle of Ewias Harold, which was evidently an important stronghold;[1] while beyond, was Ewias Lacy (now Longtown), which was said to be within the bounds of Ewias although it did not belong to the jurisdiction of the castle or to the hundred. This was apparently the farthest point of Anglo-Norman penetration into Ewias, and there Roger de Laci received 15 sesters of honey, and 15 swine 'when the men were there', and pleas over them (184). None of the holdings in Ewias was hidated.

APPENDIX: THE WELSH BORDERLAND

Although Domesday Herefordshire included parts of the present counties of Monmouth and Radnor, the Welshry thus described is not comparable in extent with that described in the Cheshire or Gloucestershire folios. In fact, the essentially Welsh districts of Ewias and Archenfield have long been an integral part of Herefordshire and, as such, have already been discussed. There remains, however, a certain amount of information about Monmouthshire and Radnorshire which has still to be mentioned.

The Gloucestershire folios also include a number of places which are now in Monmouthshire, but there are also two entries relating to

---

[1] See p. 105 above.

Monmouthshire among the Herefordshire folios. Under the heading *In Wirecestre Scire* is an entry for Monmouth Castle (180b) which gives a concise account of that place:

In MONMOUTH CASTLE the King has on the demesne 4 ploughs. William son of Baderon has the custody of them (*custodit eas*). What the King has in this castle (*castello*) is worth 100*s*. William has 8 ploughs on the demesne there, and there can be more (*et plures possunt esse*). There are Welshmen having 24 ploughs. They pay 33 sesters of honey and 2*s*. There [are] 15 serfs and bondwomen together and 3 mills worth (*de*) 20*s*. The knights of this William have 7 ploughs. What William holds is worth £30.

The church of this castle and all the tithe with 2 carucates of land St Florence of Saumur holds.

The Domesday record of Caerleon does not appear to be complete. The revenue (*redditionem de Carleion*) appears in the Gloucestershire folios (162) as appurtenant to Chepstow Castle, but no details are given of Caerleon Castle itself. The Herefordshire description (185 b) refers only to the 8 carucates of land in the jurisdiction of Caerleon Castle, which were not assigned to any hundred. The full entry is given below:

William de Scohies holds 8 carucates of land in the jurisdiction of Caerleon Castle, and Turstin holds of him. There he has on the demesne 1 plough and 3 Welshmen living under Welsh law (*lege Walensi viventes*) with 3 ploughs and 2 bordars with half a plough. They render 4 sesters of honey. There, 2 serfs and 1 bondwoman.

This land was waste T.R.E. and when William received it. Now it is worth 40*s*.

At neither of these places do we find any mention of hides; both reckoned in carucates from which we may conclude that they were relatively recent English acquisitions.[1]

When, however, we come to that part of Radnorshire which was surveyed under the Herefordshire folios, we find evidence, not of Welsh, but of English settlement, assessed in hides. J. H. Round noted that the diocese of Hereford drives a wedge into that of St David's at this point, and that this wedge is crossed by Offa's Dyke; he concluded, therefore, that this might have been English soil perhaps even from Offa's day.[1] Eight places now in Radnorshire are thus recorded in the Herefordshire folios. They were waste in 1086, and apparently had been so earlier. Their names and assessments were: Harpton, 3 hides; Clatterbrune,

[1] *V.C.H. Herefordshire*, I, p. 264.

2 hides; Querentune, 1 hide; Discoed, 3 hides; Cascob, ½ hide (186b); Burlingjobb, 2 hides (181); Pilleth on Lugg, 2 hides (183b); and Old Radnor, 15 hides (181). The first five of these places are described in a composite entry that also includes six places within modern Hereford-shire; the remarks applicable to all eleven places are as follows: 'land for 36 ploughs, but it was and is waste. It never paid geld. It lies on the marches of Wales.'[1] We are also told that 'on these waste lands there have grown up woods in which Osbern hunts, and thence he has whatever he can take. Nothing else.'[2] Cascob, now in Radnorshire, is described partly under Herefordshire (186b) and partly under Shropshire (260).

BIBLIOGRAPHICAL NOTE

(1) Remarkably little work appears to have been done on the Domesday folios for Herefordshire. There is a list of Domesday names of places and of landed proprietors given in John Duncumb's *Collections towards the History and Antiquities of the county of Hereford* (Hereford, 1804), 1, pp. 59–65.

(2) *The Victoria County History, Herefordshire* (London, 1908), 1, contains a valuable introduction to the Herefordshire Domesday by J. H. Round, (pp. 263–307), together with the only available translation (309–45).

(3) The unique transcript of the Herefordshire folios (Balliol MS. 350) probably made between 1160 and 1170 is not only of great interest in itself, but throws much light on the identification of many Domesday place-names: V. H. Galbraith and J. Tait, *Herefordshire Domesday, circa 1160–1170* (Pipe Roll Society, London, 1950).

(4) The *Transactions of the Woolhope Naturalists' Field Club* (Hereford, 1904) contain a paper (pp. 318–23) on 'The Herefordshire Domesday' by A. T. Bannister.

See also Lynn H. Nelson, *The Normans in South Wales, 1070–1171* (University of Texas, Austin, 1966).

(5) The following by Lord Rennell of Rodd deal with the north-west of the county:

'Aids to the Domesday Geography of North-west Hereford', *Geog. Journ.* (1954), cxx, pp. 458–467.

'The Domesday manors in the hundreds of Hezetre and Elsedune in Here-fordshire', being chapter xi (pp. 130–58) of *Centenary of Woolhope Club* (Hereford, 1954).

---

[1] The holding at Pilleth and another of 2 hides at Harpton, together with two places in modern Herefordshire, are described in another composite entry as *wastae in marcha de Wales* (183b). [2] See p. 84 above.

Valley on the March (London, 1958).

'The Land of Lene', being chapter 12 (pp. 303–26) of I. Ll. Foster and A. Alcock (eds), *Culture and Environment* (London, 1963).

(6) In the absence of an English Place-Name Society volume, useful reference may be made to A. T. Bannister, *The Place-Names of Herefordshire* (Hereford, 1916).

CHAPTER III

SHROPSHIRE

BY V. A. SAUNDERS, B.A.

The Domesday folios for Shropshire, like those for Herefordshire, are of especial interest because they throw light on the unsettled condition of the Welsh borderland in the years immediately following the Norman Conquest. The evidence of the activities of the Welsh gives some colour to the Shropshire folios, and the enumeration of Welsh districts dependent on Shropshire in 1086 illustrates how successfully the traditional enemies of the county had been defeated.

The extent of the Domesday county of Shropshire was very different from that of its modern counterpart, in terms of which this study is written. On the northern side of the county, Tittenley, which was surveyed in the Cheshire folios, is now in Shropshire. In the east, Cheswardine, Chipnall, Sheriff Hales and Brockton Grange have been transferred from, and Tyrley to, Staffordshire. In the south-east, a substantial area was surveyed in the folios for Staffordshire and Warwickshire.[1] On the other hand, there have been losses in the south-west and west. In the south-west, the hundred of Wigmore in modern Herefordshire occupies what was once the greater part of the Shropshire hundred of Lenteurde, and, further north, on the western border, much of Witentreu hundred was placed in Montgomeryshire in 1536. Furthermore, the whole Welsh border must have been in an extremely fluid condition in the eleventh century, and we cannot be certain that the western half of Purslow hundred (Clun Forest) was part of the Domesday hundred of Rinlau.

During the Norman period, perhaps in the reign of Henry I, the hundreds as they existed in 1086 were drastically re-organised, and there have also been later changes, so that the present-day hundreds coincide in neither name nor area with the Domesday hundreds (Fig. 37). An important feature of the Shropshire folios is the care with which the Domesday scribes rubricated the hundreds, but even so, the hundred rubrics do sometimes appear to be misplaced. Sleap Magna (257b), for

[1] Alveley, Claverley, Kingsnordley and Worfield from the Staffordshire folios. Quatt, Romsley, Rudge and Shipley from the Warwickshire folios.

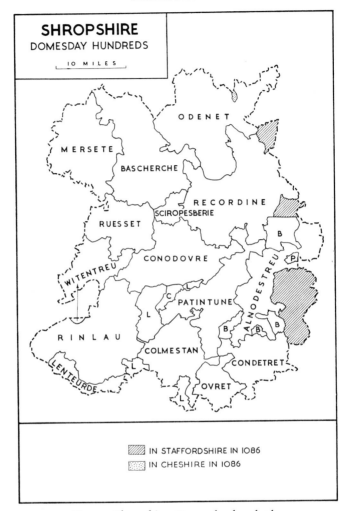

Fig. 37. Shropshire: Domesday hundreds.
Detached portions of hundreds are indicated as follows: B, Bascherche;
C, Colmestan; L, Lenteurde; P, Patintune.

example, was entered under Colmestan, yet it seems to have been in
Bascherche; and for several places in Odenet the rubric was omitted.
Perhaps the best example of the result of what may have been lack of local
knowledge on the part of the scribes is the entry in which they appear to
confuse a place-name with a landholder's name: 'The same Ralf holds
Mawley for (*de*) 1 hide and Lel for 1 virgate and Fech for 1 virgate' (260).

There was no place Fech, but there was a man called Fech who held one virgate at Baveney. Another slip can be detected in the mistaken entry for the lands of St Mary of Bromfield under those of St Mary of Shrewsbury (252b).

Other difficulties which arise when the Shropshire material is analysed are the presence of unidentified place-names, and the apparently incomplete character of some entries. There are also a number of duplicate entries relating to the same holding—for Sheinton (256b and 260b), for Overs (254 and 259b) and for Brockton in Longford (257 and 259b). In some counties such duplicate entries disagree one with another. Fortunately the problem does not arise in Shropshire, although for both Sheinton and Overs one of the entries is more detailed than the other. Composite entries, which present problems for some counties, are not important in the Shropshire folios.

SETTLEMENTS AND THEIR DISTRIBUTION

The total number of separate places mentioned in the Domesday Book for the area now included in the modern county of Shropshire seems to be 437, including the two boroughs of Shrewsbury and Quatford. This figure, however, may not be quite accurate because there are some instances of two or more adjoining villages bearing the same surname today, and it is not always clear whether more than one unit existed in the eleventh century. There is no indication that, say, the Market Drayton and Little Drayton of today existed as separate villages; the Domesday information about them is entered under only one name (*Draitune*) though there may well have been separate settlements in the eleventh century. The distinction between the two names appears later in time; the name Little Drayton, so far as the evidence goes, dates from 1327. For some counties the Domesday text occasionally differentiates between the related units of such groups by designating one unit as *alia* or *parva*, but none of the Shropshire groups is distinguished in this way.

The total of 437 includes about half a dozen places for which very little information is given; the record may be incomplete, or the details may have been included with those of neighbouring villages. Thus we are told nothing of the population of Poston beyond the fact that one man rendered a bundle of box (*fascem buxi*) on Palm Sunday (252b). The entry for *Sudtelch* in Colmestan, on the other hand, tells us that it answered

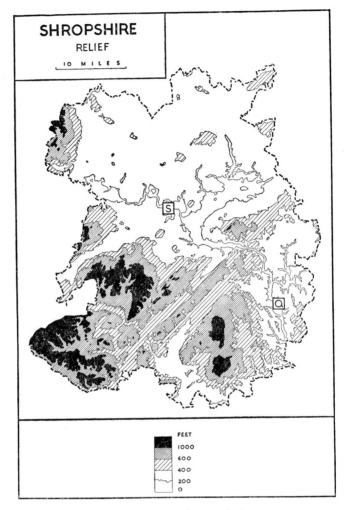

Fig. 38. Shropshire: Relief.
Domesday boroughs are indicated by initials: Q, Quatford; S, Shrewsbury.

for half a hide, that there was land for 2 ploughs and that its value had
decreased from 5*s.* to 2*s.*; but we are told nothing about its other resources
or about its population (257b). Or again, at an unnamed manor of one
virgate (Eye?), described on fo. 256b, there was land for one plough, and
one villein and 2 serfs lived there; and yet, apparently, they had nothing
(*nil habent*) and we are given no indication of how they supported them-

selves. In rather a different category are the waste vills, for which little or nothing was entered.[1] The total of 437 places also includes three which are described as berewicks—Drayton (257), Milson (260) and Hopton (256b).[2] But it is a striking fact that in addition to these three, a large number of unnamed berewicks are also mentioned. We are told that there were 57 on the twelve manors formerly held by King Edward (254). Presumably these refer to the groups separately entered, e.g. 10 berewicks for Condover (253) and 8½ for Whittington (253b) but the total of these amounts only to 53, a discrepancy which could be due to error or to omission. Another 82 unnamed berewicks are recorded for other manors e.g. Ford with 14 (253b), Worthen with 13 (255b) and other places with fewer. Was there then a total of, say, 135 localities in addition to the 437 named? Or were all or many of the unnamed berewicks in places named elsewhere in the text?

Not all the Domesday names appear on the present-day map of Shropshire villages. Some are represented by hamlets, by individual houses, or by the names of topographical features. Thus *Sudtone* is now the two hamlets of Sutton in the parish of Diddlebury; *Netelie* survives as Netley Hall in Stapleton, while *Melicope* is represented by Millichope Park on Wenlock Edge. Some Domesday names have disappeared entirely. The Domesday *Cesdille* or *Cestulle* does not appear on the modern map, although its site is known to lie in the parish of Sutton upon Tern. Another example is *Slacheberie* in the parish of Ellesmere Rural, of which nothing is known after the thirteenth century.[3] Finally, there are a number of names which remain unidentified, and which can be assigned to no site.[4] Whether they will yet be located or whether the places they represent have completely disappeared, leaving no record or trace behind, we cannot say. In any case, some of these places were either waste or only very small settlements.

On the other hand, some villages on the modern map are not mentioned in the Domesday Book. Their names do not appear until the twelfth and

[1] See p. 144 below.
[2] A fourth berewick, the unidentified *Calvestone* is said to be 'in Worcestershire' (253)—see *V.C.H. Shropshire* (London, 1908), I, p. 286n.
[3] *V.C.H. Shropshire*, I, p. 341n.
[4] The complete list is: *Aitone* (254b), *Benehale* (254), *Caurtune* (258), *Cheneltone* (259), *Chinbaldescote* (258), *Cleu* (260b), *Hibrihteselle* (258b), *Humet* (260), *Newetone* (259b), *Petelie* (252b), *Staurecote* (254), *Sudtelch* (257b), *Tibetune* (254b), *Tumbelawe* (260), *Tunestan* (259b), *Udeford* (257b), *Wlferesforde* (259b).

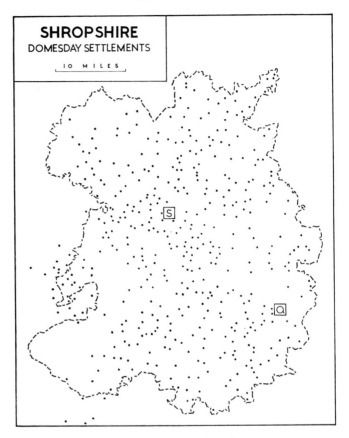

Fig. 39. Shropshire: Domesday place-names.

Domesday boroughs are indicated by initials: Q, Quatford; S, Shrewsbury. The 18 settlements to the west of the modern county are described in the folios for the Domesday county. So are the 5 settlements to the south, but one of these also appears in the Herefordshire folios (Cascob, 186b). The settlements and their resources have not been included in our totals for the county.

thirteenth centuries, and, presumably, if they existed in 1086, they are accounted for under the statistics of neighbouring settlements. Thus, as far as record goes, Coalbrookdale was first mentioned in 1250, and Cruckton in 1272, while Ironbridge is a very late name.[1] The town of Ludlow is not mentioned in the Domesday Book. It was once believed

---

[1] The dates in this paragraph are from E. Ekwall, *The Concise Oxford Dictionary of English Place-Names*, 3rd ed. (Oxford, 1947).

that the entry under *Lude* (186b) in the Herefordshire folios referred to the site of the present-day town, but this has now been disproved.[1]

Despite the complications of extinct and additional villages, the fact remains that the distribution of Domesday names shows a general similarity to that of the present-day settlements (Fig. 39). The vills were more or less evenly scattered over much of the country. Settlements were most numerous in the fertile Severn Valley from the Welsh border to Ironbridge Gorge. In the southern upland region where Boulder Clay fills the valley bottoms, the villages are strung out along the upper limits of the glacial clay. This is particularly noticeable on the lower slopes of the Longmynd, and in Ape Dale, Hope Dale and Corve Dale below the limestone escarpments, e.g. Wenlock Edge. Lines of villages also encircle the infertile sandstone and volcanic rocks of the Brown and Titterstone Clee massifs—but the hills themselves stand out as empty areas. In the northern plain, the Triassic sandstones are masked by an undulating layer of Boulder Clay and glacial sands, with which are associated numerous heaths and marshes. Prees Heath is an example of such an infertile area, where settlements were sparsely distributed. Other districts with few vills were Clun Forest, the Wrekin, and the Shelve and Stiperstones mountains in the west.

THE DISTRIBUTION OF PROSPERITY AND POPULATION

Some idea of the information in the Domesday folios for Shropshire, and of the form in which it was presented, may be obtained from the account of the village of Stoke upon Tern in the hundred of Recordine (256 b). The village was held entirely by Earl Roger, and so it is described in a single entry:

> The same Roger holds Stoke. Edmund held [it]. There 7 hides. In demesne are 3 ploughs and 6 serfs and 3 bondwomen. A church and a priest and 11 villeins and 3 radmen and one Frenchman with 10 ploughs among them (*inter omnes*). And still 5 more ploughs might be there (*et adhuc v carucae plus possent esse*). There a mill of (*de*) 12s., and a third part of one league of wood. T.R.E. it was worth £6, and afterwards it was waste. Now it is worth £7.

This entry does not include all the kinds of information that appear elsewhere in the folios for the county. It does not mention, for example,

---

[1] See *V.C.H. Herefordshire*, I, p. 305, and V. H. Galbraith and J. Tait, *Herefordshire Domesday, circa 1160–1170* (Pipe Roll Society, London, 1950), pp. 66 and 120.

the categories of population known as bordars, oxmen and freemen. There is also no mention of a hay or a hawk's eyry. Moreover, as we shall see, the reference to the possibility of further plough-teams is not characteristic of all entries. But although not comprehensive, it is a fairly representative and straightforward entry, and it does set out the recurring standard items that are found for most villages. These are five in number: (1) hides; (2) plough-lands; (3) plough-teams; (4) population; (5) values. The bearing of these five items of information upon regional variations in prosperity must now be considered.

(1) *Hides*

The Shropshire assessment is stated in terms of hides and virgates, and, very rarely, of acres. The normal formula runs simply: 'there n hides paying geld' (*ibi n hidae geldantes*), or merely 'there n hides' (*ibi n hidae*). Evidence for the equation of the hide with four virgates can be found in the entry for Baveney (257), where there was half a hide, and 'one virgate of these two lay in Cleobury Mortimer' (*Ibi dimidia hida. Una virgata ex his duabus jacebat in Claiberie*). Acres are mentioned only twice—at Stanton Lacy (260b),[1] and at Brockton in South Maddock (257b). At the latter a virgate of land and 8 acres are recorded (*Ibi i virgata terrae et viii acrae*); the use of the word *terra* in connection with the assessment is exceptional. Other unusual entries are those for Clunbury (258) and Neen Solars (260) which had never paid geld nor been assessed in hides.

A glance through the Shropshire folios reveals a number of examples of the five-hide unit. Twenty places were assessed at exactly 5 hides each, three at 10, four at 20, one at 30, and Shrewsbury itself at 100 hides; there were also eleven places assessed at 2½ hides. It is possible that some villages were grouped in blocks, for the purpose of assessment, as in Leicestershire.[2] The method of apportioning hides in fractions or multiples of five may be observed in the following groups taken from Colmestan hundred:[3]

Stanton Lacy (260b)	20½ hides	
Clee Stanton (252b)	2 hides	25 hides
Aldon (260b)	2½ hides	

---

[1] The Stanton Lacy entry also mentions a ferling.
[2] See p. 325 below.
[3] *V.C.H. Shropshire*, I, pp. 284–5.

Sutton (257b, 258b)	$2\frac{1}{2}$ hides	
Ledwych (257b)	2 hides	5 hides
Poston (252b, 258b)	$\frac{1}{2}$ hide	

The artificiality of the hidage can also be seen in the round totals (e.g. 100 hides) assigned to many of the Domesday hundreds.

Taking the county as a whole the assessment was low. There were only about 1,400 hides as compared with over 3,000 plough-lands. One reason for the disparity between hides and arable capacity is the fact that a reduction in assessment had taken place. In the eleventh-century document known as the County Hidage,[1] Shropshire answered for 2,400 hides, and beneficial hidation may have taken place before the Conquest. Thus on fo. 252b we are informed that 4 of the 20 hides at Much Wenlock were quit of geld in the time of King Canute (*Ex his iiii erant quietae a geldo Tempore Regis Chnut*). The church of St Milburga held some half-dozen places in Patintune hundred, including Much Wenlock, and beneficial hidation may be inferred at several of them. At Ticklerton there were 7 hides paying geld and '3 other hides quit of geld' (252b); and at Shipton, there was 'half a hide not paying geld and 3 other hides paying geld' (252b); At neither place is there any indication that the reduction took place before the Conquest. On the other hand, there is some evidence that the general reduction took place between 1070 and 1086. But whatever the date of the reduction, it is clear, as Tait says, that 'Shropshire had certainly suffered ravages which could have justified some abatement'.[2] Beneficial hidation may to some extent reflect the wasting of the county, but the operation of the five-hide unit introduces an element of artificiality which renders the assessment quite unreliable as an index of prosperity in 1086.

The total assessment (including non-gelding hides) amounted to $1402\frac{1}{2}$ hides and 8 acres, but it must be remembered that this refers to the area included in the modern county.[3] Maitland estimated the number of hides in the Domesday county at 1,245.[4] Tait's figure of 1,438 likewise refers to the Domesday county.[5] Eyton[6] did not produce a county total,

---

[1] F. W. Maitland, *Domesday Book and Beyond* (Cambridge, 1897), p. 456.

[2] *V.C.H. Shropshire*, I, p. 281. The problem of the waste is discussed on pp. 144–6 below.

[3] This figure includes the 100 hides at which the city was assessed in the time of King Edward (252).

[4] F. W. Maitland, *op. cit.* p. 400.   [5] *V.C.H. Shropshire*, I, p. 281.

[6] R. W. Eyton, *Antiquities of Shropshire*, 12 vols. (London, 1854).

but his hundred totals agree with those of Tait. In any case, a count of
Domesday hides can rarely be definitive. All the present estimate can do
is to indicate the order of magnitude involved.

(2) *Plough-lands*

Information about plough-lands seems to be given in two quite different
ways.[1] In the first half of the Shropshire folios (252–7) we can only infer
the number of plough-lands on a holding from other information that is
given. With very few exceptions, entry after entry states the number of
plough-teams at work, and then goes on to say that there could be others
(*et adhuc n carucae plus possent esse*). The exact form of this additional
information varies. Thus at Much Wenlock (252b) there were 'in
demesne $9\frac{1}{2}$ ploughs and 9 villeins and 3 radmen and 46 bordars. Among
them all they have 17 ploughs and another 17 could be there' (*et aliae
xvii ibi possent esse*). From this it seems that there were $43\frac{1}{2}$ plough-lands
at Much Wenlock, and that only $26\frac{1}{2}$ of these were cultivated. Occasionally in these earlier folios the number of plough-teams alone is recorded,
and presumably on such holdings the arable was fully cultivated, and there
were no potential teams to be added.

In the second half of the Shropshire folios (257–60) the plough-land
entry consists of the common formula: 'land for *n* ploughs' (*terra est n
carucis*). At Wolverley (257), for example, there was land for 4 ploughs.
In a few entries no plough-lands are indicated—but there is little information of any kind about these places, and the omissions are not sufficiently
numerous to affect the general picture. Even where the land was waste,
plough-lands are usually recorded. Both methods of recording plough-
lands are employed in every hundred of the county, so that we cannot
assume that the jurors of some hundreds employed one formula, and that
those of other hundreds employed the other. A total of $2,664\frac{3}{8}$ plough-
lands can be inferred for the modern county. If, however, we assume that
plough-lands equalled teams in those entries where a figure for the former
is not given, as many as $468\frac{1}{2}$ teams must be added, and the total becomes
$3,132\frac{7}{8}$ plough-lands. Tait's figure of 3,173 plough-lands, refers to the
Domesday county.[2] Clearly such totals involve many assumptions and

---

[1] The 1066 teams at Clunbury, Onibury and Ovret hundred have also been counted
as plough-lands (see p. 126 below).

[2] *V.C.H. Shropshire*, I, p. 281.

Maitland may have been wise in not giving a total.[1] The varying relations of hides, plough-lands and teams is shown below:

Some Representative Shropshire Villages

	Hides	Plough-lands	Plough-teams
Oxenbold (258b)	1	4	2
Whitton (255b)	$1\frac{1}{2}$	4	2
Merrington (258)[2]	2	5	3
Wem (257)	4	8	2
Upton Magna (254b)	5	23	15
Clungunford (255, 258b)	8	23	14
Worthen (255b)	$14\frac{1}{2}$	45	21
Lydham (253b)	15	30	14

The general impression is that nearly one-half of the Shropshire plough-lands were not cultivated, but there were many variations. Kinlet, for example, seems to have been tilled to capacity (260), whereas at Myddle (255) there was just one team on 20 plough-lands. More striking still was the discrepancy on the manor of Lydbury North (252); 28 teams were entered but there seem to have been 120 plough-lands there (*In hoc manerio possent esse quater xx et xii carucae plusquam sunt*). Over the county as a whole there was a deficiency of teams in as many as 88% of the entries which mention teams that could be added or which record both plough-lands and teams.

Five entries record more teams than plough-lands, but whether this was due to overstocking or was the result of clerical error we cannot say.[3] The details are as follows:

	Plough-lands	Plough-teams
Wigwig (258)	2	$2\frac{1}{2}$
Pulley (260b)	5	7
Stanton Lacy (260b)	50	$50\frac{1}{2}$
Quatt (239)	12	14
Worfield (248b)	30	34

The Quatt entry comes from the Warwickshire folios, and the Worfield entry from those for Staffordshire.

---

[1] F. W. Maitland, *op. cit.* p. 400.
[2] The Domesday *Gellidone*.

[3] See *V.C.H. Shropshire*, I, p. 282n.

(3) *Plough-teams*

The Shropshire entries, like those of other counties, usually draw a distinction between the teams on the demesne and those held by the peasantry. But at times the distinction is not made, and no demesne ploughs are recorded, especially on the very small manors. Furthermore, as we have seen, some entries appear to be defective and make no reference to teams. Oxen are occasionally mentioned: thus at Wattlesborough (255b) there were 3 ploughs on the demesne and 5 oxen with the peasantry. Some other variations should also be noted. The two holders at Prees (252) had men who ploughed and paid a rent: '*In dominio habent ii carucas et ii villanos cum i caruca et alii iii homines ibi arantes reddunt x solidos.*' This is a unique entry in the Shropshire folios. In three entries we are given information about plough-teams in 1066, or at any rate for some year previous to 1086:

> Onibury in Colmestan (252): In demesne there is 1 plough and 4 whole villeins and 6 half-villeins (*iiii villani integri et vi dimidii*) and a priest and 1 coscet with 3 ploughs. One serf [is] there. A knight holds 1 hide there and has 1 plough and 5 villeins. T.R.E. there were in this manor 9 ploughs.
>
> Clunbury in Rinlau (258): This manor never paid geld nor was it assessed in hides. T.R.E. it was worth £4. There were 6 ploughs (*Ibi erant vi carucae*).
>
> Unnamed holding in Ovret hundred (252): There were and are 2 ploughs (*ibi fuerunt et sunt ii carucae*).

A total of 1,833¾ teams is recorded for the area included in the modern county. Tait[1] arrived at a total of 1,784 and Maitland's[2] figure is 1,755—but both of these refer to the Domesday county.

(4) *Population*

The main bulk of the population was comprised in the five categories of villeins, bordars, serfs, oxmen (*bovarii*) and radmen. In addition to these were the burgesses together with a large miscellaneous group that included Welshmen, priests, *francigenae*, freemen and others. The details of the groups are summarised on p. 129. There are three other estimates

---

[1] *V.C.H. Shropshire*, I, p. 307.   [2] F. W. Maitland, *op. cit.* p. 401.

of population, by Ellis,[1] Eyton,[2] and Tait,[3] but none is comparable with the present estimate which has been made in terms of the modern county, and which excludes landholders, tenants and bondwomen (*ancillae*). Eyton's total of 5,080 is precisely the same as that of Ellis. Both interpreted the formula *n inter servos et ancillas* to mean the number of ploughs held jointly by the serfs and bondwomen, whereas it refers to the serfs and bondwomen themselves. Tait's estimate, which was also in terms of the Domesday county, amounted to 5,162. Definitive accuracy rarely belongs to a count of Domesday population, and all that can be claimed for the present figures is that they indicate the order of magnitude involved. The figures are those of recorded population, and must be multiplied by some factor, say 4 or 5, in order to obtain the actual population; but this does not affect the relative density as between one area and another.[4] That is all that a map, such as Fig. 42, can show.

It is impossible for us to say how complete were these Domesday statistics. As we have seen, a few entries contain no reference to population, but we cannot be certain about the significance of these omissions. Another source of inaccuracy is the use of a single formula for two categories of persons. On fo. 252b, at Sutton, the phrase '8 men between freemen and villeins' (*viii homines inter francos et villanos*) is employed, and it is not clear how many of each there were. Similarly, at Stanton Lacy (260b), there were '28 serfs and bondwomen altogether' (*xxviii inter servos et ancillas*). There are a number of entries of this kind and the total in each has been divided equally between the categories.

Villeins constituted the most important element in the population, and amounted to about 40% of the total, while bordars take second place with nearly 25%. Half-villeins are entered for Stanton Lacy (260b) and Onibury (252); they were presumably men who each held only half a villein tenement, or those whose land and services were divided between two lords. The statistics relating to serfs (19%) present no special feature for our purpose. Those relating to oxmen ($7\frac{1}{2}$%) are interesting, because this category of the population is confined to the counties of the Welsh border, more particularly Herefordshire, Shropshire and Cheshire. Tait,

---

[1] Sir Henry Ellis, *A General Introduction to Domesday Book* (London, 1833), II, p. 481.

[2] R. W. Eyton, *op. cit.* does not give a total for the county as a whole, but sets out hundred totals.

[3] *V.C.H. Shropshire*, I, p. 307.

[4] But see p. 430 below for the complication of serfs.

in his introduction to the Cheshire text writes: 'Mr. Round has shown from later cartularies that these oxmen had charge of the oxen of the demesne ploughs and took part in the ploughing; also that two of them were apparently assigned to each plough. This proportion is found to obtain in a large majority of the cases where they are mentioned in Domesday. As the "servi" often bear the same ratio to the demesne ploughs, Mr. Round concludes that it is highly probable that Domesday uses the terms alternatively, the one expressing legal status, the other occupation.'[1] The mention of a *liber bovarius* at Upton Cresset (255) suggests that the other *bovarii* were not free, and gives added weight to Round's thesis.[2] Tait himself, however, saw many difficulties in accepting Round's view. On the other hand, Mr W. J. Slack has suggested that the expression *liber bovarius* may be employed to distinguish the oxman from the *servi* also recorded at Upton Cresset.[3] Vinogradoff was also of the view that the *bovarii* and the *servi* were distinct groups, and points out that the expression '*inter bovarios et ancillas*' never occurs while '*inter servos et ancillas*' occurs frequently.[4]

The radmen or 'riding-men' take fifth place numerically, and account for $3\frac{1}{2}\%$ of the population. We are not told precisely what these men did, but it is known that one of their chief functions was to act as escort when the lord went on a journey, and to perform the duties of reeve and messenger between the various scattered estates which a lord normally possessed. In return for these services the radmen received a share in the open fields, and a few had villeins and bordars of their own.[5]

The miscellaneous category is a very varied one. Most important from a numerical point of view were the Welshmen,[6] priests and Frenchmen (*francigenae*). Only a few freemen are specifically mentioned. Another interesting group is that of the coliberts who were in some way intermediate between serfs and villeins. Among the smaller groups there were the knights (*milites*) who were no doubt connected in some way with the defence of the Welsh border, and the French serjeants (*francigenae servientes*). *Hospites* or settlers were a small group invited on to the land

---

[1] J. Tait, *The Domesday Survey of Cheshire*, Chetham Society (Manchester, 1916), N.S. LXXV, pp. 67–8.

[2] For *liber bovarius* in Herefordshire, see p. 74 above.

[3] W. J. Slack, 'The Shropshire Ploughmen of Domesday Book', *Trans. Shrops. Archaeol. Soc.* (Shrewsbury, 1939), L, pp. 31–5.

[4] For a discussion, see *V.C.H. Shropshire*, I, p. 302.

[5] J. Tait, *op. cit.* p. 65.   [6] See p. 158 below for Welshmen.

by the lord of a manor, and they are mentioned in three Shropshire entries:[1]

Hatton (259): *et unus hospes reddit ii solidos.*
Colemere (259): *et iiii hospites ibi reddunt xl denarios.*
Leaton (259 b): *et ibi sunt ii hospites reddentes iiii solidos et viii denarios.*

In addition to all these groups there were a few smiths, cottars, coscets, reeves and 'men', together with a miller and a bee-keeper. Finally a number of women were mentioned. Apart from the bondwomen (*ancillae*), who have already been mentioned, there were 2 widows (*ii viduae feminae*) at Besford (259) and 9 female cottars (*ix feminae cotarii*) at Stokesay (260b).

Recorded Population of Shropshire in 1086

A. Rural Population

Villeins	.	.	.	.	.	1,985
Bordars	.	.	.	.	.	1,198
Serfs .	.	.	.	.	.	922
Oxmen	.	.	.	.	.	361
Radmen	.	.	.	.	.	171
Miscellaneous	.	.	.	.	.	269
Total	.	.	.	.	.	4,906

There were also 88 bondwomen (*ancillae*), 9 female cottars and 2 widows who have not been included in the above total.

Details of Miscellaneous Rural Population

Welshmen	64	Smiths	.	.	.	8
Priests	55	Reeves	.	.	.	7
Francigenae	34	*Hospites*	.	.	.	7
Freemen	20	*Servientes*	.	.		6
Cottars	15	*Angli*	.	.	.	3
Men	14	*Francigenae servientes*				3
Coliberts	13	Bee-keeper	.	.		1
Coscets	9	Miller	.	.	.	1
Knights	9	Total	.	.	.	269

B. Urban Population

QUATFORD   See p. 152 below.
SHREWSBURY   See pp. 152–3 below.

[1] See p. 74 above for *hospites* in Herefordshire.

(5) *Values*

The value of an estate is most frequently given for two dates, 1066 and 1086, but in about one-third of the entries a return is also entered for the year in which the Norman tenant took possession; more often than not the holding was waste at this intermediate date. The entry for High Hatton (254) is typical: 'T.R.E. it was worth 60*s.*, afterwards (*post*) waste, now 10*s.*' There are also twenty-five entries in which a value for 1086 alone is recorded. Of course there were some exceptions. For example, we are told of an estate at Eyton on Severn (252b) that it was worth £21 T.R.E. and yielded £14 when the earl gave it to the church (St Peter's at Shrewsbury)—but no value is given for 1086. Sometimes a distinction was made between the value and the amount of money rendered. Of Marton (253) we read that 'T.R.E. it yielded 8*s.* Now it is worth 10*s.*, but pays no more than 6*s.* 2*d.*' A similar distinction occurs in an entry for Bausley in Ruesset (255b), a place now located in Wales. The entry for Strefford (255) is unusual: *Non sunt homines ibi et tamen [Rainaldus] habet xx solidos, T.R.E. valebat xxx solidos et post wasta fuit.*

There is a distinct tendency for the values to be entered to the nearest five or ten shillings, and this suggests that the valuation of some estates was little more than an estimate.[1] For example, English Frankton (255) was worth 10*s.* in 1066 and 15*s.* in 1086. On the other hand, some figures suggest a greater degree of precision. The three figures (*T.R.E., post* and *modo*) for Hopton Wafer (260b) were 10*s.*, 12*s.* and 9*s.* 2*d.* Or again, the 1066 figure for Shipton (252b) was 30*s.*, but that for 1086 was 4*d.* more (*Modo iiii denarii plus*). The 1086 figure for Waters Upton (256b) was as precise as 30*s.* 2¼*d.* (*i ferding*).

Generally speaking, the greater the number of plough-teams and men on an estate, the higher its value, but it is impossible to discern any constant relationships as shown by the figures for four vills on p. 131. It is true that the variations in the values, as between one holding and another, may not reflect variations in the amount of arable, but even taking the other resources into account, the figures are not easy to explain.

[1] The frequent appearance in values of Shropshire holdings of sums such as 1*s.* 4*d.*, 2*s.* 8*d.*, 5*s.* 4*d.*, 6*s.* 8*d.*, etc. is evidence that a Scandinavian monetary system was in use, as well as the older English system of pounds. The *ora* or ounce was equivalent to 16 silver pennies, and the mark to 10*s.* 8*d.*

Fitz and Forton for example had similar numbers of ploughs and men; but although Forton had wood enough to feed 100 swine, its value was only about one-fifth that of Fitz, where there was no wood.

	Teams	Population	Value 1086	Other resources
Forton (256b)	3½	16	25s.	Wood
Fitz (258)	4	15	£6	Nil
Welshampton (255)	3	13	30s.	Nil
Colemere (259)	1½	8	30s.	Nil

The most significant feature of the Domesday values for Shropshire is the frequent decrease in value that occurred after 1066 and the subsequent recovery by 1086. The state of affairs at Hatton (259) was typical: 'T.R.E. it was worth 12s., later it was waste, now it is worth 11s.' Many vills were waste or partially waste when the new tenants took over, and the values of many others were much reduced.[1] By 1086 the values of the great majority of places had recovered almost to their former amounts. In some places, however, there had been no recovery at all, and indeed even a decline. Bratton in Recordine (257b) was worth 24s. in 1066, and in 1086 it was almost waste (*Modo pene wasta est*). Similarly the Domesday *Corselle* (Cross Hill?) in Recordine was worth 20s. in 1066, 40s. later, and only 1s. in 1086 (257). Myddle in Bascherche (255) also declined in value from £6 (1066) to £4 later, and finally to £3. 10s. in 1086.

Conclusion

For the purpose of calculating densities the fifteen Domesday hundreds have been adopted, but many of these have been subdivided in the light of variations in surface geology and soils. The limits of the resulting twenty-seven districts are very artificial for they coincide with the boundaries of civil parishes, in order to make possible a statistical analysis of the information (Figs. 40 and 42). Although this does not give as perfect a division as a geographer could wish, it provides a useful basis for distinguishing the degree of variation over the face of the county.

Of the five standard formulae, those relating to population and to plough-teams are most likely to reflect something of the distribution of wealth and prosperity throughout the county. Taken together, they

[1] See pp. 144–6 below for discussion of waste.

5-2

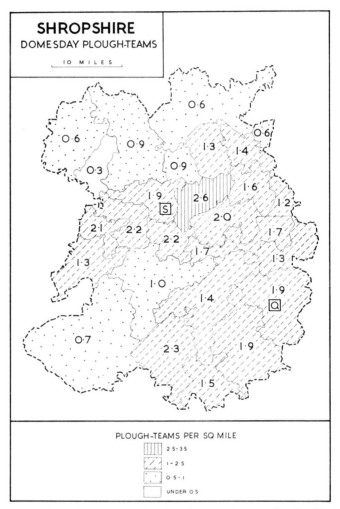

Fig. 40. Shropshire: Domesday plough-teams in 1086 (by densities).
Domesday boroughs are indicated by initials: Q, Quatford; S, Shrewsbury.

supplement one another to provide a general picture. The most striking
feature of Figs. 40 and 42 is the concentration of population in a belt
extending across the Boulder Clay region of loamy soils in the centre of
the county. Thus along the upper Severn valley the density of recorded
population was 6 per square mile, and a relatively high number of ploughs
per square mile (2) was to be found in the same district. Only in the Corve

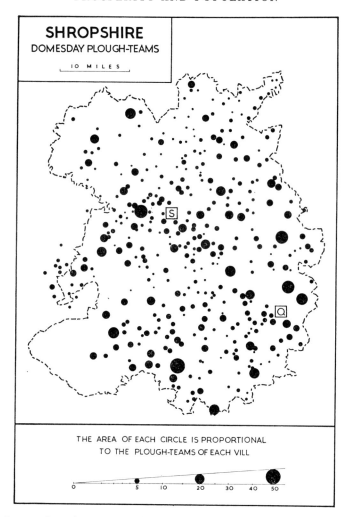

SHROPSHIRE
DOMESDAY PLOUGH-TEAMS

10 MILES

S

Q

THE AREA OF EACH CIRCLE IS PROPORTIONAL
TO THE PLOUGH-TEAMS OF EACH VILL

0   5   10   20   30   40   50

Fig. 41. Shropshire: Domesday plough-teams in 1086 (by settlements).
Domesday boroughs are indicated by initials: Q, Quatford; S, Shrewsbury.

valley in the south are comparable figures to be found—there were 6 recorded persons and over 2 teams per square mile in this well-populated valley.

In contrast to this there were other districts with low densities per square mile. These negative areas were of two kinds. In the first place were the upland districts of the south-west and north-west where the land

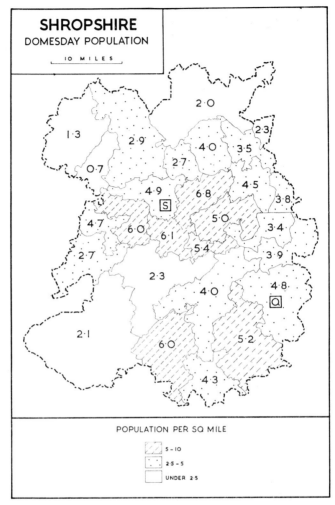

Fig. 42. Shropshire: Domesday population in 1086 (by densities).
Domesday boroughs are indicated by initials: Q, Quatford; S, Shrewsbury.

rises in places to over 1,000 ft. above sea-level. Not only was there much infertile land but also a number of villages lay waste in 1086. The result was only between about 1·0 and 2·0 people and less than 1·0 team per square mile. In the second place there was the northern plain with low undulating relief, for the most part below 300 ft. above sea-level, and covered by glacial deposits of various kinds. The Boulder Clay, stiff and

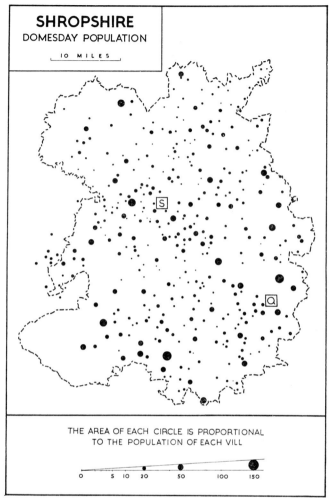

SHROPSHIRE
DOMESDAY POPULATION

10 MILES

THE AREA OF EACH CIRCLE IS PROPORTIONAL
TO THE POPULATION OF EACH VILL

0   5 10 20   50   100   150

Fig. 43. Shropshire: Domesday population in 1086 (by settlements).
Domesday boroughs are indicated by initials: Q, Quatford; S, Shrewsbury.

impervious in places, the stretches of sands and gravels, the marshy areas
on post-glacial deposits of peat—all were infertile in the context of
eleventh century conditions. Over much of this northern plain there were
only between 2·0 and 3·0 people per square mile and less than 1·0 team.

Elsewhere, along the eastern border of the county, over a variety of
terrain, densities were intermediate in character, ranging between 3·5
and 5·0 or so for population and less than 2·0 for teams.

Figs. 41 and 43 are supplementary to the density maps, but it is necessary to make one reservation about them. As we have seen on p. 117, it
is possible that some Domesday names may have covered more than one
settlement, e.g. the present-day villages of Dawley and Dawley Magna
are represented in the Domesday Book by only one name. A few of the
symbols should therefore appear as two or more smaller symbols, but
this limitation does not affect the main pattern of the maps. Generally
speaking they confirm and amplify the information of the density maps.
In view of the doubtful nature of the plough-land entries, the implications
of Fig. 44 are uncertain, but the map has been included for comparison
with Fig. 40.

WOODLAND

Types of entries

The amount of woodland on a holding in Shropshire was normally
indicated in one of two ways. Sometimes it was measured by the number
of swine it could feed, and a typical entry is that for Betton in Hales (259),
where there was 'wood for fattening 60 swine' (*Silva lx porcis incrassandis*). The phrase *incrassandis* or *ad incrassandos* is omitted at Sheinton
(260b) where the entry states 'wood there for 100 swine' (*Ibi silva c
porcis*). Figures for swine range from 6 to 600 in these entries.

The second type of entry records wood in terms of dimensions, usually
single dimensions; thus at Cardington (255) there were *ii leuuae silvae*.
Curiously, the Shropshire folios themselves never use furlongs and
perches. Sometimes, two dimensions are given, as is more usual for other
counties; thus at Donington (253b) there was *silvae i leuuae longa et
dimidia lata*. The exact significance of these linear dimensions is far from
clear—whether they are extreme lengths or averages, or whether they
imply some idea of area. It may well be that single dimensions were
sometimes used when length and breadth were the same, but they have
been plotted as single lines on Fig. 45 to indicate their frequency. In any
case, we cannot hope to convert any of these measurements into modern
acreages.[1] Nor can we relate swine units to linear measurements. Fig. 45
must inevitably be diagrammatic. We are given a glimpse into the relationship between the two types of entry in the account of Worthen which
states (255b): 'Wood 2 leagues long, in which are 4 hays, and it is

[1] See p. 437 below.

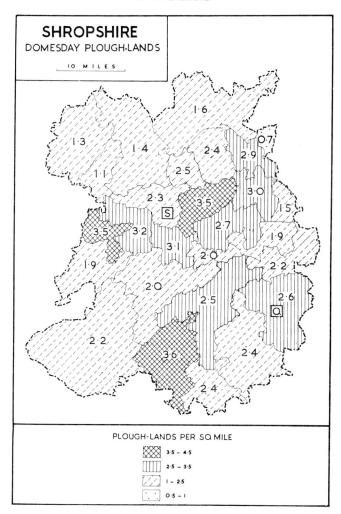

Fig. 44. Shropshire: Domesday plough-lands (by densities).
Domesday boroughs are indicated by initials: Q, Quatford; S, Shrewsbury.

sufficient for fattening 200 swine' (*Silva ii leuuis longa in qua sunt iiii haiae et cc porcis incrassandis sufficit*).[1]

In addition to the two main types of entries there are eight miscellaneous entries that refer to wood in other ways:

[1] For a similar equation of swine and wood at Leighton (*Lestune*), a place now in Wales, see p. 161 below. See also pp. 439–40 below.

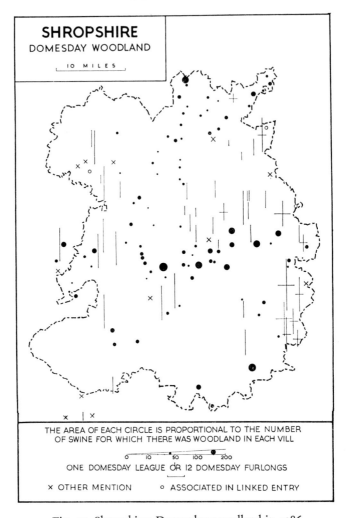

SHROPSHIRE
DOMESDAY WOODLAND

|____ 10 MILES ____|

THE AREA OF EACH CIRCLE IS PROPORTIONAL TO THE NUMBER
OF SWINE FOR WHICH THERE WAS WOODLAND IN EACH VILL

0   10   50   100   200

ONE DOMESDAY LEAGUE OR 12 DOMESDAY FURLONGS

× OTHER MENTION   ○ ASSOCIATED IN LINKED ENTRY

Fig. 45. Shropshire: Domesday woodland in 1086.

Maesbury (253b): There a small wood rendering nothing (*Ibi silva parva
nil reddens*).

Wykey (257b): Wood in which there is 1 hay (*Silva in qua est una haia*).

Chetwynd (257b): A small wood there (*Ibi silva parva*).

Stretton (254): in a wood 5 hays (*in silva v haiae*).

Eaton Constantine (254b): The underwood renders 5*d*. (*Silva modica reddit
v denarios*).

Hodnet (253): There a small wood rendering nothing (*Ibi Silva parva nil reddens*).

Morton and *Aitone* (254b)[1]: There a small tract of wood (*Ibi parva landa silvae*).

Wlferesforde (259b): There a small wood (*Ibi parva silva*).

Furthermore, the two more normal types of entry sometimes give additional detail. A number of entries mention the presence of hays for catching roe-deer,[2] while at Little Wenlock (252b) there was also a hawk's nest (*Silva ccc porcis incrassandis in qua sunt ii haiae et aira Accipitris*), and likewise another at Wem (257). Oak trees are mentioned for Shipley (239): *una quarentena quercuum in longitudine et latitudine*. There was a park at Marsley (252).[3]

Distribution of woodland

When plotted on a map the wood entries are seen to be disposed in four main groups (Fig. 45). One of these stretches over the central part of the county from the north-facing slopes of the upland region south of the Severn, across the Severn valley to the farther side of the Wrekin. The clay soils of this district are ideal for the support of woodland, and the entries relating to the district indicate the presence of considerable pannage for swine. A second well-wooded area was the hundred of Odenet, in the north-eastern part of the northern plain. The soils here are a mixture of glacial clays, sands and morainic debris. Similarly, the glacial gravels covering both sides of the mid-Severn valley, and the valleys of the Severn tributaries (e.g. the river Corfe) seem to have carried much wood. The only other area with a substantial amount of wood lay in the west of the county, on the slopes of Long Mountain and in the Rea-Brook valley, and especially on the east-facing hill slopes around Oswestry.

It is not surprising that certain parts of the county should be lacking in wood. The higher parts of the hilly region in the south, where the thin acid soils are unsuitable, the peaty Weald Moors, and the waterlogged plain of Baschurch, in the north, supported little or no wood.

The Domesday folios contain no mention of the royal forests which feature in later documents.[4] We are, however, told that King Edward had

---

[1] This is one of the few composite entries for Shropshire wood. For each such entry the wood has been plotted for the first-named place only, but associated places have also been indicated.

[2] See p. 140 below.   [3] See p. 154 below.

[4] *V.C.H. Shropshire*, I, pp. 485–6.

hunted from Shrewsbury (252), and there is one entry that may indicate
an extension of forest rights in King William's time; the wood which fed
100 swine at Albrighton, near Shifnal (259), had been taken into the
king's hands (*Ibi Silva c porcis incrassandis. Sed modo est in manu regis*).
The term forest does not necessarily mean woodland, but it may well
be that these Shropshire forests were well-wooded areas, suitable for
hunting. It is possible therefore that Shropshire, in 1086, was more
heavily wooded than is indicated by Fig. 45.

Hays

Hays (*haiae*) or enclosures in the wood are recorded in connection
with thirty-six places, two of which are unidentified—*Chinbaldescote* (258)
and *Petelie* (252b). Not included in this total is Clunton where there
seems to have been 5 hays at an earlier date—*Ibi v haiae fuerunt* (258);
the entry makes no reference to hays in 1086 nor does it mention wood.
The number on a single holding was usually one or two and never above 5.
More often than not hays were mentioned in connection with wood,
whether the wood was measured in terms of swine or of linear dimensions.
The following entries are typical:

> Uppington (258b): Wood one league in length and there one hay (*Silva i
> leuua longa et ibi i haia*).
> Ightfield (259): There wood for fattening 60 swine and 2 hays (*Ibi Silva lx
> porcis incrassandis et ii haiae*).

But at thirteen places we read of hays where there is no record of woodland, and this seems to suggest that the record of wood is incomplete.
We are merely told at Clungunford (255), for example, that there were
3 hays (*Ibi sunt iii Haiae*). The entry for Adderley is even more terse—
Ibi ii haiae (259). Some of the entries are more explicit. The 3 hays connected with the wood at Lee in Leebotwood (254b) were 'fixed hays'
(*haiae firmae*); while at Corfton (256b), where there was no wood, we
are told that the hay was for catching roe-deer (*haia capreolis capiendis*).
Fig. 46 shows the distribution of hays, but no well-marked pattern can
be observed.

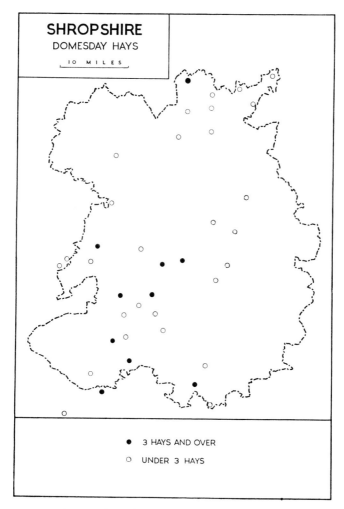

Fig. 46. Shropshire: Domesday 'hays' in 1086.

MEADOW

There are no references to meadow in the Shropshire folios, but the modern county includes places which are surveyed in the Staffordshire and Warwickshire sections of the Domesday Book. The entry for Quatt (239) is typical: 'There one acre of meadow' (*Ibi i acra prati*). The complete list of entries is as follows:

	Acres
Alveley (248)	6
Cheswardine and Chipnall (248b)	1
Claverley (248)	12
Quatt (239)	1
Sheriff Hales (248)	8
Worfield (248b)	16

With the exception of Sheriff Hales, all these places are located near the river Severn where it leaves the county. It seems reasonable to suppose that the other river valleys of the county also carried meadow. There must have been meadow, for example, along the Tern and Corve valleys, but for some reason unknown to us this was not entered in the Domesday Book as we now know it.

FISHERIES

Fisheries (*piscariae*) are specifically recorded in connection with at least sixteen places in Shropshire, one of which (*Udeford*) cannot be identified. It is difficult to be sure of the exact number of places because of a joint entry for *Udeford* and Ruyton of the Eleven Towns (257b) where there were five fisheries. In each entry the number of fisheries is stated, usually with their yield in money or kind. The number of fisheries varied from five to one-half at Montford; but there is no indication of who held the other half at Montford. The most frequent type of entry is illustrated by that for Cressage (256b) where there was a fishery yielding 8s. (*Ibi piscaria de viii solidis*). On the other hand, at Ercall Magna (253b), the yield is stated in eels, for there was a fishery rendering 1,502 large eels (*piscaria de mille et quingentis et ii anguillis magnis*). Several entries are less informative. All we are told for Upton Magna (254b), for example, is that its fishery yielded what it could (*piscaria reddit quod potest*). Frequently a village possessed both fishery and mill, but the record is often a composite one: thus at Chetwynd (257b) there was 'a mill with 2 fisheries yielding 5s. and 64 sticks of eels' (*molinum cum ii piscariis reddit v solidos et lxiiii stichas anguillarum*).[1] At Huntington (256b) and Little Ness (255) eels are recorded in connection with mills making a total of 18 fisheries mentioned or implied. Some additional information is occasionally given. The five fisheries at *Udeford* and Ruyton of the Eleven Towns (257b) were 'rented by the villeins' (*in censu villanorum*); while

---

[1] A stick consisted of 25 eels.

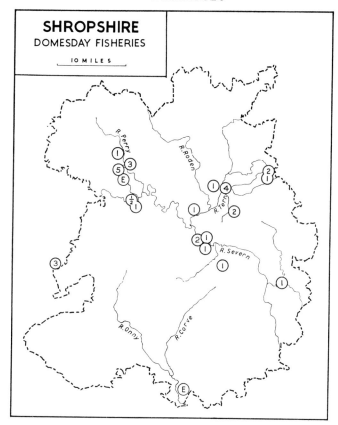

Fig. 47. Shropshire: Domesday fisheries in 1086.

The figure in each circle indicates the number of fisheries. Fisheries are not specifically
recorded for Ness and Huntington, but only mills rendering eels (E).

one of the two fisheries at Eyton on Severn (252b) was 'for the sustenance
of the monks' (*ii piscariae. Una reddit xvi solidos, alia est ad victum
monachorum*). The fishery at Eaton Constantine (254b) and that at
Wykey (257b) each rendered nothing; the former was said to be in the
Severn (*in Sauerna*).

Fig. 47 shows that most of the fisheries lay along the river Perry, and
on those tributaries of the Severn which drain the north-east of the county,
i.e. the Tern and Roden. Other fisheries were located along the Severn
itself, and near Much Wenlock and Worfield. No fisheries are specifically
mentioned for the streams of the southern upland apart from the eel
render at Huntington on the Teme.

WASTE

It is clear that a number of places in the west of the county were waste in 1066, presumably as a result of Welsh raiding. Very many more became wasted in the years following the Norman Conquest, and a large number of these still lay waste in 1086.

There is no doubt about the date implied by the phrase *wasta est*; it must refer to 1086. The phrases *wastam invenit* or *post fuit wasta* imply an intermediate date between 1066 and 1086; such phrases are frequent in the Shropshire folios whereas they occur only twice in those for Herefordshire.[1] The phrase *T.R.E. wasta fuit* is also clear in that it must refer to 1066. But the phrase *wasta fuit* (without *T.R.E.*) is not so clear. Does it refer to 1066 or to an intermediate date, or to both dates? We cannot be certain and on Figs. 48 and 49 it has been shown as a separate category. Referring to this problem, Tait wrote that 'if to the manors expressly stated to have been waste *T.R.E.* we could safely add all those where the formula "wasta fuit et est" appears without a valebat, lands assessed at nearby 200 hides must have been waste in 1066'.[2] The following representative entries indicate the ambiguities involved:

> *Udeford* and Ruyton (257b): *T.R.E. wasta fuit et post valuit xiii solidos. Modo xx solidos.*
> Maesbury (253b): *T.R.E. erat wasta. Modo valet xl solidos.*
> Eyton in Baschurch (256): *Wasta fuit. Modo valet v solidos.*
> Audley (256): *Valuit vi solidos. Modo ii solidos. Wasta fuit.*
> Edgeley (256): *T.R.E. valebat xl solidos. Modo xii solidos. Wastam invenit.*
> Marchamley (254): *T.R.E. valebat c solidos et post fuit wasta. Modo valet xlvi solidos et iiii denarios.*
> Weston Rhyn (254b): *Wasta fuit et Wastam invenit. Valet modo x solidos.*
> West Felton (255): *Wasta fuit et est.*
> Spoonley (259): *Wasta fuit et est. T.R.E. valebat xx solidos.*
> Netley (259b): *Wasta est et fuit. Valebat xii solidos* (presumably *T.R.E.*).
> *Slacheberie* in Ellesmere (259): *Valebat v solidos T.R.E. Modo wasta est.*

That some vills were only partly waste we may judge from two types of evidence. In the first place, there were those vills comprising two or more holdings, of which at least one was waste. Thus of two holdings at

---

[1] See p. 95 above.   [2] *V.C.H. Shropshire*, I, p. 281 n.

Chesthill, one was waste (252) and the other was tilled (258b). In the second place, single holdings were themselves partly waste, as the following examples show:

Lydbury North (252): 53 hides. *De hac terra sunt wastae xxxii hidae et dimidia.*
Rowton (259b): 2 hides. *T.R.E. valebat ix solidos. Modo iii solidos. Wasta fuerunt et adhuc satis sunt.*
Womerton (256): 2½ hides. *Ibi sunt ii villani cum dimidia caruca. Maxima pars hujus manerii est wasta.*
Cause (*Alretone*, 253b): 40 plough-lands. *Wasta terra est ibi ad xxxi carucas.*

Even when a holding was waste, it was not necessarily completely devoid of value or resources. Here are four representative entries relating to such holdings:

Hawkesley (259): *Wasta fuit et est. Ibi silva xl porcis incrassandis. Hoc manerium est ad firmam pro vi denarios.*
Wollaston (255b): *Wasta fuit et est, et tamen reddit xii denarios.*
Horton (257b): *Wasta est. Ibi est dimidia leuua silvae et una haia.*
Minton (258): *Wasta fuit et est. Ibi i haia.*

These and similar holdings have been regarded not as partly waste, but as entirely waste in the sense that all their arable seems to have been devastated and that they had no population. They do not necessarily appear as 'wholly waste' on Figs. 48–50 because some vills had fully stocked holdings in addition to their devastated land.

Fig. 48 shows the places that were waste and possibly waste (*wasta fuit*) in 1066. As might be expected, the greater number lay in the western part of the county. By 1070, the wasted areas extended all over the county and were as frequent in the east as in the west (Fig. 49). 'Manor after manor which yielded a good income in 1066 is entered as "waste" when its new holder received it.'[1] Considerable recovery had taken place by 1086 but, as Fig. 50 shows, many places were in a devastated condition. The table below gives the number of places, including unidentified vills, affected in the area covered by the modern county.

	Wholly waste	Partly waste
1066 (minimum)	9	2
1066 (possible total)	47	7
c. 1070 (minimum)	80	5
c. 1070 (possible total)	118	10
1086	36	15

[1] *V.C.H. Shropshire*, I, p. 281.

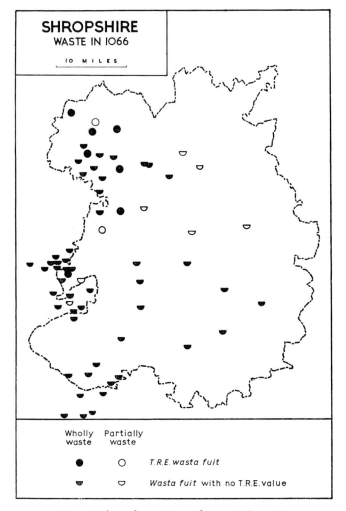

Fig. 48. Shropshire: Domesday waste in 1066.

The extent of the recovery can be seen more clearly on Fig. 51 which shows the number of teams at work in 1086 on holdings which had been waste at some earlier date.

It may well be that Fig. 50 does not give a complete picture of the empty villages of the county in 1086. A few holdings not specifically described as waste had no population or teams entered for them, e.g. Broome (259b) with two plough-lands but no teams or population.

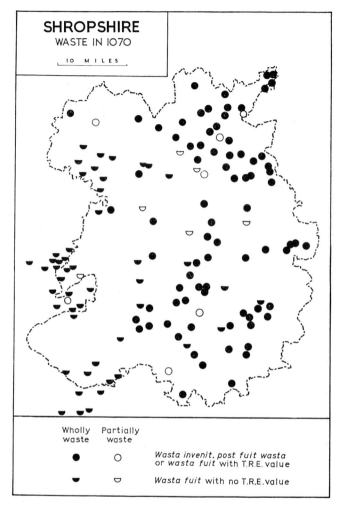

Fig. 49. Shropshire: Domesday waste in 1070.

MILLS

Mills are mentioned in connection with eighty-eight of the 437 Domesday settlements within the area covered by modern Shropshire. Some entries state the presence of a mill and give its annual value. Thus at Sheinton (256b) there was a mill yielding 10s. (*Ibi molinum de x solidis*). Other entries omit the value; thus the entry for Ellesmere (253b) merely states

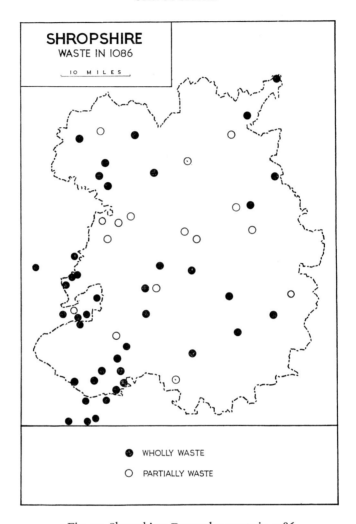

Fig. 50. Shropshire: Domesday waste in 1086.

that there was a mill there (*Ibi molinum*). Occasionally, the yield is stated not in terms of money but in terms of eels. At Huntington (256b) we read of 'a mill yielding (*de*) 400 eels', while at Little Ness (255) there was a mill yielding 20*s*. and 600 eels. At Aston Eyre (255) and at Lilleshall (253) there were mills that rendered nothing (*nil reddit*). The following entries illustrate some other variations:

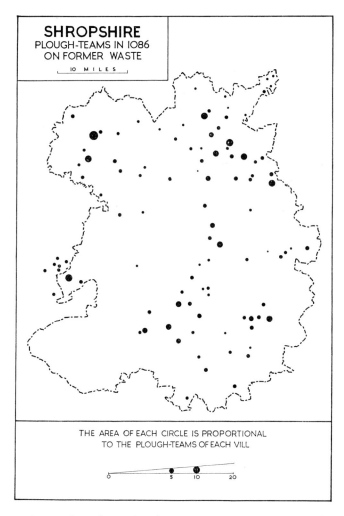

Fig. 51. Shropshire: Plough-teams in 1086 on former waste.

Bunford (260): Osbern has there two mills yielding 12 loads of grain (*Ibi habet Osbern ii molinos [sic] reddentes xii summas Annonae*).

Ford (253b): There a mill yielding 3½ ora (*Ibi molinum reddit iii ores et dimidiam*).

Lydham (253b): A mill yields 1 pig (*molinum reddit i porcum*).

Yockleton (255b): There a mill yields 1 load of malt (*Ibi molinum reddit unam summam brasii*).

Ryton (257b): There a mill yields 8 sesters of rye (*Ibi molinum reddit viii sextarios siliginis*).

There is one mention of a seasonal mill at Welbatch (255b) where there was 'a mill for winter, but not for summer' (*Ibi molinum hiemale non aestivum*), a restriction due to the seasonal nature of the stream. Only once do we read of a miller; this unique entry is for Stokesay (260b) and it reads thus: 'There a mill yields 9 loads of grain, and there [is] a miller' (*Ibi molinum reddit ix summas frumenti et molinarius ibi*). At Chetton (254) there was a new mill (*molinum novum*) for which no render is mentioned. No fractions complicate the Shropshire record of mills.

Domesday Mills in Shropshire in 1086

1 mill	80 settlements	3 mills	2 settlements
2 mills	6 settlements		

Fig. 52 shows the alignment of the mill sites along the main streams. A large number of mills were to be found in the Tern valley, and in the tributary valleys that drain the Weald Moors in the north-east. Another line of mills followed the river Roden from near its source to its confluence with the Tern at Rodington. Mills were frequent in the whole of this north-eastern district. A number of mills were located on the tributaries of the Severn, and a few along the Severn itself. Another noticeable feature of the distribution is the grouping of mills on the lower slopes of the Clun and Longmynd Uplands, and in the scarp and dale region farther east. A similar group encircles the Clee platforms, where the mills are situated on the lower reaches of the tributaries of the rivers Onny, Corve, Rea and Severn.

CHURCHES

Apart from Shrewsbury which possessed five churches, there are twenty-two places for which churches are recorded, within the area of the modern county. In each of these places a single church is mentioned, except at Chirbury where there were two. Altogether, twenty-eight churches are mentioned. At Much Wenlock we are informed that Earl Roger had made the church of St Milburga into an abbey (252b). In five places we are merely told that there was 'a church there', but in the remaining localities mention is made of one or more priests. Wroxeter, for example, had a church with four priests.

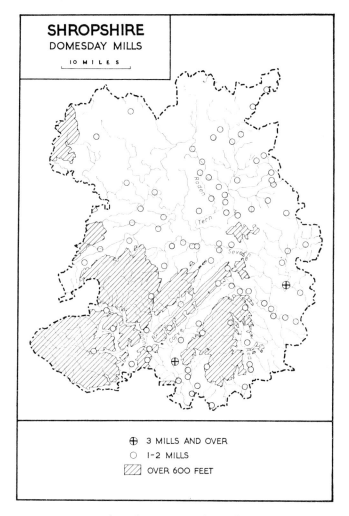

Fig. 52. Shropshire: Domesday mills in 1086.

The usual formula employed is 'a church and a priest there'; but we are sometimes given additional information. At Aldon we are told, 'the church holds half a hide, and a priest [has] one plough with one coscet' (260b). Again at Morville we read: 'the church of this manor is in the honour of St Gregory and used to hold T.R.E. 8 hides of this land, and 8 canons served there. The Church of St Peter [now] holds this church from the earl, with 5 hides of land.... There [are] 9 villeins and 1 bordar and 3 priests with 9 ploughs' (253). In this entry the priests are grouped with the rest of the population.

In addition to the places for which churches are recorded, there are a further twenty-seven places where priests alone are mentioned. Two places each had two priests—Ellesmere (253) and Westbury (255 b). It is possible that the mention of a priest implies the existence of a church; if this is so, then there were fifty places with churches out of a total of 437 places in the county. Clearly the Domesday enumeration of churches was very incomplete.

URBAN LIFE

Two boroughs are mentioned in the Shropshire folios. On fo. 254 there is a solitary reference to the borough of Quatford; the entry for Earding-ton (*Ardintone*) tells of a new house and of a borough called Quatford yielding nothing (*nova domus et burgus Quatford dictus nil reddit*). No other details are given about the place.

The other borough was Shrewsbury. The Shropshire folios begin with an account of the city (*Civitas Sciropesberie*), which is the only place for which burgesses are specifically recorded. The greater part of the descrip-tion is taken up with an account of the customs of the city, and we are given only a few unsatisfactory details of the state of the city itself. For the year 1066, we are told that 'in the city of Shrewsbury, in the time of King Edward there were 252 houses (*domus*) and the same number of burgesses' (fo. 252). In addition we are told that the bishop of Chester had 16 *masurae* and the same number of burgesses in the city T.R.E., and 16 canons—but whether these were in addition to the 252 houses mentioned earlier, is not clear.

By 1086, the Normans had wrought many changes, and it appears that the burgesses had fared badly under the new regime. The English-born burgesses complained that:

it is very hard on them (*multum grave sibi esse*) that they now pay the same geld as they used to pay in the time of King Edward, although the Earl's castle has occupied the place of 51 burgages (*masuras*) and another 50 are waste, and 43 French-born burgesses occupy burgages geldable in the time of King Edward, and the Earl has granted to the abbey which he is forming there, 39 burgesses formerly geldable in like manner with the others. In all there are 193 burgages which do not now pay geld (fo. 252).

But if the details are added, the total of non-gelding burgages is 183 and not 193.[1] If from the total of 252 in 1066 we take the 51 burgages destroyed for the castle, and the 50 waste burgages, we are left with 151 burgages of various kinds (French and other) in 1086, but these figures may well be wrong. Burgesses are also mentioned in seven other connections, viz.:

Fo. 252. The bishop of Chester held in Shrewsbury 16 burgages, and the same number of burgesses paying geld with the rest of the burgesses. Now 10 of these burgages are waste, and the other 6 render 4s. 7d.

Fo. 252b. In Shrewsbury city Earl Roger is making an abbey, and has given to it the church (*monasterium*) of St Peter, where there was a parish in the city; and as many of his burgesses and mills as yield £12 to the monks.

Fo. 252b. The same church [St Remigius] holds Emstrey....In the city [there is] 1 burgage property (*masura terrae*) yielding 2s.

Fo. 253. The church of St Almund holds in Shrewsbury 21 burgesses besides 12 houses for the canons. These burgesses render 8s. 8d.

Fo. 253. The church of St Juliana holds half a hide, and it has there 1 plough, and 2 burgesses labouring on the land (*in hac terra laborantes*) render 3s.

Fo. 256. The same Robert holds Woodcote (*Udecote*)...and 1 burgess renders 8s.

Fo. 260b. The same Ralf holds Meole Brace (*Mela*),...To this manor pertain 9 burgesses in the city.

No other reference to contributory burgesses is to be found in the Shropshire folios. It is difficult to see what relation this total of 40 burgesses and the unknown number belonging to the abbey, bears to the 151 burgesses that may have been in the city in 1086.[2] But it seems reasonable to assume that in 1086 Shrewsbury was a settlement of at least a thousand people, and probably of considerably more. Other indications of the life of the town are found in the presence of moneyers who worked the mint, and in the five churches. Domesday Book leaves us with only a vague picture of the border fortress but we are reminded of its strategic importance in relation to the Welsh borderland by the remains of a castle within the present-day city.

[1] See *V.C.H. Shropshire*, I, p. 294.

[2] There is also the further complication of the '12 houses for the canons' held by St Almund.

MISCELLANEOUS INFORMATION

(1) *Castles*. Three castles are mentioned in the area covered by modern Shropshire. The relevant entries are as follows:

Stanton (258b): *Ibi habet Helgot castellum*.[1]
Maesbury (253b): *Ibi fecit Rainald castellum Luvre* (Oswestry).
Shrewsbury (252): *castellum comitatis occupaverit li masuras*.

There was also a castle at Montgomery in Wales (254).[2]

(2) *Salt*. Three references are made to salt in the Shropshire folios:

Caynham (256b): *iiii summae salis de Wich*.
Ditton Priors (253b): *In Wich una salina reddit ii solidos*.
Donington (253b): *In Wich v salinae reddunt xx solidos*.

In all three entries *Wich* presumably refers to Droitwich in Worcestershire.[3]

(3) A park is entered for *Marsetelie* (252) in the entry for Shrewsbury.[4]

REGIONAL SUMMARY

Shropshire may be broadly divided into a southern hilly region and a northern lowland region, but the varied nature of the lowland with its heath, moorlands, and the Wrekin mass, makes further sub-division necessary. The hills of south Shropshire are formed of ancient rocks with many volcanic intrusions. Infertile grits and flagstones are predominant, although limestones occur around Wenlock, and marls and sandstones on the Clee platform. Within this area, the population was confined mainly to the clay-filled valleys. In the northern lowland, the Triassic sandstones are thickly covered with Boulder Clay and glacial sands; the natural drainage has been disorganised by glacial action, and a good deal of marshland still remains. For purposes of regional description the county has been divided into eight districts (Fig. 53).

---

[1] This part of Stanton later became Holdgate from the name of the Domesday holder.
[2] See p. 158 below.　　　　　　　　　[3] See p. 257 below.
[4] Identified with Marsley in Habberley—R. W. Eyton, *Antiquities of Shropshire*, VII, p. 45.

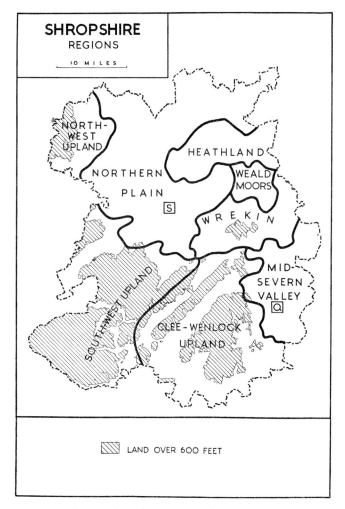

Fig. 53. Shropshire: Regional subdivisions.
Domesday boroughs are indicated by initials: Q, Quatford; S, Shrewsbury.

(1) *North-west Upland*

This region rises from 250 ft. above sea-level in the east, to over 1,000 ft. in the west. Apart from the outcrop of limestones and sandstones in the extreme west, the surface is covered by Boulder Clay. The region supported a few small settlements; the figures for population (1·3) per square mile and plough-teams (0·6) per square mile indicate that the general prosperity was not great. Very little wood was recorded for the area.

(2) *The Northern Plain*

This plain lies at between 250 ft. and 350 ft. above sea-level. A complicated distribution of glacial drift and fluvio-glacial deposits has produced soils varying from light sands to heavy loams obscured in many places by peat and marshland. A belt of alluvium is found in the Severn valley, and it is here, where the soil is a medium loam, that the greatest number of settlements and the highest density of population were to be found. The contrast between the valley district and the sandy and marshy district to the north may be seen from the population figures which range from about six per square mile in the south, to two per square mile in the north. The density of plough-teams was about two per square mile in the Severn valley, decreasing to under one in the north. The north-east section of the plain supported a little woodland which was used for pannage. Almost one-half of the fisheries recorded in Shropshire were in the northern plain, but mills were few. The southern part of this region was the most prosperous part of Shropshire in the eleventh century.

(3) *The Heathland*

The soils of this region are light sands or loams. The sandy soils support heathland, but towards the eastern part of the region these sands pass gradually into more fertile light loams. The average of one plough per square mile on the loams is evidence of their arable nature in 1086. None of the settlements had a recorded population of more than twenty-five, and the density of population was about three to four per square mile. Some mills were located in the area, but there was very little wood in the heathland region as a whole. There were no fisheries.

(4) *The Weald Moors*

Physically, the area of the Weald Moors is a basin in which peat has accumulated since the disappearance of the Pleistocene ice-sheet. The rim of this basin lies at 250 ft. above sea-level, whereas the centre is only about 180 ft. In some places there are patches of light loam or loamy sands. A few of the vills on the eastern rim were quite large. These large settlements resulted in a population density of 4·5 per square mile, though the waterlogged nature of the region as a whole is shown by the fact that the plough-team average was only 1·6 per square mile. Even now, special drainage arrangements are needed to make farming profitable. There was some wood in the south of the area, and two isolated fisheries in the north, together with a few scattered mills.

(5) *The Wrekin Area*

The ancient volcanic mass of the Wrekin reaching 1,000 ft. above sea-level forms the core of this region, much of which is above the 400 ft. contour. The greater part of the area was devoid of settlements in 1086, although near the river Severn, which flows through its south-western corner, the population density was over five per square mile. There were up to two plough-teams per square mile from which it can be inferred that some parts of the region must have been relatively fertile. The area was well wooded, and several fisheries were situated on the river Severn.

(6) *The Mid-Severn Valley*

The Triassic sandstones of this region are deeply dissected by streams, and are partly covered by glacial pebble beds and gravels. On the whole, the settlements were in the valleys. The population figure of four to five per square mile provides evidence of moderate prosperity, due, no doubt, to the nature of the soil—a fertile light or medium loam. There were one to two teams per square mile. Some woodland flanked the valleys, and there were some mills and one fishery in the area.

(7) *The Clee-Wenlock Upland*

This region displays considerable variety of surface relief, and its settlements were located mainly in the clay-filled valleys. The Clee hills themselves, which rise to about 2,000 ft. and which are composed of hard volcanic rocks, were almost completely avoided. Taken as a whole, the area was relatively well developed: there were between one and two teams per square mile, and four to six recorded persons. Small amounts of wood were to be found, mainly in the north of the region. A considerable number of mills was to be found in the valleys, but there were only two fisheries. Many of the holdings were wasted after the Conquest, but these had largely recovered by 1086.

(8) *The South-west Upland*

Almost all of this region lies at over 600 ft. and much of it at over 1,000 ft. It includes the hard grits, shales and volcanic rocks of the Longmynd, Stiperstones and Clun massifs. Settlement was confined chiefly to the east of the area; Clun Forest was an empty district. These physical conditions are reflected in the low densities of much of the region—rarely more than 2·5 people and one team per square mile.

Some wood was recorded, particularly in the north, and a number of hays. Mills were located in the valleys, but there were no fisheries. Some holdings in the extreme south were waste, a reminder of the proximity of this region to the unsettled Welsh border.

APPENDIX: THE WELSH BORDERLAND

The Domesday evidence relating to Welsh influence in the area included in modern Shropshire shows two things quite clearly. The first, as we have seen, is that the Welsh were responsible for the wasting of many areas both before and after the Norman Conquest.[1] The second fact is that sixty-four Welshmen were recorded for vills in the extreme west of the modern county (Fig. 54).

But the Domesday county extended westward to include places that now lie in Wales itself, and it is clear that the boundary between Shropshire and Wales was in an extremely fluid condition at the time of the Domesday Survey.

Among the customs of Shrewsbury, we are told that 'when the sheriff decided to march into Wales, anyone who after being warned by him did not proceed forfeited 40s.'. The Normans made a deliberate attempt to stabilise the frontier by the construction of castles at Montgomery (254), at *Luvre* or Oswestry (253b), and at Shrewsbury (252), with the result that in 1086 some unhidated Welsh districts were paying tribute to the Norman conquerors. On fo. 253b we are told that 'Tuder a certain Welshman, holds of Earl [Roger of Shrewsbury] a certain district of Welsh land (*fines terrae Walensis*) and renders therefrom £4. 5s.'. The exact location of this district is not known, but it has been suggested that it formed part of Maelor Saesneg (Flintshire).[2] In the north, the tributary and

---

[1] See p. 144 above.
[2] See (i) R. W. Eyton, *op. cit.* XI, pp. 48–9; (ii) *V.C.H. Shropshire*, I, p. 287.

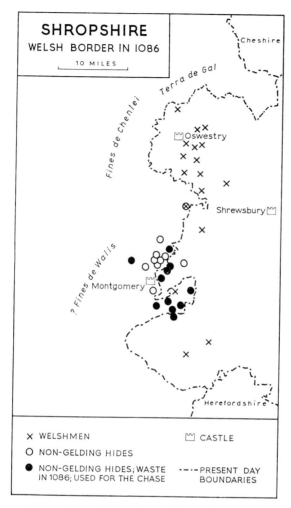

Fig. 54. Shropshire: The Welsh Border in 1086.

There were also non-gelding hides at the unidentified *Horseforde, Staurecote, Benehale* and *Goseford* (254); the two later places were also waste in 1086 and were used for the chase (*ad venandum*).

unhidated Welsh districts of Cynllaith and Edeyrnion (*Chenlei* and *Derniou*) had been joined to the hundred of Mersete, and were held by Rainald the Sheriff. Of these places we read that 'the same Rainald has in Wales two districts (*fines*), *Chenlei* and *Derniou*. From the one he has 60s. by way of rent, and from the other, 8 cows from the Welshmen' (255).

Another unhidated district in Wales was held by Earl Hugh: 'Earl Hugh holds of Earl Roger in Wales the land of Gal (*terra de Gal*).[1] This land extends 5 leagues in length and 1½ leagues in breadth. T.R.E. it was waste and likewise when Hugh received it. In demesne are 3 ploughs and 2 priests and 33 men having between them 8 ploughs and 1 plough more could be employed. A mill there yields nothing. The whole is now worth 40s.' (254). Farther south, at the castle of Montgomery, we read that 'the Earl himself has 4 ploughs and he has 6 pounds of pence from a certain district of Wales (*de uno fine de Walis*) belonging to the same castellany. Roger Corbet has there 2 ploughs and from Wales, he and his brother have 40s.' (253 b). This nameless district of Wales was possibly the hundred of Arvester, which is mentioned in the Domesday Survey of Cheshire, and which has been identified as the 'cantref' of Arwystli adjoining Kerry in the west; the hundred lay in the debatable area between Gwynedd and Powys. Although in the possession of Roger of Montgomery we are told in the Cheshire folios that this land was claimed by Robert of Rhuddlan and that the Welsh themselves bore witness that it belonged to North Wales (269).

The Normans also held a number of hidated but geld-free settlements in Wales. Most of them were attached to the castle of Montgomery, and were located in the countryside around, mainly to the west of the later boundary of Shropshire. They were entered as two groups (254). The first comprised holdings at 13 places; they had been, and still were, waste, and were quit of all geld and used for the chase (*ad venandum*).[2] The second group consisted of holdings at 9 places: they were inhabited and their resources are summarised together as follows:

In these, there are in demesne 9½ ploughs, and 15 villeins and 14 bordars with 3 radmen and 3 serfs have 12½ ploughs. They were waste. Now they are worth 100s. In Hem are 3 fisheries and wood with one hay. In Edderton, wood for fattening 60 swine. In Ackley 1 hay.

Nine other entries relating to places now in Wales are scattered over folios 255b–260b. Only one holding is said to pay geld; two are specifically described as non-gelding; and for the remainder, the hidage is

[1] Yale, the district about Llangollen in Denbighshire.
[2] One of these holdings was at Mellington where there was a second holding of one virgate with 2 serfs and a team (259b).

entered without comment. The details of the nine entries are summarised below:

Mellington (259b). One gelding virgate; Land for 1 plough, which is in demesne with 2 serfs; It was and is worth 5s.

Bausley (255b). 1 non-gelding hide; 2 Welshmen with 1 plough; It was and is worth 2s., nevertheless is at farm for 6s. 8d.

Leighton (255b). 1 non-gelding hide; In demesne 1 plough and 2 oxmen and 1 radman; Wood 2 leagues long sufficient for fattening 200 swine; It was and is worth 5s.

Church Stoke (259b). 5 hides; Land for 7 ploughs; One Welshman has 1 plough; Wood for fattening 100 swine; It was worth 10s. Now 64d.

Stanage (260): 6 hides; Land for 15 ploughs; It was and is waste; 3 hays.

Cascob (260): ½ hide; Land for 2 ploughs; It was and is waste; A wood and 1 hay.[1]

Ackhill (260): 3 hides; It is and was waste; ½ league of wood.

Knighton (260b): 5 hides; Land for 12 ploughs; It was and is waste: A large wood (*Silva magna*).

Norton (260b): 5 hides; Land for 12 ploughs; It was and is waste; A large wood (*Silva magna*).

Taken together, these various entries serve to indicate something of the character of the Welsh Marches.

BIBLIOGRAPHICAL NOTE

(1) The standard text of the Shropshire folios is that prepared by C. H. Drinkwater in *V.C.H. Shropshire* (London, 1908), I, pp. 309–49. This contains a valuable series of identifications and footnotes and is preceded by an introduction by J. Tait (pp. 279–308).

(2) R. W. EYTON, *Antiquities of Shropshire*, in twelve volumes (London, 1854), contains many Domesday references, and includes a table and a map for each Domesday hundred.

J. C. ANDERSON, *Shropshire: its early history and antiquities* (London, 1864). This also contains tables of Domesday hundreds.

(3) The following should also be noted:

W. J. SLACK, 'The Shropshire Ploughmen of Domesday Book', *Trans. Shrops. Archaeol. Soc.* (Shrewsbury, 1939), L, pp. 31–5.

D. SYLVESTER, 'Rural Settlement in Domesday Shropshire', *Sociological Review* (London, 1933), XXV, pp. 244–57.

[1] Another half-hide for Cascob is entered in the Herefordshire folios (186b)—
see p. 113 above.

LYNN H. NELSON, *The Normans in South Wales, 1070–1171* (University of Texas, Austin, 1966).

D. SYLVESTER, *The rural landscape of the Welsh Borderland* (London, 1969).

(4) There are two maps showing Domesday woodland and Domesday arable land, together with some discussion of them, in E. J. Howell, *Shropshire* (London, 1941), pp. 278–82, being Part 66 of *The Land of Britain*, ed. L. Dudley Stamp.

CHAPTER IV

STAFFORDSHIRE

BY P. WHEATLEY, M.A.

Although Sampson Erdeswick in the late sixteenth century had shown the value to Staffordshire historians of Domesday Book as a source for genealogical studies,[1] and Stebbing Shaw had included a Latin transcription of the text in his history of the county,[2] it was not until 1881 that a systematic study of the Staffordshire folios appeared. This was the third of R. W. Eyton's county studies in which he applied to Staffordshire the principles of interpretation that he had already developed in monographs on Dorset and Somerset.[3] Although his deductions and calculations can no longer be accepted as valid, the tables in which he set out and totalled the items recorded for all the holdings in each hundred, have been the quarry for most subsequent investigators. Eyton was also the first to comment on that feature which characterises the Staffordshire record, that is, the high proportion of waste villages and the low standard of prosperity over the county as a whole.

The county of Stafford, in terms of which this study is written, is far from being the same as the Domesday county, for the intervening centuries have seen frequent boundary changes. In the west and south, Staffordshire has on the whole lost territory, but has gained a small area in the east. These changes have not been confined to any one period. Alveley, Kingsnordley, Claverley and Worfield were included in the Domesday county, but had been transferred into Shropshire before the death of Henry I, while Cheswardine and Chipnall followed during the reign of Henry II. The parish of Sheriff Hales, and with it Brockton Grange, was also transferred into Shropshire during the administrative reorganisation of the last century, and it was during the same period that

[1] S. Erdeswick, *A Survey of Staffordshire, containing the antiquities of that county, with a description of Beeston-Castle in Cheshire; publish'd from Sir W. Dugdale's transcript of the author's original copy* (London, 1717).
[2] Stebbing Shaw, *The History and Antiquities of Staffordshire* (London, 1798–1801), Appendix.
[3] R. W. Eyton, *Domesday Studies: an Analysis and Digest of the Staffordshire Survey* (London, 1881).

Upper Arley and, presumably, *alia Ernlege*, were transferred to Worcestershire. During the past century, too, the concentration of population in and around Birmingham necessitated the creation of new parishes and the reconstitution of old boundaries, with the net result that the settlements of Perry Barr, Handsworth and Harborne, all situated within Staffordshire in 1086, are now in Warwickshire. Tamworth is mentioned only incidentally in the folios for both Staffordshire and Warwickshire, and it was not wholly included within the former county until as late as 1890.

Against these losses must be set some gains. Tyrley, surveyed in Domesday Shropshire, was transferred into Staffordshire during the reign of Henry I. Edingale, Croxall, Stapenhill and Winshill of the Derbyshire folios have been brought within the bounds of Staffordshire during the course of the last seventy years.[1] Only in the north has the boundary remained stable.

There is another complication in that several villages in Domesday Staffordshire are described in the folios of other counties, and holdings in other counties appear in the account of Staffordshire. Of the first type are Lapley, Marston, and West Bromwich which are all recorded in the Northamptonshire folios, while Chillington, although in Domesday Staffordshire, was surveyed in Warwickshire. *Sibeford* on fo. 250 of the Staffordshire record seems to have been a holding in Sibford Gower, situated as far away as Oxfordshire. Similar to this is the entry for Drayton which, apart from one or two slight variations in wording and the substitution of Turchil for Turstin as tenant, is repeated on fo. 160b of the Oxfordshire Domesday. Charlotte S. Burn, who brought this gloss to the notice of Eyton, comments, 'It is clear from this, that the entry of Draitone, like that of Sibeford, which it immediately follows, was inserted in an almost blank column, after the fair copy of the Survey was completed, by a scribe who had confused his materials, and did not know where these two entries ought to go'.[2] Another duplication, but this time of a Staffordshire village (250), is to be found on fo. 243 of the Warwickshire Survey, where the entry for Essington is repeated word for word.[3]

[1] For these boundary changes, see R. W. Eyton, *op. cit.* pp. 6–7. Other changes in the county boundary have occurred, but have proved to be only temporary. Tardebigge, Clent and Broom, for example, were for a time annexed to Staffordshire, but are now again incorporated in Worcestershire.

[2] Quoted by R. W. Eyton, *op. cit.* p. 133.

[3] R. W. Eyton, *op. cit.* p. 4. Eyton believed that the four manors of Quatt, Romsley,

These anomalies in the county headings are relatively few. More numerous and more confusing are the errors, and especially the omissions, in the hundred rubrics. Fifteen consecutive waste holdings belonging to the King, for example, were assigned to Cuttlestone hundred (246b), whereas it is clear that they must have formed part of Pirehill; and on the same folio a further block of seventeen was falsely rubricated under Pirehill instead of Totmanslow. The fiefs of Westminster Abbey and Burton Abbey, and of the clergy of Wolverhampton, all on fo. 247b, were wholly without hundred rubrics, and in a number of other instances it is difficult or impossible to say to how many entries a hundred heading was intended to apply. The hundred boundaries as reconstructed by Eyton are shown on Fig. 55.

In addition to these errors in rubrication there is a number of omissions which will be noted in the following pages. One of the more interesting is the lack of a name for one of Robert de Stafford's holdings on fo. 249b, but Eyton has identified this anonymous village as Weston Jones. Usually the omissions concern only the figures for one of the items in an entry. The figures for plough-lands in the entries for Chartley (246b) and Cotwalton (248), for example, were never inserted, while a space follows the figure *ii* at the end of the list of peasantry in the village of Salt (248b), so that we have no means of deciding to which group these two men belonged. Other entries omit, or seem to omit, any reference at all to plough-lands, plough-teams or population, but on the whole, these omissions are exceptional, and the Staffordshire folios seem to be fairly complete.

Another difficulty in handling the Staffordshire material from a geo-graphical point of view is the fact that the information about two or more places is sometimes combined into one statement (Figs 56 and 57). Altogether, twenty-seven holdings are listed as having dependencies or 'members'.[1] In many entries, such as those for Wolstanton (246b), Penkhull (246b), Yarlet (248) and Norton in the Moors (249), the dependencies are not named, and their appurtenances are presumably included with those of the main village. But sometimes the resources of

Rudge and Shipley were actually in Staffordshire although they were rubricated as being in Stoneleigh hundred in Warwickshire; but whether this were so or not, all four have since been transferred into Shropshire (*op. cit.* p. 2).

[1] In addition there were eleven places to which single holdings belonged (*pertinet*) but which were separately described. There are also a number of straight-forward joint entries covering two or more places.

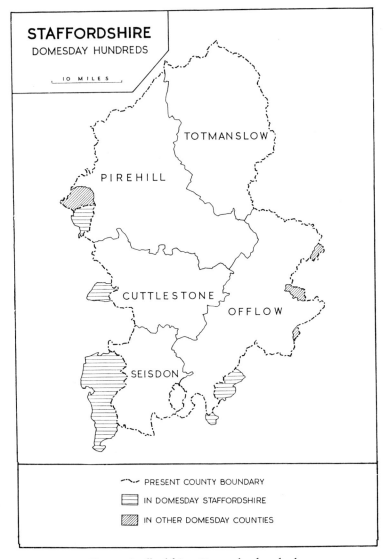

STAFFORDSHIRE
DOMESDAY HUNDREDS

10 MILES

TOTMANSLOW

PIREHILL

CUTTLESTONE

OFFLOW

SEISDON

PRESENT COUNTY BOUNDARY
IN DOMESDAY STAFFORDSHIRE
IN OTHER DOMESDAY COUNTIES

Fig. 55. Staffordshire: Domesday hundreds.
From R. W. Eyton's *Domesday Studies: an Analysis and Digest of the Staffordshire Survey* (London, 1881).

the head of a manor are described separately, followed by those of its members combined into one account. The manor of Sugnall with its nine berewicks, which are named, is treated in this way on fo. 247. The entry for Penkridge (246) is also on this pattern but a little more complicated: the hidages and plough-lands of its six members are given separately, but their other resources are described collectively.

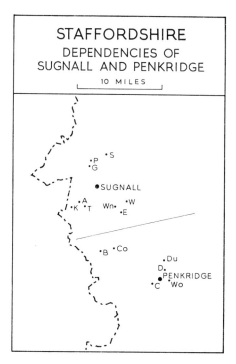

Fig. 56. Staffordshire: Dependencies of Sugnall and Penkridge.

(a) *Sugnall:* A, Adbaston; E, Ellenhall; G, Gerrard's Bromley; K, Knighton; P, Podmore; S, Swinchurch; T, Tunstall; W, Walton; Wn, Wootton.

(b) *Penkridge:* B, Beffcote; C, Congreve; Co, Cowley; D, Drayton; Du, Dunston; Wo, Wolgarston.

Bradley and its eleven dependencies (248b, 249b) constitute another composite manor where the hidage is entered for each village, but where plough-lands, values and woodland are recorded for the unit as a whole; the rest of the manor's resources are given under two totals, one for Bradley itself, and one for all the dependencies (*inter omnes berewichas*). This entry has an additional interest for the component villages are

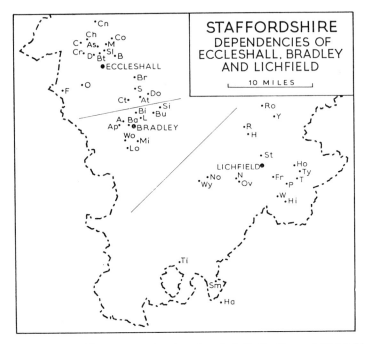

Fig. 57. Staffordshire: Dependencies of Eccleshall, Bradley and Lichfield.

(a) *Eccleshall:* As, Aspley; At, Aston; B, Baden; Br, Bridgeford; Bt, Brockton; C, Charnes; Ch, Chatcull; Cn, Chorlton; Co, Cotes; Ct, Coton; Cr, Croxton; D, Dorsley; Do, Doxey; F, Flashbrook; M, Meece (Cold and Mill); O, Offley; S, Seighford; Sl, Slindon.

(b) *Bradley:* A, Alstone; Ap, Apeton; Ba, Barton; Bi, Billington; Bu, Burton; L, Littywood; Lo, Longnor; Mi, Mitton; Si, Silkmore; Wo, Woolaston.

(c) *Lichfield:* Fr, Freeford; H, Handsacre; Ha, Harborne; Hi, Hints; Ho, Horton; N, Nether Hammerwich; No, Norton Canes; Ov, Over Hammerwich; P, Packington; R, Ridware; Ro, Rowley; Sm, Smethwick; St, Stychbrook; T, Tamhorn; Ti, Tipton; Ty, Tymmore; W, Weeford; Wy, Wyrley; Y, Yoxall. *Burouestone* and *Litelbech*, dependencies of Lichfield, cannot be identified; Harborne lies outside the county.

referred to in no less than four different ways: as dependencies (*appendicii*), berewicks (*berewichae*), members (*membra*), and lands (*terrae*). Lichfield and Eccleshall, with twenty-two and twenty dependent holdings respectively, were two of the largest composite manors in Staffordshire, and their entries are among the most complicated. In the first place the survey of each manor and its members is given in two distinct groups of

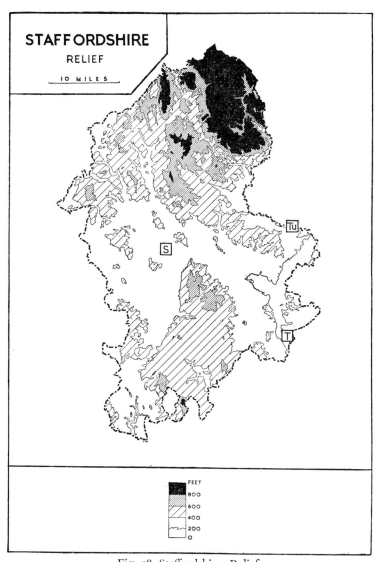

Fig. 58. Staffordshire: Relief.
Domesday boroughs are indicated by initials: S, Stafford;
T, Tamworth; Tu, Tutbury.

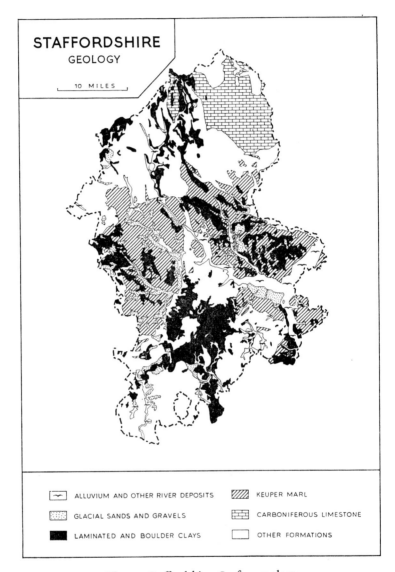

Fig. 59. Staffordshire: Surface geology.

Based on Geological Survey One-Inch Sheets. (New Series) 110, 123, 139–141, 153, 154, 167 and 168; and D. P. Bickmore and M. A. Shaw (eds), *Atlas of Britain and Northern Ireland* (Oxford, 1963), p. 59.

entries on fo. 247, separated by entries for other places. The first group of entries for each manor consists of a general statement of the resources of the head of the manor itself, followed by a list of its waste members, with their plough-lands (or sometimes carucates). The second group of entries gives a list of the inhabited members of each manor followed by a summary of their resources. This arrangement is further complicated by the fact that three of the Eccleshall holdings are described separately in detail, and by the presence of entries for Ellastone and Broughton which seem to belong to neither manor. For the purpose of constructing distribution maps the details of these combined entries have been divided exactly among all the villages concerned. But as the woodland is measured in leagues and furlongs, it cannot be so divided, and it has therefore been plotted for the first-named village only, or, in the case of a manor and its dependencies, for the head of the manor. Mills have been treated likewise.[1]

SETTLEMENTS AND THEIR DISTRIBUTION

The total number of separate places mentioned in the Domesday Book for the area included in the modern county of Stafford seems to be 334, including the three places for which burgesses are recorded—Stafford, Tamworth and Tutbury.[2] This figure may not be quite accurate for two reasons. In the first place, each of a dozen or so entries mentions an unspecified number of dependencies which may or may not refer to localities named elsewhere in the folios.[3] The second source of possible inaccuracy is that when two or more adjoining villages bear the same name today, it is not always clear whether more than one unit existed in the eleventh century. There is no indication that, say, the Upper and Lower Tean of today existed as separate villages; the Domesday information is entered under only one name, *Tene*. The same applies, for example, to Stanton Mayfield and Church Mayfield, to Sugnall and Little Sugnall, and to Upper and Lower Hatton. The distinction between the respective units of each of these groups appears later in time. In one entry we are told specifically that there were two villages of the same name. On fo. 247 the text refers to 'the two Hammerwiches' (*Duae Humerwich*). For some

---

[1] See pp. 194 and 203.

[2] T. Harwood counted '334 towns, villages and hamlets' in the Domesday county (*A Survey of Staffordshire* (Westminster, 1820), p. 421).

[3] For similar entries in Shropshire, see p. 119 above.

counties the Domesday text occasionally differentiates between the related units of a group by such designations as *alia* or *parva*. The only example of this in Staffordshire is the distinction between Sandon and Little Sandon, which appear as *Scandone* (246) and *parva Sandone* (248 b), but the two Arleys, *Ernlege* and *alia Ernlege* (247 b), now in Worcestershire, were also described in the Staffordshire folios.

Apart from those vills which were waste,[1] the total of 334 includes a number of places for which very little information is given; the record may be incomplete, or the details may have been included with those of neighbouring villages. The hidage for Compton is embedded in the entry for Tettenhall (246), together with the information that it belonged to the manor (*pertinens ad Totehala*), but there is no other mention of it. Agardsley, *Colt*, Walton and Brockhurst (*Ruscote*) are similar places whose details are almost certainly included respectively with those of Marchington (248 b), Colton (248), Baswich (247) and Blymhill (249 b). For a holding in Onn (250 b) only hides and plough-lands are recorded; for Marston in Church Eaton (222 b) only hides, plough-lands and its value.

Not all the Domesday names appear on the present-day map of Staffordshire villages. Some are represented by hamlets; about a score survive only as names of farms, and another thirty or so as country seats. Thus *Cuneshala* is now the hamlet of Consall in the parish of Cheddleton, and *Sceotestan* is that of Shushions in Church Eaton. *Mortone* is the hamlet of Moreton in Colwich, and *Morve* is now the name (Morfe) of four farms in Enville. Penkridge is an interesting example of a parish in which no less than twelve Domesday names are represented by present-day hamlets,[2] while Eccleshall parish, with seventeen such names, is even more noteworthy.[3] *Pendeford* is preserved unaltered as the names of Pendeford Hall and Pendeford Farm, while *Chenwardestone*, *Musedene* and *Turvoldesfeld* have become respectively Kinvaston Hall, Musden Grange and Thursfield Lodge. *Fricescote* is now represented by Syerscote Barn and Manor, and *Cocretone* by Crockington Lane in Trysull and

---

[1] As many as one fifth of the villages in the county were wholly waste in 1086— see p. 202 below.

[2] Bickford, Congreve, Drayton, Gailey, Levedale, Mitton, Otherton, Pillaton (*Beddintone*), Rodbaston, Water Eaton, Whiston and Wolgarston.

[3] Aspley, Baden, Brockton, Broughton, Charnes, Chatcull, Cotes, Croxton, Dorsley, Gerrard's Bromley, Meece (Cold and Mill), Podmore, Slindon, Sugnall, Swinchurch, Walton and Wootton.

Seisdon.[1] Still other villages, such as *Haswic*, have disappeared altogether from the landscape. Another is *Dorueslau* (247), which continued to appear in records for more than five centuries after it was mentioned in Domesday Book. Shaw[2] and Harwood[3] identified it as Dodsley, while Eyton[4] and Wedgwood[5] concluded correctly that it was an obsolete settlement, but its site was unknown until C. G. O. Bridgeman found two pastures in Little Sugnall which as recently as 1900 were known as Doesley or Dorsley.[6] These are but a few of the vicissitudes of the Staffordshire villages. To them must be added a number of unidentified names: *Burouestone* (247), *Cippemore* (249b), *Cobintone* (248), *Colt* (248), *Litelbech* (247) and *Monetvile* (249b). It is impossible to say whether they will yet be located or whether the places they represent have disappeared without trace, but each of them may be fairly confidently assigned to one or other of the regions on Figs. 62 and 64, so that the densities are unaffected by these uncertainties. *Colt* (248), for example, was almost certainly situated very near to the modern village of Colton; *Burouestone* and *Litelbech* (247) were in the vicinity of Weeford; while *Monetvile* (249b) seems to have been near Stafford.[7] Although there is room for a difference of opinion as to whether *Cippemore* (249b) is to be identified with Combere, near Kinver,[8] or Great Moor in the parish of Pattingham,[9] it seems reasonably certain that the village was situated somewhere in the region entitled South-western Lowland.

On the other hand, some villages on the modern map are not mentioned in the Domesday Book. Their names do not appear until later, and, presumably, if they existed in 1086, they are accounted for under the statistics of neighbouring settlements. A few of the places not mentioned in the Domesday Book (e.g. Walsall) must have existed, or at any rate been named, in Domesday times because they appear in pre-Domesday

[1] J. C. Wedgwood, 'Early Staffordshire History (from the Map and from Domesday)', *Staffordshire Record Society: Collections for a History of Staffordshire* (London, 1916), p. 170.

[2] S. Shaw, *op. cit.* Appendix, p. xi.   [3] T. Harwood, *op. cit.* p. 411.

[4] R. W. Eyton, *op. cit.* p. 36.   [5] J. C. Wedgwood, *op. cit.* p. 168.

[6] C. G. O. Bridgeman, 'Some Unidentified Domesday Vills', *Staffordshire Record Society: Collections for a History of Staffordshire* (London, 1923), p. 33.

[7] W. F. Carter, 'On the Identification of the Domesday Monetvile', *Staffordshire Record Society: Collections for a History of Staffordshire* (London, 1908), pp. 227–30.

[8] (i) R. W. Eyton, *op. cit.* p. 35; (ii) J. C. Wedgwood, *op. cit.* p. 170.

[9] C. G. O. Bridgeman and G. P. Mander, 'The Staffordshire Hidation', *Staffordshire Record Society: Collections for a History of Staffordshire* (London, 1919), p. 165.

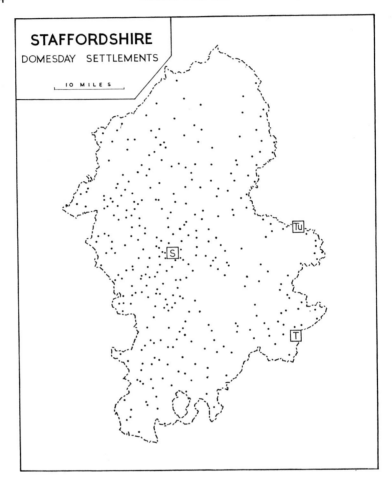

Fig. 60. Staffordshire: Domesday place-names.
Domesday boroughs are indicated by initials: S, Stafford;
T, Tamworth; Tu, Tutbury.

documents. Sometimes, a modern parish which is not named in the Domesday Book, contains hamlets that are mentioned. A noteworthy example is the parish of Stone. This is not represented by a Domesday name but the parish includes eight Domesday vills.[1] From this account, it is clear that there have been many changes in the village geography of

---

[1] Aston, Cotwalton, Darlaston, Little Stoke, Meaford, Moddershall, Tittensor and Walton.

the county, and that the list of Domesday names differs considerably from that of present-day parishes.

Fig. 60 shows that vills lay thickest where soils were most fertile, that is, on the strong soils of the central claylands. Over stretches of Bunter Sandstone and Coal Measures, which, when drift free, usually give rise to poor soils, villages were much less frequent. These areas are conspicuous by their emptiness. In the Pennine area of the north-east there were relatively few villages above 600 ft., and practically none above 800 ft. The village pattern of this upland was fitted into the valleys, leaving the ridges bare of settlement. The other two areas noticeably without villages were the Cannock Hills and the broad upland between the Dove and Trent rivers, and it may not be without significance that here, at a later date, were the sites respectively of Cannock Chase and Needwood Forest. Although the Domesday Book makes no reference to 'afforestation' in these areas, it is possible that they already lay under forest law, and that no returns were made for them.

THE DISTRIBUTION OF PROSPERITY AND POPULATION

Some idea of the nature of the information in the Domesday folios for Staffordshire, and of the form in which it is presented may be obtained from the entry relating to Penkhull (246b), a village situated on a spur overlooking the Upper Trent Valley, and now a suburb of Stoke on Trent. In this village the king was the only land-holder, so it is covered by a single entry:

The king holds Penkhull. Earl Algar held [it]. There [are] 2 hides with their dependencies. There is land for 11 plough-teams. In demesne are 2 [plough-teams] and 17 villeins and 6 bordars with 8 plough-teams. There [are] 2 acres of meadow. Wood one league long and 2 furlongs broad. It is worth £6.

By no means every entry includes all the items listed for this village, and neither does this entry include all the kinds of information that appear elsewhere in the folios for the county. It does not mention serfs, for example, or priests, or freemen. It makes no reference to a mill, to underwood or to waste and its value is given for only one date, 1086, whereas some holdings are valued for both 1086 and 1066. The Domesday folios of some counties record churches but only two are mentioned in Staffordshire, and there is no reference to pasture, vineyards, salt-pans, or other agricultural and mineral products, which feature prominently in some other counties. But although the entry for Penkhull is not comprehensive, it is fairly representative and straightforward, and it does set out the recurring standard items that are found for most villages. These are five in number: (1) hides; (2) plough-lands; (3) plough-teams; (4) population; (5) values. The bearing of these five items of information upon regional variations in the prosperity of the county must now be considered.

(1) *Hides*

The Staffordshire assessment is normally stated in terms of hides and virgates. Part of a hide is expressed either as a simple fraction, or as so many virgates:

> Crakemarsh (246b): There [is] half a hide (*Ibi dimidia hida*).
> Bucknall (246b): a third of a hide (*tercia pars hidae*).
> Loxley (248): There [is] a quarter of a hide (*Ibi iiiita pars hidae*).
> Shelton under Harley (246b): one virgate of land (*una virgata terrae*).

Three-quarters of a hide is written in one of two ways:

> Milwich (249): *iiies partes unius hidae*.
> Lichfield (247): *xxv hidae et dimidia et una virgata terrae*.

Nowhere in the Staffordshire folios is there any reference to geld acres, a fact which is the more remarkable in view of the low assessment of the county. Precise assessments are rare, but when they do occur the scribe seems to have been at some pains to avoid using acres to express even the more complicated fractions. The following entries give some idea of the circumlocutions to which he resorted:

> Balterley (250b): half a virgate of land (*dimidia virgata terrae*).
> Himley (249b): 2 hides less half a virgate of land (*ii hidas dimidia virgata terrae minus*).
> Hanchurch (250b): three-quarters of half a hide (*iii partes dimidiae hidae*).

The hidage given in the entries for Eccleshall and Lichfield (247) includes those of their dependencies, but in other combined entries the hidage is stated separately for each member of the manor. No holding in Staffordshire was specifically exempted from assessment, but occasionally the hidage of a holding seems to have been omitted. This is most conspicuous for a group of fourteen waste estates on fo. 246b. Presumably these were unhidated or it was impracticable to ascertain their earlier assessments. Spaces were left in the three entries for Acton Trussell, Hixon and Fradswell (247), as if the scribe had intended to fill in the assessment after some further enquiry. Indeed, he had made a note in the margin against Hixon to enquire how much land was there (*require quantum terrae*).

The first published attempt to discover how far the five-hide principle was applicable to Staffordshire was that of J. C. Wedgwood in 1916.[1] In the southern parts of the county he found several examples of vills each assessed at 5 hides, but more often the total assessment of a group of vills made up a five-hide unit. Oaken (249), Kingswinford (246), and Morfe (249b), for example, were each assessed at 5 hides; while 2 hides at Oakley (249) and 3 at Elford (246b) made up a similar unit. In addition, Wedgwood showed that the hides of several groups of villages could be combined to form multiples of 5 hides, e.g. Wombourn with 7 hides and Orton with 3 together formed a unit of 10 hides (249b). North of Stafford town, Wedgwood could find but one such unit of 5 hides, so he concluded that the south was hidated in five-hide units, the north was not. In 1919 C. G. O. Bridgeman continued this study and showed that it was possible to make up not only many more groups of 5 hides, but also a considerable number of other groups which gave totals very close to 5 or a multiple of that figure (see table on p. 178).[2]

Shortly after this Bridgeman collaborated with G. P. Mander in a more intensive analysis of the Staffordshire hidage.[3] These authors agreed with Wedgwood that the clearest evidence for the five-hide principle came from the southern hundreds of Seisdon and Cuttlestone, but they also found traces of such a grouping in the northern hundreds of Pirehill and Totmanslow. In the south-eastern hundred of Offlow, too, there were many indications of five-hide units, but there they seemed to be competing with three-hide units. In all, Bridgeman and Mander counted sixteen three-hide units in Offlow hundred, together with six units of a hide and a half. Yet it does seem fairly well established that the five-hide principle had been applied to the whole of Staffordshire, but that it had been complicated perhaps in the eastern part of the county by some Danish influence.

[1] J. C. Wedgwood, *op. cit.* pp. 175–8.
[2] C. G. O. Bridgeman, 'Notes on the Contents of the Volume for 1916; the Five-Hide Unit in Staffordshire', *Staffordshire Record Society: Collections for a History of Staffordshire* (London, 1919), pp. 134–51.
[3] C. G. O. Bridgeman and G. P. Mander, *op. cit.* pp. 154–81.

	Hides	Virgates		Hides	Virgates
Biddulph (246b)	1	0	Heighley (246b)		0½
Thursfield (250b)		1	Knutton (250b)		1
Talke (250b)		1	Whitmore (250b)		2
Audley (250b)		2	Hanchurch (250b)		1½
Balterley (250b)		0½	Madeley (249)	1	0
Balterley (250b)		0½	Shelton under		1
Betley (250b)		2	Harley (246b)		
Dimsdale (250b)		1	Hatton (246b)		2
Wolstanton (246b)	2	0	Swynnerton (249)	2	0
Total	5	0	Total	5	0

	Hides	Virgates
Bushbury (247b)		1
Bushbury (250)	2	2½
Bushbury (250)		1
Moseley (250)	1	0
Oxley (249b)	1	0
Total	5	0½

These same authors also came to the conclusion that the hidage of Staffordshire was probably apportioned as follows:

Seisdon hundred   180 hides (a long hundred and a half).
Cuttlestone hundred 120 hides (a long hundred).
Offlow hundred   120 hides (a long hundred).
Pirehill hundred   90 hides (three-quarters of a long hundred).
Totmanslow hundred   30 hides (one-quarter of a long hundred).

The use of the long hundred might be interpreted as further evidence of Danish influence in Staffordshire.

Whether the five-hide unit can be demonstrated or not, it is clear that the assessment was largely artificial in character, and bore no constant relation to the agricultural resources of a vill. In a very general way, it is true that the larger the number of hides, the greater the resources of a vill, but there are many exceptions. The variation among a representative selection of two-hide vills speaks for itself:

	Teams	Population
Bradley in the Moors (250b)	3	10
Ashley (248)	1	4
Swynnerton (249)	7	15
Marchington (248b)	5	28
Wolstanton (246b)	10	17

This relatively straightforward picture of the Staffordshire assessment is complicated by occasional references to carucates (*carucatae terrae*). Some of these are easily explained. In the entries for the Staffordshire villages which were recorded in the Derbyshire folios, the hide is replaced by the carucate, but this is the normal unit of assessment in that county. In an entry for Edingale (278b), for example, we read of 2 carucates of land for geld (*ii carucatae terrae ad geldum*), and under Croxall (274) of 3 carucates of land for geld. As is usual in the carucated shires, fractions of carucates are expressed in terms of bovates. Three-quarters of a caru-cate which comprised one of the holdings at Stapenhill (278) is recorded as 6 bovates of land for geld (*vi bovatae terrae ad geldum*), and four and a quarter carucates on the other holding in the same village (273) as 4 carucates of land and 2 bovates for geld (*iiii carucatae terrae et ii bovatae ad geldum*). But *carucatae terrae* are also sometimes mentioned in the Staffordshire folios. The holdings which appear to be assessed in this way are set out below (Fig. 61):

Bescot (246): *una carucata terrae vasta*.
Aston (246b): *una carucata terrae et una virgata terrae*.
Consall (246b): *terrae una carucata*.
Cheadle (246b): *una carucata terrae*.
Newton (246b): *ii carucatae terrae*.
Broughton (247): *dimidia carucata terrae*.
Coley (247): *ii carucatae terrae*.
Moreton (247): *ii carucatae terrae*.
Bishton (250b): *i carucata terrae*.
Cannock (250b): *i carucata terrae*.
Stramshall (250b): *ii carucatae et una virgata terrae*.

In each of seven other entries the carucated land is only a subsidiary part of a larger tenement which is assessed in hides, but in the present count all carucates have been totalled separately:

Hammerwich, Nether and Over (247): *terrae v carucatae*.
Norton Canes and Little Wyrley (247): *iiii carucatae terrae*.
Rowley (247): *i carucata terrae*.
Walton (248b): *una carucata terrae*.
Stretton (249b): *una carucata terrae*.
Madeley in Checkley (249, 2nd entry): *i carucata terrae et dimidia*.
Caverswall (249): *cum dimidia carucata terrae*.

The entries for Aston and Stramshall seem to suggest, but not necessarily, that a virgate was a fraction of a carucate, but this possibility has been ignored in the count of carucates.

These Staffordshire carucates have given rise to a considerable amount of speculation. Eyton thought they were 'conceptional hides', applied by the commissioners to 'land which had never been hidated, or whose hidation they could not fix or discover'.[1] Wedgwood assumed a Danish origin for at least some of the carucates, e.g. for those at Bishton and Cannock, and possibly for those of the waste vills of the northern uplands.[2] Bridgeman noted that more than half the carucated land in Staffordshire was waste, and suggested that for these holdings the jurors had made an entirely new estimate.[3] As this was based not on households but on teams, the jurors would naturally express the new assessment in terms of carucates instead of hides. The rest of the carucated land was in the hands either of sub-tenants or of king's thegns, and for these holdings the carucate was a supplementary estimate dating from the creation of the tenancy. None of these interpretations is completely satisfactory.

The assessment of the area covered by the modern county amounted to $440\frac{23}{120}$ hides, $28\frac{1}{2}$ Staffordshire carucates and 13 Derbyshire carucates. Eyton counted 454 hides $2\frac{13}{30}$ virgates for the Staffordshire of his day, but as many as 499 hides $2\frac{13}{30}$ virgates for the Domesday county. In a second total, for the Domesday county, he included $32\frac{1}{2}$ carucates as conceptional hides, and the 35 plough-lands recorded for non-hidated estates as quasi-hides, to make a total of $567\frac{13}{20}$ hides and quasi-hides.[4] Maitland's total of 505 hides for the Domesday county falls at about the average of these estimates.[5] C. F. Slade's figure for the Domesday county was $513\frac{13}{120}$ hides.[6] The fact is that the nature of some of the composite entries makes exact calculation exceedingly difficult. All the figures can do is to indicate the order of magnitude involved.

Whatever be the exact total, one thing is clear and that is the very low assessment of the county as compared with its area (similar to that of Gloucestershire) and its numbers of plough-lands and teams. The County Hidage of the early eleventh century records a similar assessment of

---

[1] R. W. Eyton, *op. cit.* p. 25.

[2] J. C. Wedgwood, *op. cit.* p. 150.

[3] C. G. O. Bridgeman, *op. cit.* p. 147.

[4] R. W. Eyton, *op. cit.* pp. 110 and 132.

[5] F. W. Maitland, *Domesday Book and Beyond* (Cambridge, 1897), p. 400.

[6] *V.C.H. Staffordshire* (London, 1958), IV, p. 2.

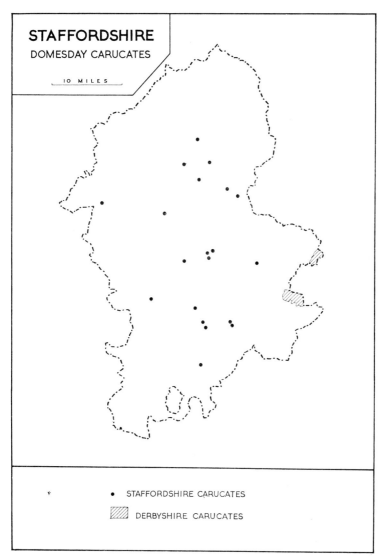

STAFFORDSHIRE

DOMESDAY CARUCATES

10 MILES

• STAFFORDSHIRE CARUCATES

▨ DERBYSHIRE CARUCATES

Fig. 61. Staffordshire: Places assessed in carucates.

The shaded areas are described in the Derbyshire folios, and were assessed in the carucates normal to that county.

500 hides,[1] so that there had been no reduction since then for Stafford-shire; the Shropshire hidage, on the other hand, was considerably reduced.[2]

(2) *Plough-lands*

The normal formula runs: 'There is land for *n* ploughs' (*Terra est n carucis*). Only once does the phrasing vary, and the form of this entry is so unusual as to indicate that it is a scribal error. At Offley (247) there was 'land for 3 ploughs in demesne' (*In dominio habet terram iii carucis*), but *terram* is inserted above, with all the appearance of an afterthought. Before this alteration the entry would have resembled the normal plough-team formula: 'In demesne he has 3 teams' (*In dominio habet iii carucas*). There are a few entries, such as those for Wolstanton (246b) and Sedgley (249b), in which there is no mention of plough-lands, and we are left in doubt as to whether the omission was accidental or intentional. In no less than five entries, although the usual formula is included, a lacuna replaces the figure.[3] Presumably the scribe was unable to obtain any estimate of plough-lands, a not unlikely event in a countryside which had suffered systematic devastation during the past half-century. In fact, there is positive evidence that the Commissioners could do no more than specu-late about the number of plough-lands on six of King William's waste estates (246b). In the five entries for Sheen, Stanshope, Farley, Endon and Rudyard, 'or 2' (*vel ii*) is written above the phrase, 'There is land for one plough', while in the entry for Wootton under Weaver, 'or 3' (*vel iii*) is interlined above '2'. The lower figures have been counted.

Fractional parts of a plough-land sometimes occur e.g. both at Dorsley (247) and at Hilderstone (246b)—*terra est dimidiae carucae*; and at Coton Clanford (247) there was land for 2 oxen (*terra est ii bobus*).

There has been considerable discussion as to the exact meaning of the term plough-land. Some scholars have regarded it as a statement of the number of teams at work in 1066. Others have believed it was an estimate of the potential arable land of a manor; yet others have seen in it a large conventional element. The Staffordshire evidence is not conclusive on any of these points. The entry for Burton (248b)[4] would seem to confirm

---

[1] F. W. Maitland, *op. cit.* p. 456.   [2] See p. 123 above.

[3] The five entries are those for Drayton (246), Chartley (246b), Eccleshall (247), Cotwalton (248) and Warslow (248). The *est terra* for Cotwalton (*Codeuualle*) was deleted in the MS—see *V.C.H. Staffordshire*, IV, p. 47 n.   [4] See p. 210 n. below.

the first of these interpretations, for we read: 'In the time of King Edward there were 12 plough-teams. There are now 4 plough-teams in demesne' (*T.R.E. erant xii carucae. Ibi sunt modo iv carucae in dominio*). No tenants' teams are mentioned. Thus, in this entry it seems that the scribe has deliberately inserted the number of teams in 1066 instead of the number of plough-lands. When this happens frequently, as in the Leicestershire folios, it is natural to ask whether the two expressions were synonymous.[1] No plough-lands are likewise entered for Syerscote (249) but the entry says: *Ibi possent esse carucae*, and then goes on to record the presence of 3½ teams.

The description of Wolverhampton (247b) fits this interpretation less happily, for there is a statement of the teams at work both in 1086 and in 1066, as well as an estimate of plough-lands (*Terra est iii carucis. T.R.E. fuerunt ibi viii carucae, modo sunt x et xiiii servi et vi villani et xxx bordarii cum ix carucis*). The entry for Weston, Beighterton and Brockton Grange (250b) takes a similar form: *Ibi fuerunt xi carucae. Terra est vi carucis. In dominio sunt iii carucae et ii servi et x villani cum ii carucis*. In these entries where plough-lands and teams in 1066 occur side by side, it would seem that the plough-lands must be read as an index of agricultural efficiency. Yet against this conclusion must be urged the fact that on many of the Staffordshire holdings plough-teams out-numbered plough-lands. The entry above shows that 19 teams were at work at Wolverhampton, where the Commissioners, on this hypothesis, considered 3 to be adequate. Equally anomalous is the entry for Clifton Campville (246b) where there were 13 plough-teams and only 4 plough-lands, but these are only two examples from the many that might be quoted. Thus, each of these interpretations is far from satisfactory, but it is possible that the plough-land was neither an estimate of past nor of potential cultivation. There are indications in the folios for other counties that the plough-land was in part an artificial estimate,[2] but on this point the Staffordshire evidence does not seem to be conclusive. Of the entries which record both plough-lands and teams, there was a deficiency of teams on 56% and an excess on 22%.

The total number of plough-lands recorded for the area covered by the modern county is 1,270¼. Eyton counted 1,223¼ for the Staffordshire of 1881,[3] and Maitland 1,398 for the Domesday county,[4] but both these

---

[1] See p. 329 below.   [2] See pp. 329, 369 and 394 below.
[3] R. W. Eyton, *op. cit.* p. 132. This table is a correction of that on p. 15.
[4] F. W. Maitland, *op. cit.* p. 401.

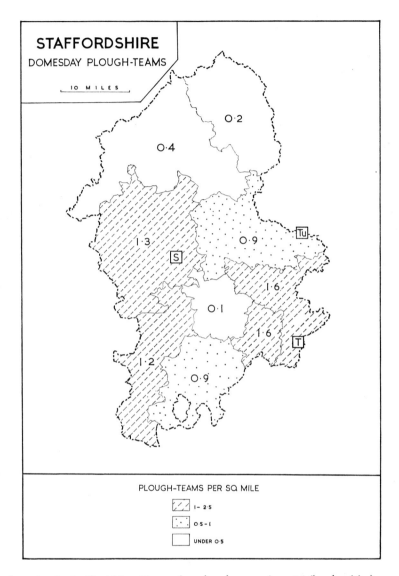

Fig. 62. Staffordshire: Domesday plough-teams in 1086 (by densities).
Domesday boroughs are indicated by initials: S, Stafford;
T, Tamworth; Tu, Tutbury.

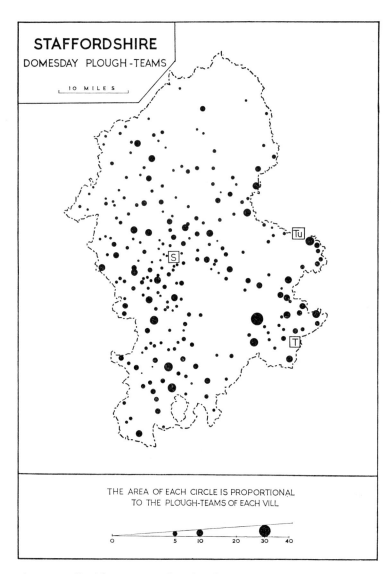

Fig. 63. Staffordshire: Domesday plough-teams in 1086 (by settlements).
Domesday boroughs are indicated by initials: S, Stafford;
T, Tamworth; Tu, Tutbury.

totals were on a rather different basis from the present estimate. In those entries where there seemed to be no mention of plough-lands, and where there were lacunae in those items, Eyton and Maitland both assumed that there were plough-lands equal in number to the recorded plough-teams. A comparable total for the present study is 1,316¾ plough-lands.

(3) Plough-teams

The Staffordshire entries, like those of other counties, draw a distinction between the teams on the demesne and those held by the peasantry. There are frequent variations in the formula, and occasional entries seem to be defective. There is, for example, no information about demesne teams at Sugnall (247) and Derrington (248). At Hamstall Ridware (248), on the other hand, it was the peasantry who had no ploughs. But the record of teams is usually complete and fairly straightforward, and it is possible that the absence of teams may well have been but a transitory vicissitude which, caught in the cross-section of Domesday Book, has been preserved for all time. Half-teams occur occasionally. At Wightwick (246), for example, there was half a team, at Drayton Bassett (246b) 4½ teams, and similar moieties are recorded at Drointon (247), Branston (247b) and Hamstall Ridware (248). At Coton near Stafford (248) there was one plough and 6 oxen (i caruca et vi boves). Just once Domesday Book supplements its usual formula with an item of information unique in the Staffordshire folios. At Rolleston (248b) where there was land for 8 ploughs and where 18 teams were at work, we are told that the arable was 2 leagues long by one wide (Terra arabilis ii leuuis longa et una lata). The exact significance of this statement is obscure, and the notional character of the leagues does not allow of any correlation in terms of modern measure with the 18 teams at work on the holding.

The total number of plough-teams recorded for the area covered by the modern county is 979¼. For the Domesday county, Maitland counted 951; and Eyton counted 990½ for the Staffordshire of 1881.[1]

(4) Population

The main bulk of the population was comprised in the three categories of villeins, bordars and serfs. In addition to these were the burgesses,

---

[1] (i) F. W. Maitland, op. cit. p. 401; (ii) R. W. Eyton, op. cit. p. 132 (this is a correction of his original table on p. 15).

together with a small miscellaneous group that included freemen, priests and others. The details of these groups are summarised on p. 190. There are three other estimates of population, by Ellis,[1] Eyton[2] and Slade,[3] but they are comparable neither with one another nor with the present estimate. Ellis's grand total for the Domesday county included tenants-in-chief, under-tenants, burgesses and the solitary *ancilla*, who are all omitted from the present calculation. Eyton's total referred to the Stafford-shire of his day, but several villages have been transferred into other counties since that time. Furthermore, he counted only the three main groups of peasantry, the villeins, bordars and serfs, and so ignored the miscellaneous category. Slade's totals are for the Domesday county. All that can be claimed for the present figures is that they indicate the order of magnitude involved. These figures are those of recorded population, and must be multiplied by some factor, perhaps 4 or 5, in order to obtain the actual population; but this does not affect the relative density as between one area and another.[4] That is all a map such as Fig. 64 can roughly indicate.

It is impossible to say how complete were these Domesday statistics. As we have seen, a few entries contain no reference to population, but we cannot be certain about the significance of this. Where no peasants are recorded, as at Marston in Church Eaton (222b), it may be that the under-tenants worked the land themselves. This would seem also to be true of Kingsley (250b) and of Normacot (250b), estates where a plough was at work although no labourers were recorded. Again, at Alstonfield (248) a single villein apparently managed two teams. In the entry for Salt (248b) a lacuna follows a *ii* at the end of the list of peasantry, that is, at the point where interpolation is difficult; possibly serfs were meant.

Villeins constituted the most important element in the population of Staffordshire, and amounted to almost 60% of the total. Bordars comprised about 30% of the peasantry, but present no special features for our purpose. The category of serfs formed about 8% of the total, a proportion considerably greater than is found in most of the counties to the east, but much smaller than occurs in parts of the western Midlands. The proportion of serfs to the rest of the population was not uniform throughout the county. In south-western Staffordshire one man in every six or so

[1] Sir Henry Ellis, *A General Introduction to Domesday Book* (London, 1833), II, pp. 486–7.
[2] R. W. Eyton, *op. cit.* p. 132. This table is a correction of that on p. 15.
[3] *V.C.H. Staffordshire*, IV, p. 20.
[4] But see p. 430 below for the complication of serfs.

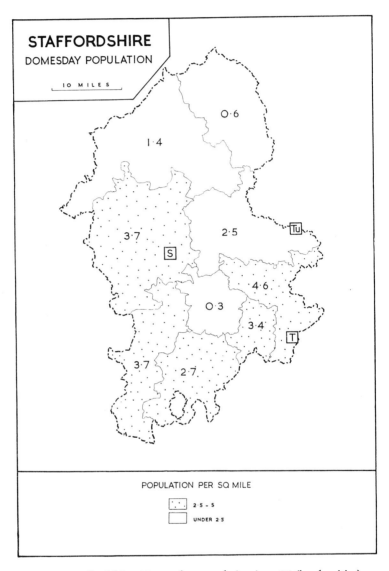

Fig. 64. Staffordshire: Domesday population in 1086 (by densities).
Domesday boroughs are indicated by initials: S, Stafford;
T, Tamworth; Tu, Tutbury.

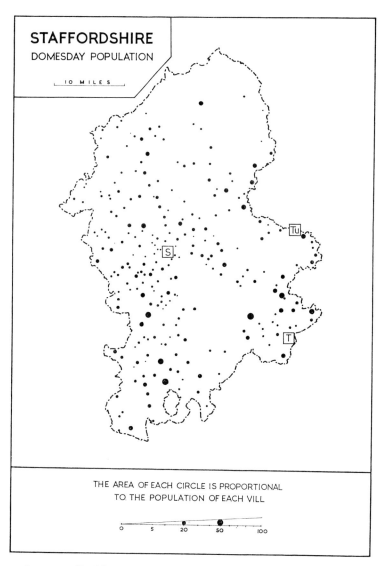

STAFFORDSHIRE

DOMESDAY POPULATION

| 10 MILES |

THE AREA OF EACH CIRCLE IS PROPORTIONAL
TO THE POPULATION OF EACH VILL

0   5   20   50   100

Fig. 65. Staffordshire: Domesday population in 1086 (by settlements).
Domesday boroughs are indicated by initials: S, Stafford;
T, Tamworth; Tu, Tutbury.

was a serf, but in the south-east barely one man in twenty-five. Yet in both districts serfs were to be found in about half the total number of villages. It is interesting to note that there was but a single serf among all the settlements of the north-eastern valleys. At Rolleston (248b) there was a bondwoman (*ancilla*). These were fairly common in some counties, but this is the only one recorded in the Staffordshire folios.

Recorded Population of Staffordshire in 1086

A. Rural Population

Villeins	1,669
Bordars	886
Serfs	229
Miscellaneous	59
Total	2,843

There was also one bondwoman (*ancilla*) who has not been included in the above total.

Details of Miscellaneous Rural Population

Priests	30
Freemen	13
Sokemen	6
Knights	3
Angli	3
Servientes	2
Thegn	1
Prepositus	1
Total	59

B. *Urban Population*

Villeins, bordars and serfs are also included in the table above.

STAFFORD   36 burgesses; 110 *mansiones hospitatae*; 13 canons; 13 villeins; 8 bordars; 4 serfs; 51 *mansiones vastae*; 1 *vasta masura*.

TAMWORTH   22 burgesses.

TUTBURY   42 *homines*.

The miscellaneous category is a varied one with no prominent group. There were a few freemen (*liberi homines*), distributed singly throughout

the county except for two each at Bilbrook (247b) and *Cobintone* (248);[1]
six sokemen 'belonging to Repton' were actually at Winshill (273), a
village in Domesday Derbyshire but in modern Staffordshire. It is difficult
to draw a comprehensive distinction between these two categories of men,
but together they comprised the free element in the rural population.
A number of priests (*presbyteri*) are mentioned, and it is clear from the
phrasing of such references that they often joined with others of the
peasantry to provide teams of plough-oxen. At Walton in Stone (248b),
for instance, 7 villeins, 2 bordars, 5 serfs and a priest had 4 teams, and at
Alrewas (246) 20 villeins, 6 bordars and a priest had 6 teams. The rest
of this miscellaneous category was composed of small groups of men:
3 knights (*milites*),[2] 3 Englishmen (*anglici*),[3] 2 serjeants (*servientes*),[4]
a thegn (*tainus*),[5] and a reeve (*prepositus*).[6] It is difficult to say why these
individuals were recorded, for there must have been others in the same or
similar groups. In particular, the complete absence in the Domesday
folios of craftsmen, such as smiths and carpenters, and of such agri-
culturalists as swineherds, is a surprising feature. It may well be that the
craftsmen in general were usually reckoned among the villein tenants.

Fifty-four thegns and 21 freemen are recorded for 1066, e.g. 5 thegns
at Brineton (249b) and 3 freemen at Coppenhall (249b). Included in the
total of thegns are 4 free thegns (*teini liberi*) at Saredon (249b).

(5) *Values*

The value of an estate is normally recorded for only one date, 1086, and
is given as a plain statement of money, under the formula, 'It is worth
n pounds or shillings'. Rarely a little further information is added.
Burton (248b), for example, was worth 24s. a year (*per annum*).[7] Sometimes the rents of a number of holdings are included in a single total.
The value of Warslow seems to be included with that of Alstonfield (248),
and of Basford with that of Cheddleton (248b), while the whole manor of
Bradley (248b) with its ten members was worth £7 (*Totum Manerium
cum membris valet vii libras*). A notable example of such a single total
occurs on fo. 247b, where we read that the total value of the thirteen
estates of the canons of Wolverhampton was £6 a year (*Tota haec*

---

[1] Unidentified, but see *V.C.H. Staffordshire*, IV, p. 46 n.
[2] Hatherton (247b); Moreton in Gnosall (248); Alstone (248b).
[3] Moreton in Gnosall (248); Syerscote (249).   [4] Seisdon (250).
[5] Members of Penkridge (246).   [6] Trentham (246).   [7] See p. 210 n.

canonicorum terra valet per annum vi libras). About the value of Codsall, Saredon and Bickford (250b) Domesday Book is explicit, for we read that 3 men paid 12s. a year to the sheriff (*Hi iii homines reddunt vicecomiti xii solidos per singulos annos*). Six holdings are without 1086 values: Moreton in Hanbury parish (248b); Onn (250b); Moreton in Gnosall (248); Burton, near Stafford (248b); Mayfield (246b); and Wednesbury (246). At first the scribe omitted the values for Wigginton (246), both in 1066 and 1086, but later he added them in the margin. Other places without values were waste or included elsewhere.

The values of holdings at earlier periods were recorded only sporadically, and then usually for 1066. In 1066 and in 1086, Sedgley (249b) was worth £10 (*T.R.E. valebat x libras. Modo similiter*), while Meertown (246b), valued at £4 in 1066, was worth 10s. more by 1086 (*T.R.E. valebat iiii libras. Modo x solidos plus*). Nine berewicks of the manor of Sugnall (247) were worth 62s. in 1086 as in 1066 (*Totum T.R.E. valebat lxii solidos. Modo similiter inter omnes*). This earlier value is recorded most consistently for those manors forming the Ancient Demesne of the Crown and for the estates of Burton Abbey, but it is also noted for many of the larger holdings throughout the county. In addition, three of the Burton manors are valued for an intermediate period. Stretton (247b) was worth 60s. in 1066, later (*post*) 20s., and 40s. in 1086. During the same three periods the values of Wetmoor and Darlaston likewise fell steeply (by 75% for Wetmoor) to a nadir at the time known as *post*, but by 1086 they had markedly increased although none of the three had attained its initial value (247b). This decline in intermediate values may well reflect the harryings of 1069. For Mayfield (246b) the value for 1066 only is recorded, but although it comes at the end of the entry, it is followed by an unusually large gap, and it is just possible that this was intended for the value in 1086.

Generally speaking, the greater the number of plough-teams and men on an estate, the higher its value, but the table on p. 193 shows that there is no constant ratio between these two sets of items. Each of the manors was valued at £1 in 1086.

It is true that the variations in the arable, as between one manor and another, did not necessarily reflect variations in total resources, but even taking the other resources into account the figures are not easy to explain. Moreover, in the absence of information about conditions in 1066, we cannot explain such changes in value as the Domesday Book does record.

	Teams	Population	Other resources
Acton Trussell (247)	5	18	Mill, wood, meadow
Abbots Bromley (247b)	2	3	Wood
Church Eaton (249b)	4	20	Wood, meadow
Derrington (248)	1	3	Wood, meadow
Blymhill and Brockhurst (249b)	7	12	Wood, meadow

Conclusion

The large size and confused rubrication of the Domesday hundreds do not make them convenient units for the purpose of calculating densities. The county has therefore been divided into nine districts, each characterised by a fairly uniform physical background.

Of the five standard formulae, those relating to plough-teams and population are most likely to reflect something of the distribution of wealth and prosperity throughout the county in the eleventh century. Taken together, they supplement one another to provide a general picture (Figs. 62 and 64). The outstanding feature of Staffordshire is the paucity of teams and population when contrasted with, say, the East Anglian or South Midland counties. Nowhere in the county are there averages of more than two teams or five men per square mile. But it was not a uniform poverty which lay over the countryside. The upland area of the north was a poor area with not more than 1·5 people and 0·5 teams per square mile. Even more empty were the Cannock Hills with their poor sandy soil, and with only 0·3 people per square mile and practically no teams. The upland south of Cannock was more prosperous with 2 to 3 people and about one team per square mile. The figures for what was later Needwood Forest were somewhat similar. The rest of the county comprised the lowlands of the centre of the Peak basin and of the Trent-Tame area, all relatively prosperous with 3·4 to 4·6 people and 1·2 to 1·6 teams per square mile.

Figs. 63 and 65 are supplementary to the density maps, but it is necessary to make one reservation about them. As we have seen on p. 171, it is possible that some Domesday names may have covered two or more settlements, e.g. the present-day villages of Church and Stanton Mayfield are represented in the Domesday Book by only one name. A few of the symbols should therefore appear as two or more smaller symbols, but this limitation does not affect the main pattern of the maps. Generally speaking, they confirm and amplify the information of the density maps. In view of the doubtful nature of the plough-land entries, the implications of Fig. 66 are uncertain, but the map has been included for comparison with Fig. 62.

WOODLAND AND FOREST

Types of entries

About one half of the Staffordshire vills laid claim to a tract of woodland, whose extent was usually given in terms of leagues and furlongs. The phrasing of the entry for Penkridge (246) is typical: 'There is wood one league long and one broad' (*Silva habet i leuuam longa et unam lata*). The exact wording of the formula often varies slightly; occasionally a little more detail is provided, but the substance is always the same:

> Stafford (247b): *Silva dimidia leuua longa et tantundem lata.*
> Uttoxeter (246b): *Silva ii leuuis longa et totidem lata.*
> Wetmoor (247b): *Silva una leuua longa et alia lata.*
> Maer (249): *Silva i leuua longa et altera lata.*
> Weston under Lizard, Beighterton, Brockton Grange (250b): *Silva harum terrarum habet i leuuam longitudine et dimidiam leuuam latitudine.*

The usual term used in the Staffordshire folios to denote woodland is *silva*. This is true also of the Northamptonshire entries, but in the account of Lapley (222b), in Domesday Northamptonshire but in modern Staffordshire, *nemus* is used: *Nemus iii quarentenis longum et totidem latum.*

Despite this apparent abundance of information it is difficult to be sure of the exact significance of these measurements, and we cannot hope to convert them into modern acreages.[1] Fig. 67 must inevitably be diagrammatic. It is interesting to note that where several villages formed a single holding, their woodland was often recorded in a composite entry. The wood entered under Lichfield (247), for example, must have included those tracts pertaining possibly to as many as 22 places—*membra* or *berewichae* (Fig. 57). In the record of the king's manor of Penkridge (246) the arrangement was slightly different. Here the woodland belonging to the caput was entered separately, while that of its six members (*membra*) was given as a combined entry: *Inter omnes...silva dimidia leuua longa et iii quarentenis lata.* This type of entry seems to imply, although not necessarily, some process of addition whereby the dimensions of separate

---

[1] See p. 437 below.

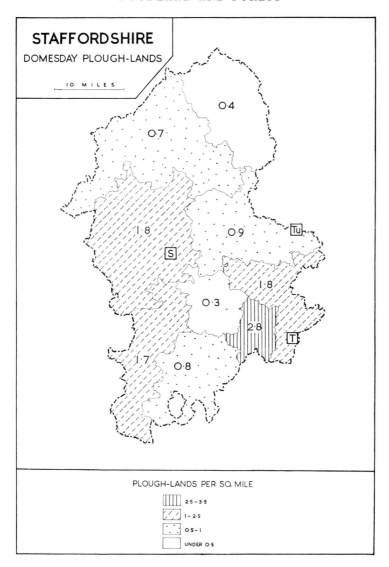

Fig. 66. Staffordshire: Domesday plough-lands (by densities).
Domesday boroughs are indicated by initials: S, Stafford;
T, Tamworth; Tu, Tutbury.

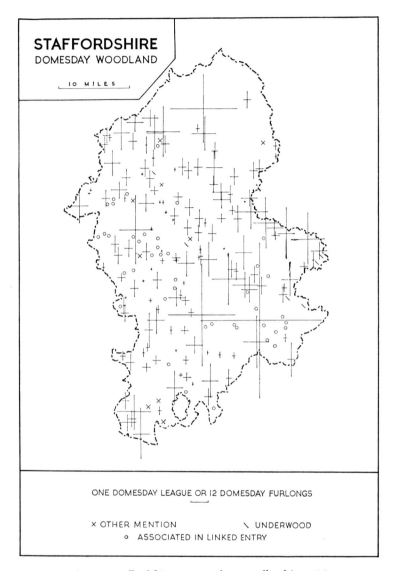

STAFFORDSHIRE
DOMESDAY WOODLAND

10 MILES

ONE DOMESDAY LEAGUE OR 12 DOMESDAY FURLONGS

× OTHER MENTION   \ UNDERWOOD
        ○ ASSOCIATED IN LINKED ENTRY

Fig. 67. Staffordshire: Domesday woodland in 1086.

Where the wood of a village is entered partly in linear dimensions
and partly in some other way, only the dimensions are shown.

tracts of woodland could be consolidated into one set of measurements. On Fig. 67 these dimensions have been plotted for the caput of the manor or for the first-named village in a composite entry. This means, in the above examples, for Lichfield itself and for both Penkridge and the first-named of its members. The associated villages in all such entries have also been indicated.

Five entries are phrased somewhat differently from the above:

Moddershall and Cotwalton (248): *Silva dimidia leuua longitudine et latitudine.*
Tettenhall (246): *Silva ibi dimidia leuua in longitudine et latitudine.*
Ettingshall (250): *Silva habet iii quarentenas in longitudine et latitudine.*
Cippemore (249b): *Silva i leuua in longitudine et latitudine.*
Hanford (250b): *Silva modica xx perticae in longitudine et latitudine.*

C. S. Taylor and R. W. Eyton believed that similar entries in the folios for Gloucester and Dorset referred not to dimensions of length and breadth, but to areal leagues of 120 acres. If this were so these measurements thus differed from those in the normal formula represented by that for Penkridge given on p. 194; but on Fig. 67 these entries have been plotted in the usual way on the assumption that the phrasing merely represents scribal variations.[1] Whatever the true interpretation may be, these five entries cannot greatly distort the pattern of woodland over the county as a whole. There is one example of a wood for which only one measurement is given. On the nameless manor on fo. 249b there was one furlong of wood (*una quarentena silvae*).[2]

While leagues and furlongs form the usual units of measurement, there are a number of exceptional entries that refer to wood in other ways:

Brough Hall (248b): *ii acrae silvae.*
Salt (248b): *iiii acrae silvae.*
Barlaston (249): *iii acrae silvae.*
Himley (249b): *et Silva.*
Amblecote (249b): *et Silva.*

In the Staffordshire folios themselves underwood (*silva modica*) is mentioned only twice, at Hopton (248b) and Hanford (250b); but the holdings at Edingale (274, 278b), Winshill (273), Stapenhill (278), and Croxall (274), villages in Domesday Derbyshire but in modern Staffordshire, are credited with quantities under the term *silva minuta*. In addition

[1] See p. 437 below.
[2] Identified as Weston Jones—see p. 165 above.

there are ten references to pasturable wood (*silva pastilis*), which was often of considerable extent. At Rolleston (248b), for example, there was a tract of 3 leagues by 2, at Marchington (248b) of 3 leagues by 1½, and at Fauld (248b) of 3 leagues by 1.[1] Finally, at Blore (249) there were 2 furlongs of spinney (*ii quarentenae Spineti*); at Burslem (249) there were 2 acres of alder wood (*ii acrae alneti*); and at Standon and The Rudge (248b) there was a grove (*xiiii acrae gravae*).

Distribution of woodland

Eyton, by translating the leagues of Domesday Book into modern measures, and converting them to acres, concluded that, 'Considerably more than half of the whole registered territory was woodland.'[2] These calculations can no longer be accepted, but it is clear from Fig. 67 that, despite the centuries of Anglo-Saxon occupation, Staffordshire was still a well-wooded county in 1086. Soils derived from clays and sands alike seem to have carried a heavy woodland cover. The one notable exception to the wooded nature of the landscape was the north-eastern upland, whose height and waterlogged soils were unfavourable to tree growth. Such woodland as the Domesday Book does record for this area was found along the better-drained slopes of the upper Dove and Churnet valleys. Comparatively little woodland was recorded for the watershed between the lower Dove and the Trent, which is difficult to explain until we remember that later sources refer to this area as the site of Needwood Forest. There is no mention of this in the Domesday Book but if afforestation had already taken place no returns would have been made. The relatively empty areas around Lichfield result from the method of compilation of the map, for, as was shown above, the enormous extent of woodland returned for that manor included amounts for over a score of surrounding villages. In the neighbourhood of Wombourn in the south-west there is a group of ten villages for which no woodland was recorded. This may perhaps be connected with the afforestation around Kinver discussed below.

[1] The other entries relating to *silva pastilis* are: Chebsey (248b); Great Barr (250); Aldridge (250); Rushall (250); Wednesfield (247b); Bobbington (249); Stapenhill (273).

[2] R. W. Eyton, *op. cit.* pp. 15 and 16.

Forests

The royal forests and the chases of Staffordshire are mentioned only incidentally in the Domesday Book. There is no reference, for example, to Cannock Chase or to Needwood Forest, but we do read that the woodland belonging to Enville (249b) had been included in a forest (*Silva i leuua longa et dimidia lata. Rex tenet eam in foresta*); so had that of Cippemore (249b): *Silva i leuua in longitudine et latitudine. Rex habet in foresta*. Presumably these woodlands formed part of Kinver Forest. At *Haswic* (247b) tillage had apparently ceased since the manor had been included in a forest (*Modo est wasta propter forestam regis. Ibi pertinuit medietas silvae quae est in foresta*). The medieval *foresta* was a legal term which did not necessarily denote a continuous tree cover, but these three entries show that considerable tracts of woodland were included within the bounds of the Staffordshire forests. A holding at Chacepool was also in the king's forest (249b). Afforestation is implied by the entry for Coven (249b) near Brewood where wood half a league by one furlong was in the king's demesne (*In dominio regis est haec silva*). *Haiae*, or enclosures for catching deer, are mentioned for some counties, but not for Staffordshire.

MEADOW

Types of entries

The record of meadow is among the most consistent of the entries in the Staffordshire folios. 'There, *n* acres of meadow' (*Ibi n acrae prati*) runs the usual formula. Altogether about half the villages in the county laid claim to some meadow, in amounts varying from half-acres at Walton Grange (248) and Knighton (250b) to stretches of as much as 50 acres at Clifton Campville (246b), Fauld (248b) and Rolleston (248b). Amounts exceeding 20 acres are rare, and where they do occur are usually in combined entries for large manors with groups of dependencies. Lichfield (247) as caput of a manor, held 35 acres, while a further 52 acres were distributed among fifteen of its members. As in other counties, no attempt has been made to translate these figures into modern acreages. The Domesday acres have been treated merely as conventional units of measurement, and Fig. 68 has been plotted on that assumption.

While measurement in acres is normal, there are a number of entries that refer to meadow in other ways:

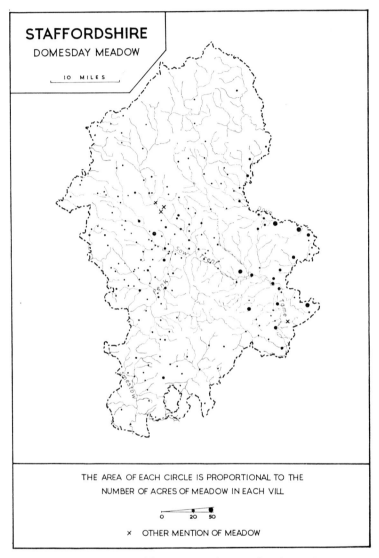

Fig. 68. Staffordshire: Domesday meadow in 1086.

Where the meadow of a village is entered partly in acres and partly
in some other way, only the acres are shown.

Stafford (247b): *ii quarentenae prati in longitudine et una quarentena latitudine.*
Wigginton (246): *vi quarentenae prati in longitudine et ii quarentenae latitudine.*

In two consecutive entries on fo. 248b it seems that the meadow was recorded almost accidentally in terms of linear measurements:

Walton: Wood 2 furlongs long and one furlong wide, and the meadow is of like extent (*et pratum similiter*).

Aston and Little Stoke: Wood 2 furlongs long and one furlong wide, and there is as much meadow (*tantundem prati*).

Can this possibly mean that for a moment the scribe forgot (or did not think it worth while) to convert the information about meadow into acres?

Distribution of meadowland

Meadow was most plentiful on the flood-plain of the Trent, flowing from north-west to south-east across the centre of the county, and along its main tributaries, the Dove, the Tame and the Sow. The larger acreages were found along the eastern stretches of these rivers where their flood-plains were best developed. Along the lower reaches of the Dove, for example, were 40 acres at Marchington (248b), 50 acres each at Fauld (248b) and at Rolleston (248b), and 24 acres at Branston (247b); while along the Tame were villages such as Elford (246b) with 24 acres, Hopwas (246b) with 30 acres, and Wigginton with its meadow of 6 furlongs by 2 (246). In this south-eastern region, comprised mainly of the vales of Trent and Tame, over 60% of the vills had some meadow, and the average size of these tracts was 20 acres. Higher up the Trent valley, and on the heavy claylands of the Sow and Penk basins, vills with meadow were still fairly numerous, but acreages were, generally speaking, smaller (Fig. 68).

Comparatively little meadow is recorded for vills above 400 ft. Much of the north-east, indeed, was entirely without, and such meadows as did occur in the narrow-floored valleys were of small extent: for example, 3 acres at Leek (246b), 1 acre at Basford (248b), 3 acres at Kingsley (250b *bis*) and 4 acres at Knutton (250b). Other areas with little or no recorded meadow were the Cannock upland and Needwood Forest.

WASTE

A feature of eleventh-century Staffordshire was the number of wholly 'waste' vills in 1086: no less than sixty-two, or about one-fifth of the total villages, fall into this category. The usual formula states simply, 'It is waste' (*Wasta est*). A group of holdings, each with its hidage and plough-lands, is sometimes classed collectively as waste. For instance, at the end of the list of King William's thirty-two holdings on fo. 246b we read: 'All this land belonging to the King is waste' (*Omnis haec terra Regis Wasta est*). And again, after enumerating the eleven waste members of the manor of Eccleshall (247) the Domesday entry adds: 'All these lands are waste' (*Hae terrae omnes sunt wastae*); and the seven waste members of Lichfield (247) are described in a similar way. All these waste entries refer to 1086, but just once we are told that a holding had been waste at some previous date: Loxley (248) 'was waste and still is' (*Vasta fuit et est*).

In addition to these waste vills, another seven were only partially waste, a condition not stated explicitly in the Domesday Book, but one which we infer when one of two or more holdings comprising a vill was entered as waste. Hilderstone is a good example of such a partially waste village, for the king's holding was waste (246b) while that of Robert de Stafford was well-stocked (249). At Bushbury, two holdings were being worked successfully (247b, 250), but one virgate on a third (250) was altogether waste (*vasta est omnino*).

The fact that a holding was described as waste did not necessarily imply that its land was entirely without profit. A waste half-hide at Loynton (249b), for example, yielded 2s. Usually there is no mention of appurtenances in the entries for waste villages, but sometimes a waste holding included a stretch of woodland or some meadow. A quarter of a hide at Loxley (248) which was waste in 1086, and earlier, returned 20s., presumably the value of its 4 acres of meadow and of its woodland. On the waste holding at Morfe (249b) there was a tract of woodland 2 leagues by 2 leagues, and a waste vill at Hanford (250b) was valued at 2s., apparently as a result of possessing some underwood 20 perches in length and breadth. Thus, it would seem that the description 'waste' referred only to land that had gone out of cultivation. It is noteworthy, too, that the entries for these waste holdings invariably state either the number of hides or the number of plough-lands, or, more usually, both.

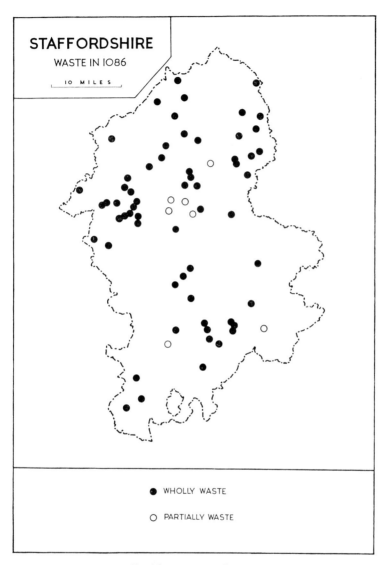

Fig. 69. Staffordshire: Domesday waste in 1086.

From the fact that about half of these waste villages were royal hold-ings, and that most of the others were church lands, J. C. Wedgwood inferred that at least some of these waste entries reflected the hazards of farming.[1] But it has usually been held that such entries record rather the effects of war, notably the widespread violence which followed upon the rebellion of 1069, when William instigated 'the most terrible visitation that had ever fallen on any large part of England since the Danish wars of Alfred's time'.[2] Fig. 69 shows that half the total number of waste villages were situated in the valleys and hollows of the northern uplands, that is, in the agriculturally marginal lands which would both be likely to carry longest the imprint of the harrying of 1069, and which would also be the first to revert in a period of lean years. Another dozen or so waste hold-ings were to be found on the South Staffordshire uplands. These, too, may reflect spoliation, or, perhaps, farming failures, but we may suspect that some of the waste villages ringing the western and southern borders of the present Cannock Chase were due to Norman afforestation. The Domesday Book affords a glimpse of this policy at work in the extreme south-west of the county, where the entries for two waste holdings add a phrase or two by way of explanation. On a hide at Chacepool (249b) tillage had ceased when the holding had been incorporated in the king's forest (*In foresta regis et vasta est*), while *Haswic* (247b) was waste 'because of the king's forest' (*Modo est wasta propter forestam regis*). Elsewhere we can only assume that the waste entries were due to special local causes.

MILLS

Mills are mentioned in connection with at least 50 of the 334 Domesday place-names within the area covered by modern Staffordshire. The fifteen berewicks of Lichfield (247), sharing one mill, have been counted as one place, and the mill has been plotted at the first-named berewick. Similarly the mill entered for Standon and The Rudge (248b) has been plotted at Stanton. If all the places named in these two entries are counted, the total becomes 64 (not 50), a very artificial figure, but all the associated villages have been indicated on Fig. 70.

In each entry the number of mills is normally given and their annual value. At Gratwich (249), for example, there was a mill worth 4s., and at

[1] J. C. Wedgwood, *op. cit.* p. 174.
[2] F. M. Stenton, *Anglo-Saxon England* (Oxford, 1943), p. 597.

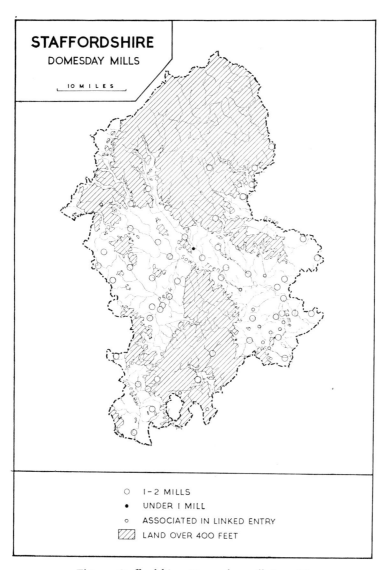

STAFFORDSHIRE
DOMESDAY MILLS

10 MILES

○   1–2 MILLS
•   UNDER 1 MILL
∘   ASSOCIATED IN LINKED ENTRY
▨   LAND OVER 400 FEET

Fig. 70. Staffordshire: Domesday mills in 1086.

Clifton Campville (246b) there were 2 mills worth 10s. The mill at Meertown (246b) paid a rent of 3s. together with 4,000 eels (*iiii millibus anguillarum*), presumably a product of Aqualate Mere. Eels were a common form of rent for mills in some counties but this reference is unique in Staffordshire. If the rents paid were a fairly accurate reflection of the quantities of grain ground, the mills of Staffordshire were small compared with those of more favoured counties. At Drayton Bassett (246b) the king had two mills paying 21s., and another two at Elford (246b) paying 20s., but one-half of the mills of the county rendered amounts between 2s. and 5s. A mill at Tittensor (249) was worth as little as 8d., and another at Rushall (250) a mere 4d. No value is entered for the mill at Okeover (247b), or for the mill shared by the fifteen berewicks of Lichfield (247).

There is one example of a half-mill. The village of Ingestre (249) is credited with half a mill worth 10d. (*de una parte molini x denarii*). In other counties where similar entries occur it is sometimes possible to assemble such fractions in an intelligible manner, and to show that one mill ground for several neighbouring settlements, but we are given no clue to the other half of the mill at Ingestre.

Domesday Mills in Staffordshire in 1086

Under 1 mill	1 settlement
1 mill	39 settlements
2 mills	10 settlements

There are a number of apparent anomalies in the location of these mills. At Hamstall Ridware (247b), with but one team, there was a mill worth 2s., and Crakemarsh (246b), where 2 teams were at work, supported a mill paying 10s. Neither was there any scarcity of larger and seemingly prosperous villages such as Sedgley (249b) and Wolverhampton (247b) which lacked mills altogether. Perhaps the unvalued mill at Okeover was no longer grinding, for the Domesday Book records no teams or peasantry on the holding.

As might be expected, mills were most frequent along those rivers which flowed through the most fertile arable lands; that is, in the south-east in the vales of Trent and Tame, along the valley of the Smestow in the south-west, and along the numerous small streams which threaded the strong claylands of central Staffordshire (Fig. 70). Both the northern and the southern uplands were all but devoid of mills. In fact there were only four mills much above 400 ft. in the whole county: at Penn (249b), Cheadle (249), Rushall (250) and Okeover (247b). The first three were worth respectively only 2s., 12d., and 4d., while there is no rent recorded for the mill at Okeover. There was no mill at all above 600 ft., owing partly to the nature of the moorland rivers, but more so to the small arable acreage in these areas.

URBAN LIFE

The information about the towns is the least satisfactory part of the Domesday evidence, and for Staffordshire it is more slender than usual. Only three places seem to have been regarded as boroughs, and the account of each is clearly incomplete. Stafford itself is described as a township (*villa*, 247b), as a borough (*burgus*, 246), and as a city (*civitas*, 247b). Tutbury is called a borough (248b); and there are references to burgesses (*burgenses*) at Tamworth (238, 246, 246b). But of the life within these settlements, of their trades and crafts and of their other activities, there is barely a mention. From the scanty evidence afforded by Domesday Book, it would seem that these boroughs were primarily agricultural or military settlements nourishing incipient urban nuclei. In view of the importance of the market and of the mint in theories of burghal origins, for example, it is surprising to find only one reference to a market, and no mention of a mint. Such meagre evidence as Domesday Book provides for the urban life of the eleventh century is summarised below.

Stafford

Stafford is described on fo. 246, at the very beginning of the survey of its county, but this main account is supplemented by further entries on fos. 247b (twice), 248 (twice), and 248b (twice). The main entry mentions 18 burgesses, and a total of 161 houses (*mansiones*) belonging to the king and his tenants-in-chief. Of these houses 110 were inhabited (*hospitatae*) and 51 were waste (*vastae*). An interesting feature which the city shares with a number of other Domesday boroughs is that some Stafford burgesses and houses are recorded as belonging to rural manors. Some of these are indicated in the main entry:

Bradley: *xiii mansiones...pertinent ad Bradelie. Ex his vi sunt vastae.*
Penn: *iiii mansiones quae pertinent ad Pennam Manerium comitis. Ex his una tantummodo est hospitata.*

Sheriff Hales: *iii mansiones quae jacent ad Halam*.
Worfield:[1] *v mansiones et pertinent ad Guruelde*.

These houses are not mentioned again in the entries for their respective villages, but another house and a number of burgesses not included in the main entry, are to be found in the descriptions of two villages:

Marston near Stafford (248): *In Stadford xviii burgenses pertinent hiuc Manerio.*
Cresswell (248): *In Stadford una vasta masura*.

In all, therefore, Domesday Book enumerates a total of 36 burgesses, and 162 houses, of which 52 were waste, together with 13 villeins, 8 bordars and 4 serfs (247b). There were also 13 canons and, by inference, a church. These figures may imply a population of at least 750.

We are told nothing of the commercial life of the city, not even of the mint, which we know to have been there during the reign of Athelstan. That the burgh of Stafford was a walled enclosure may be inferred from the statement that thirty-one of the houses listed on fo. 246 were situated 'within the wall' (*intra murum*). But despite this physical separation from the life of the countryside, there was still a strong agricultural element in the borough. From the supplementary entries on fo. 247b we learn that there were 6 teams at work, together with a mill, two parcels of meadow, and a wood half a league by half a league. We also learn from fo. 248b that King William had ordered a castle to be built in Stafford, but that it had subsequently been destroyed.[2] The returns from the borough itself had diminished since 1066 (*Tempore regis E. reddebat burgus de Stadford de omnibus consuetudinibus ix libras denariorum . . . Modo habet rex Willelmus de redditu burgi vii libras*), possibly owing to the waste state of a large proportion of the houses; but the rural parts of the settlement had increased in value from £4 to £6. 10s. (247b). This is an unsatisfactory account, but it is considerably more detailed than that for either of the other two boroughs.

Tamworth

Despite its importance in the early history of Midland England, Tamworth receives scant consideration in the Domesday Book.[3] There is no entry relating specifically to the borough, and such information as

---

[1] Now in Shropshire.
[2] See p. 210 below.   [3] See pp. 274 and 305 n. below.

we have is derived from incidental references to contributed burgesses. Wigginton (246) appears to have contributed four burgesses, and from the Warwickshire folios we learn that Coleshill (238) contributed another ten. The third entry, that for Drayton Bassett (246b), stresses the agrarian aspect of the borough, for eight burgesses in Tamworth belonged to this manor and worked there like the other villeins (*et ibi operantur sic alii villani*).

Tutbury

At Tutbury, brief though the Domesday record be, we can glimpse one of the processes by which a borough might come into being. Henry de Ferrers had founded a castle at Tutbury, and under its protection there had grown up the only trading mart recorded in the Staffordshire folios (248b). Forty-two of the inhabitants of this town, who devoted themselves wholly to trade, paid £4. 10s., which included the proceeds from their market (*In Burgo circa castellum sunt xlii homines de mercato suo tantum viventes et reddunt cum foro iv libras et x solidos*). These townsmen, it is true, are not styled 'burgesses', and it may be that legally they were not recognised as such, but both in Domesday Book and in later medieval documents there was little attempt to use technical terms with precision.

MISCELLANEOUS INFORMATION

There are no references in the Staffordshire folios to such items as pasture, vineyards or salt-pans which occur from time to time in the Domesday accounts of other counties. At Alrewas (246), on the Trent just above its confluence with the Tame, there was a fishery (*piscaria*) which yielded fifteen hundred eels (*reddens mille quingentas anguillas*). Eels also occur as part of the render of the mill at Meertown (246b).

Only two churches are mentioned and those allusively. Half of the church of Stoke (*medietas ecclesiae de Stoche*) belonged to the manor of Caverswall (249). Of the other half of the church the Domesday folios tell us nothing. The second church was at Tettenhall (247b). In addition to these there were land-owning canons, and presumably churches, at Stafford, Wolverhampton (247b) and Lichfield (247), and the abbey of Burton also appears as a land-owner (247b). There were also thirty priests recorded, for some of whom there must have been parish churches.

Three other items, all on fo. 248b, are of interest, namely the castles of Tutbury and Burton,[1] which stood guard respectively over the Dove and Trent valleys, and Stafford castle which commanded the routes to the west. About the first two Domesday Book is brief: 'Henry de Ferrers has the castle of Tutbury', around which, as we have seen, grew up a market; and in Burton he had half a hide on which his castle was situated (*in qua sedet ejus castellum*). The third reference is a little fuller and allows us a brief glimpse of the troubled times which preceded the making of the Domesday Book. The king had ordered the building of Stafford Castle on a plot of land belonging to the manor of Chebsey, but the castle had since been destroyed (*Ad hoc Manerium pertinuit terra de Stadford in qua rex praecepit fieri castellum quod modo est destructum*).

The Domesday Book rarely mentions boundaries or landmarks, but in the entry for Bramshall (249) there is a vague reference to a road which should have separated the holding of the king from that of Robert de Stafford (*unam virgatam terrae cujus virgatae medietas est regis sicut via eam dividit*).

REGIONAL SUMMARY

The evidence discussed above makes it clear that not all parts of Staffordshire had offered equal opportunities for human activity. In the north is a belt of upland country, mainly at about 550 ft. in height, but rising eastwards to the moorlands of the Southern Pennines. In the south, a sandstone upland, partly drift-covered, is flanked on east and west by lower-lying river valleys; and between these uplands there stretches a broad clayland, mainly below 400 ft. Although the nature of the record allows of only broad comparisons, there can be no doubt but that axe and plough had made considerable progress in developing regional differences. In each of these main physical divisions distinctive landscapes were coming into being. But the plough is a sensitive implement, and within these areas further subdivisions may be recognised, making six in all (Fig. 71). The geography of each of these districts, in so far as it may be deduced from the Domesday evidence, is outlined below.

(1) *The South Pennine Fringe*

This region corresponds broadly to outcrops of Carboniferous limestones, shales and grits. They are mostly over 800 ft. in height, and sometimes

[1] But Burton may be an error for Tutbury; the clerk may have misread *Burgh* (borough) as *Burt.*—see *V.C.H. Staffordshire*, IV, p. 48.

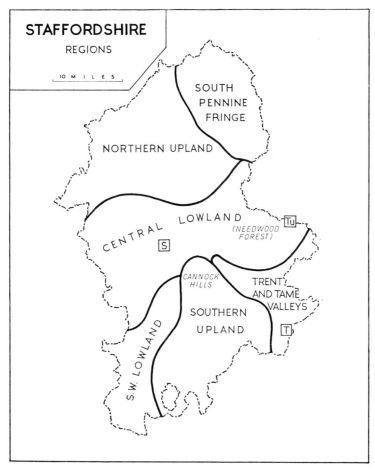

Fig. 71. Staffordshire: Regional subdivisions.
Domesday boroughs are indicated by initials: S, Stafford;
T, Tamworth; Tu, Tutbury.

rise to well over 1,000 ft. Peaty soils have developed under a heavy rainfall, and there is a considerable extent of true moorland. In such an environment it is not surprising that villages were few and small, and confined to the comparative shelter of the river valleys. Moreover, about a third of the villages were waste, and it must have been a drab and sombre landscape which met the eye among these hills. Cultivable land was scarce even in the valleys, so that there was an average of only one-fifth of a plough-team at work on each square mile of country. The density of population resulting from this pattern of settlement was less than one per square mile. It is noteworthy, too, that the ratio of population to plough-teams was the highest of any of the Staffordshire regions, which may have reflected not only a low standard of living in a harsh setting, but also a strong pastoral element in the economy of the area. As a result of the scarcity of arable land, there were only two mills in the whole of the area. Today the hills are bare of trees and in all probability were so in the eleventh century, yet over four-fifths of the cultivated manors in this region were credited with some woodland. Presumably this clothed the steep slopes of the Dove, Manifold and Dane valleys, and of their feeders. In these valley bottoms, too, were the small acreages of meadow which Domesday Book records for about half the villages.

(2) The Northern Upland

This dissected upland comprises Pennine foothills at a height of about 500 ft., and includes the basin of the Upper Trent. It is a large region in which predominantly poor soils have weathered out from widely differing geological formations. In the eastern part there is a succession of limestones, shales and grits, with clays, marls and sandstones adding to the variety in the west. Villages were fairly numerous at lower levels, especially in the Trent valley, but there was an average of less than one half a plough-team per square mile, and the density of population was also low, being only 1·4 per square mile. These low densities were due partly to the small size of the villages, and partly to the fact that nearly a quarter of them were waste. The scarcity of arable land was reflected in the lack of mills, for only one is recorded among the fifty or so villages. On the lower ground in the west of this region, and on the slopes of the valleys, woodland was fairly plentiful, but meadows were small and mostly in the valleys of the Upper Trent and its tributaries.

(3) The Central Lowland

Central Staffordshire is an undulating plain usually at about 350 ft. in height except where it rises to nearly 450 ft. along the broad watershed between the Dove and Trent rivers. Most of this lowland is floored with Keuper Marl which weathers to strong clay or clay-loam soils, of con-

siderable fertility but difficult to work. Lighter soils occur only occasionally, mainly along the western border of the county and in an attenuated
belt running northward from Cannock Chase.

Although it compared but poorly with, say, parts of East Anglia, the
central clayland was one of the most prosperous parts of Domesday
Staffordshire. Villages were spread thickly and evenly over the countryside except for an empty area on the higher ground of the east, the site
of Needwood Forest. Some of the most prosperous settlements were
situated in the wide, flat valley of the Dove, where one of them, Tutbury,
seems already to have acquired the status of a borough. The other borough
of the central plain was Stafford, situated on the river Sow, but the
Domesday information does not allow us to assess its contribution to
the prosperity of the region. The density of plough-teams in the western
half of the plain was 1·3 per square mile, that of population 3·7; the
comparable figures for the Needwood area were 0·9 and 2·5. Mills were
fairly frequent, mainly along the Trent, Sow and Penk rivers. On the
average there was one to every six villages. The heavy marls of this area
must once have carried a more or less continuous cover of wood, and
Fig. 67 shows that in Domesday times much remained. Some woodland
is recorded for over 60% of the villages. Meadows were frequent on these
low-lying claylands, especially along the Sow and Penk rivers, and along
their tributaries, but acreages were often meagre.

(4) The Trent and Tame Valleys

The Trent flows from its source among the hills of the northern upland
in a wide arc across the central clayland, but only below Rugeley does its
valley broaden sufficiently to be considered a region in itself. Below this
point the vale opens out and the river flows over the alluvium of a wide
floodplain, until just east of Alrewas it is joined by the flat-floored valley
of the Tame. Situated along the courses of these two rivers were some
of the largest and most prosperous villages of Domesday Staffordshire.
The finely-textured, fertile soils supported 1·6 plough-teams per square
mile, and the density of population was as much as 4·6 per square mile.
The relatively high proportion of arable land necessitated more numerous
and larger mills. The progress of cultivation had not entirely obliterated
the wood, and some remained, while a feature of these valleys was the
comparative abundance of meadow along the floodplains of the rivers.

(5) *The South-western Lowland*

This region comprises mainly the Smetsow and Upper Penk valleys. It
lies for the most part about 300 ft. above sea-level, rising in places to
over 400 ft. The valleys are developed chiefly on Triassic sandstones,
which weather down to light, warm soils, diversified here and there by
heavier varieties where Boulder Clay overlies the solid formations. Vills
were fairly frequent and the density of teams was 1·2 per square mile, the
density of population 3·7. Mills were not numerous. Most of the district
was well-wooded, but there was a curious group of ten villages around
Wombourn for which no woodland was recorded. Possibly this is in
some way connected with the afforestation discussed on p. 199 above. On
these free-draining soils meadow was not over-abundant. Less than half
the villages had any, and acreages were usually small.

(6) *The Southern Upland*

Within this broad category may be recognised three sub-regions: the
South Staffordshire Upland proper, the Cannock Hills and the Lichfield
area. The first of these is generally at a height of from 400 to 500 ft. above
sea-level and comprises an outcrop of mainly Coal Measure rocks, masked
extensively by Boulder Clay. The soils developed on these formations
are variable but tend on the whole to heaviness. Most of the inhabited
villages were clustered along the western edge of this region. In the
eastern half not only were settlements infrequent, but several were waste.
This almost empty eastern area brought down the density of plough-
teams to just under one per square mile, and that of population to 2·7.
There were only two mills in the whole area, and they were small. The
upland seems to have been well wooded, but meadow was scarce. Only
about half the villages had any, and several of these had but one acre,
e.g. Great Barr (250), Aldridge (250) and Bilston (246).

The South Staffordshire Upland is continued northwards in the Bunter
Sandstone and Pebble Beds of the Cannock Hills, whose light and hungry
soils, often well above 600 ft., have even yet not been brought under
cultivation. On the Domesday distribution maps this area appears as
a blank, but later medieval records show that it was maintained as a hunting preserve subject to forest law.

The third of these sub-regions is an upland only by contrast with the

low-lying Trent and Tame valleys to north and east of it. The general
level of the surface lies between 250 and 350 ft. above sea-level, but rises
gently westwards towards the South Staffordshire Upland proper. The
soils are derived from Triassic sandstones, and tend to be light, except
where they are modified by very small patches of Boulder Clay. Domesday
Book pictures this as a region of moderate prosperity, dominated by the
great manor of Lichfield with its twenty-two members. These easily-
worked soils supported 1·6 teams per square mile, while the density of
population was 3·4 per square mile. It might appear that but few of the
villages laid claim to woodland, but the unusually large tract in the entry
for Lichfield seems to have included the woodland belonging possibly to
as many as twenty-two dependencies. Some of the largest aggregate
acreages of meadow in the county were in this region, including thirty-
five acres appurtenant to Lichfield itself, and a further fifty-two acres
distributed among its berewicks. Many of these meadows were probably
along the network of small streams draining into the Trent and Tame,
but it seems likely that the larger acreages were outside the region, down
on the floodplains of the main rivers.

BIBLIOGRAPHICAL NOTE

(1) It is interesting to note that although he did not reproduce any part of
the text of the Domesday folios, Sampson Erdeswick made great use of it in
his *A Survey of Staffordshire, containing the antiquities of that county, with a
description of Beeston-Castle in Cheshire* (London, 1717).

Stebbing Shaw, in the fashion of the time, included a Latin transcription of
the Staffordshire text in an appendix to his *The History and Antiquities of
Staffordshire* (London, 1798–1801). Thomas Harwood's *A Survey of Staffordshire* (Westminster, 1820) included a list of Domesday holdings, together with
their modern names.

(2) R. W. Eyton's *Domesday Studies: an Analysis and Digest of the Staffordshire Survey* (London, 1881). Eyton's conclusions cannot now be accepted, but
this work is still of interest as an early attempt to tabulate the information of
Domesday Book, hundred by hundred.

(3) A translation of the text of the Staffordshire folios is that by H. M.
Fraser, *The Staffordshire Domesday* (Stone, 1936). An extended transcription
of the Latin is set out on the left-hand page, with a literal English translation
on the opposite page.

Another translation has been made by C. F. Slade in *V.C.H. Staffordshire*

(London, 1958), IV, pp. 37–60, accompanied by a valuable Introduction (pp. 1–36).

(4) Various aspects of Domesday Staffordshire are discussed in the following papers in the Staffordshire Record Society's *Collections for a History of Staffordshire* (London):

W. F. CARTER, 'On the Identification of the Domesday Monetvile' (1908), pp. 227–30.

J. C. WEDGWOOD, 'Early Staffordshire History (from the Map and from Domesday)' (1916), pp. 138–201.

C. G. O. BRIDGEMAN, 'Notes on the Contents of the Volume for 1916; the Five-Hide Unit in Staffordshire' (1919), pp. 134–51.

C. G. O. BRIDGEMAN, 'Some Unidentified Domesday Vills' (1923), pp. 23–44.

C. G. O. BRIDGEMAN and G. P. MANDER, 'The Staffordshire Hidation' (1919), pp. 154–81.

CHAPTER V

WORCESTERSHIRE

BY F. J. MONKHOUSE, M.A., D.SC.

The Domesday folios relating to Worcestershire are of more than usual interest. Freeman devoted an appendix to the county,[1] and Round, in his introduction to the Domesday Survey in the *Victoria County History*, declared that: 'The survey of Worcestershire in Domesday Book presents so many features of interest and historical importance that it is not easy to do them justice within the compass of a single paper.'[2] While we are given much the same details as for the other counties of the Great Domesday Book, there is considerable information about the important salt industry, so that '...it is not too much to say that Droitwich pervades the study of the shire'.[3] Furthermore, the folios are exceptionally rich in quaint and unusual entries, and reference is made to many of these in the relevant sections below.

But the usual deficiencies of the text as a source for the reconstruction of the contemporary scene must not be overlooked. There are occasional blanks or gaps in the manuscript, as in the account of Abbots Lench (173); there are also involved entries, the meaning of which must often remain a matter of conjecture; and there are also inconsistencies which sometimes emerge when it is possible to correlate scattered data. As an example of inconsistency, take two entries for Willingwick. It is not absolutely certain that they refer to the same holding, but it seems likely. As Round says, 'If they do, we have a further instance of discrepancies in Domesday Surveys'.[4] Under the entry for the king's manor of Bromsgrove, the following item appears:

Fo. 172. Of the land of this manor William Fitz Ansculf holds 3 virgates in Willingwick and Baldwin of him. Ulwinus, a thegn of Earl Eadwine, held it. There is 1 villein with half a plough, and a plough and a half more could be employed there (*plus posset esse*). It was worth 5*s.*, now 2*s.*

---

[1] E. A. Freeman, 'The condition of Worcestershire under William', in *History of the Norman Conquest* (Oxford, 1876), v, pp. 759–66.

[2] J. H. Round, in *V.C.H. Worcestershire* (London, 1901), I, p. 235.

[3] J. H. Round, *ibid.* p. 268.   [4] J. H. Round, *ibid.* p. 316.

Compare a second reference to the same holding, where the lands of William Fitz Ansculf in Came hundred are grouped together.

Fo. 177. The same William holds Willingwick and Baldwin of him. There [are] 3 virgates of land. There is [*sic*] 1 villein and 1 bordar with half a plough. There could be employed 2½ ploughs more. It was worth 5*s.*, now 3*s.*

There is a considerable measure of agreement between these two entries, but there are also some discrepancies. Rather more understandable are the inconsistencies which sometimes appear when the same holding is entered in the folios for two separate counties. Hanley Castle, for example, appears in both the Herefordshire and the Gloucestershire returns (although it was in Worcestershire). It is true that the Gloucestershire entry is given as T.R.E., but as a T.R.W. value is also stated, the whole entry may well refer to 1086.

Hanley Castle

In the Gloucestershire returns (163 b)		In the Herefordshire returns (180 b)	
Hides	4	Hides	4
Demesne plough-teams	2	Demesne plough-teams	2
Other plough-teams	Blank	Other plough-teams	21½
Villeins and bordars	40	Villeins	20
		Bordars	17
		Reeve	1
Serfs and bondwomen	8	Serfs and bondwomen	9
		Swineherds	6
A mill worth 16*d.*		A mill worth 2*s.*	
A wood in which there is a hay (*haia*)		Wood 5 leagues between length and breadth. This is placed outside the manor. There is a hawk's eyry. The forester holds half a virgate of land	
Value 1066 £15; 1086 £10		Value omitted	

The student of Domesday England is frequently able to make use of important subsidiary sources of information, contemporary or perhaps somewhat later, which may be used to amplify or corroborate the information contained in the Domesday Book. There is a number of such sources for Worcestershire.[1] The most important is a cartulary, compiled

---

[1] Translations of these are given, and their importance discussed, by J. H. Round in *V.C.H. Worcestershire*, I, pp. 324–31. See J. H. Round, *Feudal England*, pp. 169–80 (London, 1895). For a later discussion of one source, see P. H. Sawyer, 'Evesham A, a Domesday Text', *Miscellany 1, Worcs. Hist. Soc.* (Worcester, 1960), pp. 3–36.

The Domesday County and the Modern County of Worcester

A. *Transferred from the Worcestershire D.B. to modern counties*

To Gloucestershire

Blockley (173)
Church Icomb (173)
Cutsdean (173)
Daylesford (173)
Ditchford (173)
Evenlode (173)
Ildeberga (175 b)
Little Washbourne (173)
Redmarley D'Abitot (173)
Teddington (173)

To Warwickshire

Alderminster (175)
Bartley Green (177)
Bevington (175 b)

To Warwickshire

Blackwell (173 b)
King's Norton (172)
Lea Green Farm (172)
Lindsworth Farm (172)
Longdon (173 b)
Moseley (172)
Northfield (177)
Oldberrow (175 b)
Rednal (172)
Selly Oak (177 *bis*)
Shipston on Stour (173 b)
Tessall Farm (172)
Tidmington (173)
Tredington (173)
Yardley (175)

To Herefordshire

Acton Beauchamp
  (176)
Edvin Loach (176 b)
Mathon (175 b)

Note. Three parishes, not named in the Domesday Book, were also transferred from Worcestershire to Gloucestershire—Alstone in 1844, Chaceley and Staunton in 1931.

B. *Transferred from other Domesday counties to modern Worcestershire*

From Gloucestershire

Ashton under Hill (163 b)
Aston Somerville (169 b)
Beckford (164)
Bickmarsh (part) (170 b)
Broad Marston (169)
Childs Wickham (166, 168)
Cow Honeybourne (165 b)
Hinton on the Green (165 b)
Kemerton (163 b, 166 *ter*)
Pebworth (167 *bis*, 169)
Ullington (167)

From Warwickshire

Bickmarsh (part) (244)
Ipsley (244)

From Herefordshire

Rochford (186 b *bis*)
Stoke Bliss (187 b)

From Staffordshire

Upper Arley (247 b)
Alia Ernlege (247 b)

Note. Bickmarsh appears in both the Gloucestershire and Warwickshire lists. Information was returned in the Domesday folios for both these shires, although today the parish lies wholly in Worcestershire. For further complexities, see p. 220.

by Heming, a monk of the cathedral monastery at Worcester, which gives the hidage of those lands held by the Church in the hundred of Oswaldslow at a date ascribed by Round to the years between 1108 and 1118. Other subsidiary sources of information include fragments of surveys in an Evesham cartulary of the reign of Henry I, which deals with Droitwich, and another fragment from a later cartulary of the reign of Stephen. But while this material is of great importance to feudal historians, enabling them to trace changes of tenure in the period 1086 to 1118, and to disentangle some of the complicated changes in the Worcestershire hundreds, it does not add to our picture of the country-side in 1086.

Many counties do not have the same boundaries in modern as in Domesday times, but there is no greater example of disparity than that presented by the three adjacent counties of Worcestershire, Gloucestershire and Warwickshire (Fig. 2). The boundaries of these counties in 1086 interlocked in a most bewildering fashion that reflected the scattered holdings of the Church of Worcester and the confused nature of the Domesday hundreds.[1] These complicated arrangements for the most part remained stable until the nineteenth century. Successive boundary Acts in the nineteenth and twentieth centuries rearranged these involved county boundaries by transferring parishes from one county to another. The result of these changes is that the modern county of Worcestershire is very different from the county described under that name in Domesday Book (Fig. 72). The table on p. 219 summarises the changes involved. As the present study is written in terms of the modern county, it has been necessary to omit thirty-one places mentioned in the Worcestershire folios and to take in sixteen places from the folios for Gloucestershire, Herefordshire, Warwickshire and Staffordshire.

This, however, does not exhaust the complexity, for a few holdings in Domesday Worcestershire are surveyed not in the Worcestershire text, but among the Herefordshire folios.[2] At the end of the Worcestershire text there appears a reminder of this fact in two paragraphs about Esch and Doddingtree hundreds (178):

> In Esch hundred lie 10 hides in Feckenham and 3 hides in Hollow Court and they are described in the Hereford return.

[1] See p. 222 below.
[2] *Calvestone*, an unidentified berewick of Morville in Shropshire is also said to be in Worcestershire (253). See p. 119 n. above.

In Doddingtree hundred lie 13 hides of Martley and 5 hides of Suckley, which plead and geld here and which render their farm at Hereford, and are described in the King's return.

When we return to the Herefordshire folios, we do indeed find the entries in full, not only for these four places, but in addition for Bushley,

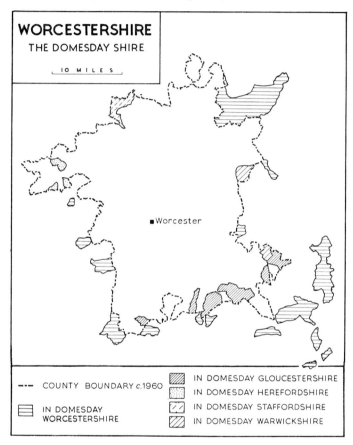

Fig. 72. Worcestershire: The Domesday shire.

Pull Court, Queenhill and Eldersfield. Most are listed under the heading *In Wirecestre Scire* (180b). Moreover, Hanley Castle, for which the Domesday folios of Worcestershire make no entry, is included in the returns both for Herefordshire (180b) and Gloucestershire (163b). Yet Hanley lies well inside both the Domesday shire and the modern county

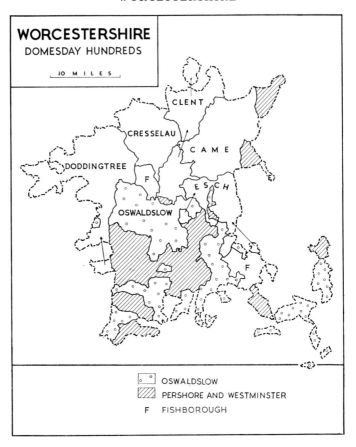

Fig. 73. Worcestershire: Domesday hundreds.
It is impossible to show the extremely complicated boundaries
between the Church hundreds of Pershore and Westminster.

of Worcester, near the banks of the Severn. It belonged, however, to
the manor of Tewkesbury in Gloucestershire.

Within the Domesday county there were twelve hundreds (Fig. 73),
and the record specifically tells us so: '*In ipso comitatu sunt xii hundreda*'
(172). As Round says, their history '...is one of much obscurity'.[1]
Seven of these hundreds were not compact territorial hundreds, but
comprised the scattered possessions of the great religious houses. The
largest of these Church holdings was the great 'triple hundred' of

[1] *V.C.H. Worcestershire*, I, p. 238.

Oswaldslow owned by the Church of Worcester.[1] Then there were the 'double hundred' of Westminster, owned by the Church of St Peter, the hundred of Pershore, owned by the Church of St Mary, and the hundred of Fishborough, owned by the Abbey of Evesham. It is impossible to distinguish on Fig. 73 between the hundreds of Westminster and Pershore, for their holdings were most complex; part of the manor of Pershore itself was in fact held by the Church of Westminster. In addition

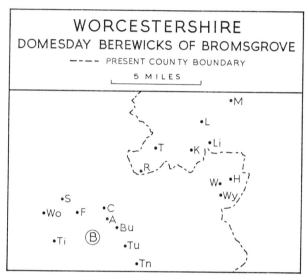

Fig. 74. Worcestershire: The Domesday berewicks of Bromsgrove (B).

Berewicks in Worcestershire: A, Ashborough; Bu, Burcot; C, Comble; F, Fockbury; H, Houndsfield; S, Suruehel; Ti, Timberhanger; Tn, Tuneslega; Tu, Tutnall; W, Warthuil; Wo, Woodcote; Wy, Wythwood.

Berewicks now in Warwickshire: K, King's Norton; L, Lea Green; Li, Lindsworth; M, Moseley; R, Rednal; T, Tessall.

to these seven ecclesiastical hundreds, there were five other hundreds corresponding to territorial divisions—Came, Clent, Cresselau, Doddingtree and Esch. Even these were not without complexities, for Clent consisted of three distinct units, while Doddingtree and Esch had several outlying portions.

One difficulty in handling the Worcestershire material from a geographical point of view is the fact that information about two or more places is sometimes combined in one statement. This is especially true

[1] See pp. 230–1.

of the very large number of named berewicks, the resources of which are described collectively under a parent manor. The manor of Bromsgrove had no less than eighteen, of which six are in the modern county of Warwick (Fig. 74); Kidderminster had sixteen; Pershore had seven;[1] Alvechurch had four; Kempsey three, and a number of manors had only one or two berewicks apiece. The resources of each of these groups of berewicks were usually included in a single composite entry covering a manor as a whole. No information, not even the hidage, is separately given for most of these named berewicks; but, sometimes, the details for a group of berewicks are given separately from those of the parent manor. Thus Kempsey had three berewicks (Mucknell, Stoulton and Wolverton), which between them answered for 7 hides and which had 7 plough-teams, 7 villeins, 7 bordars, 7 serfs and 16 acres of meadow (172 b). Only very occasionally is a berewick described separately in full; Crowle, appurtenant to Phepson (174), is one of these rare ones. In addition to the named berewicks, there were also fifteen unnamed berewicks; Chaddesley Corbett had eight of these (178), Hartlebury had six (174) and Comberton one (175). Furthermore, there are a number of composite entries covering two or more manors, e.g. that for Eardiston and Knighton (174).

For the purpose of constructing distribution maps, the details of these combined entries have been divided equally among all the villages concerned. But the item relating to wood cannot be so divided, and this has necessarily been allocated to the parent manor or to the first-named of a group of villages. Fisheries, salt-pans and mills have been treated likewise.[2]

SETTLEMENTS AND THEIR DISTRIBUTION

The total number of separate places mentioned in the Domesday Book for the area now included in the modern county of Worcester seems to be 260, including the three places for which burgesses are recorded— Worcester, Droitwich and Pershore. This figure, however, may not be quite accurate, for there are many instances of two adjoining villages bearing the same surname today, and it is not always clear whether more than one unit existed in the eleventh century. There is no indication, for example, that the Great and Little Comberton of today existed as separate villages in 1086. Similarly, the distinction between the modern Hanley

Child and Hanley William did not appear until the second half of the thirteenth century, and the Domesday Book refers only to *Hanlege*. Only where there is a specific reference to more than one village has each

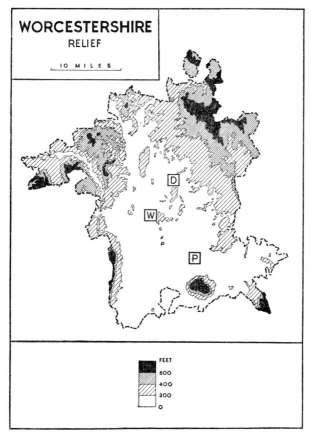

WORCESTERSHIRE
RELIEF

10 MILES

FEET
600
400
200
0

Fig. 75. Worcestershire: Relief.
Domesday boroughs are indicated by initials: D, Droitwich; P, Pershore;
W, Worcester.

been included in the total of 260. Thus there is a mention of *Frenesse et alia Frenesse* (172), so that although today there is only the single village of Franche, two Domesday settlements have been reckoned.[1] Another

[1] It is possible that only another holding is implied and that only one settlement should be counted. The same may apply to Upper Arley and *alia Ernlege* (247b). For similar instances in Essex, see *V.C.H. Essex* (London, 1903), I, p. 403.

8

fact that complicates any attempt to estimate the number of Domesday settlements is the existence of unnamed berewicks; whether these were holdings in places otherwise mentioned or whether they were separate places, we cannot say.

One interesting feature is that Worcestershire is unusually rich in examples of two or more villages with identical names, or with names containing the same element, in the Domesday record. Thus there are no less than seven Lenches and four Hamptons. Fortunately, there is rarely any problem of identification, for, as a rule, the settlements lie in different hundreds and in separate parts of the county; thus one village of Churchill (Oswaldslow) is in the south, while the other lies in the extreme north (Clent). Or again, there are two separate entries for *Crumbe*; Round has established that one entry referred to Earl's Croome, the other to Croome D'Abitot, the two being separated by the parish of Severn Stoke.[1] Hill Croome (which adjoins Earl's Croome) is distinguished as *Hilcrumbe* (173).

The total of 260 includes a number of places about which very little information is given, that is apart from the waste vills. The record may be incomplete, or the details may have been included with those of a neighbouring village. Thus an entry relating to Barley (173) states the hidage and the value in 1066, but gives no other details beyond saying that in 1086 it was in the possession of the king (*Nunc est in manu regis*). Or again, Nafford in Birlingham (175), described merely as a 'small piece of land' (*unum frustum terrae*), had a priest and was worth 5s. in 1086. All we hear of Malvern (173) is a reference to its wood which had been placed in the forest.[2] Upwich, Helpridge and Middlewich are mentioned merely as centres of salt-pans (172).[3] Broad Marston is described in the Gloucestershire folios (169), and we are told nothing about it beyond the fact that there were two hides there.

Not all the 260 Domesday names appear on the present-day parish map of Worcestershire. Some are represented by hamlets, by individual houses and farms, or by the names of topographical features. Thus *Harburgelei* is now the hamlet of Habberley in the parish of Kidderminster, and *Westmonecote* is the hamlet of Westmancot in the parish of Bredon. *Mucenhill* is represented by Mucknell Farm in the parish of Stoulton, and *Alcrintun* by Offerton Farm in the parish of Hindlip. *Estone* has survived as Aston Fields in the parish of Stoke Prior, and *Nadford* is represented

[1] *V.C.H. Worcestershire*, I, p. 292n.
[2] See p. 246 below.   [3] See p. 252 below.

by Nafford mill and hamlet in the parish of Birlingham. These are but some of the changes in the Worcestershire villages. To them must be added a number of unidentified names. Some of these were tentatively located by Round; thus he thought that *Baldehalle* lay to the north-west of Hanley Castle, that *Tonge* was between Alvechurch and Lea End, that *Cochehi* was near Wolverley in the extreme north of the county, and that *Rodeleah*, which appears in a joint entry with Whittington, was located nearby in the parish of St Peters, Worcester. Ten Domesday settlements have disappeared so completely that their names can be assigned to no locality, but their absence does not appreciably affect any general picture of the county.[1]

On the other hand, some settlements on the modern map are not mentioned in Domesday Book. Their names do not appear until the twelfth or thirteenth century, and, presumably, if they existed in 1086, they are accounted for under the statistics of neighbouring settlements. Thus, so far as record goes, Welland was first mentioned in 1182, Birtsmorton in 1241, and Redditch in 1247.[2] Some of the places not mentioned in the Domesday Book must have existed, or at any rate been named, in Domesday times because they first appear either in pre-Domesday documents or in documents that are roughly contemporary with it, and then reappear in documents of the twelfth century. Such names are Conderton, Elmley Castle, Flyford Flavell and Madresfield. Sometimes, although a modern parish is not named in the Domesday Book, it contains hamlets that are mentioned. An interesting example of this is Rock, in the north-west of the county. The earliest record of it, apparently, is from the year 1224, but the area covered by the present parish includes the Domesday settlements of Alton (176), Conningswick (176b), Hollin (177), Lindon (176), Moor (176 *bis*), Stildon (177) and Worsley (176). It is clear that the list of Domesday names differs considerably from that of present-day parishes.

The distribution of Domesday settlements can be seen from Fig. 76. While they appear to be widely scattered, this distribution is far from uniform. On the one hand, there is a marked absence of settlements in the west, that is, in the area between the Severn, the Teme and the Malvern

---

[1] The ten Domesday settlements unidentified in this analysis are as follows: *Bristitune* (172), *Broc* (176b), *Calvestone* (253), *Fastochesfelde* (172), *Halac* (176), *Hantune* (177b), *Hatete* (177b), *Ovretone* (174), *Sudwale* (172), *Tuelesberge* (172). Most of these must have been small settlements.

[2] For the names in this paragraph, see A. Mawer and F. M. Stenton, *The Place-Names of Worcestershire* (Cambridge, 1927).

Hills. There are several other empty spaces in the north-western uplands, in the Clent and Lickey Hills in the north-east, and in the east of the county in the Ridgeway Hills. Conversely, the most densely settled area forms a quadrilateral occupying the undulating Triassic sandstone country in the north, the plain of Worcester in the centre, and the Vale of Evesham in the south-east.

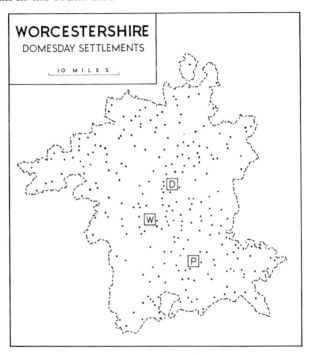

Fig. 76. Worcestershire: Domesday place-names.
Domesday boroughs are indicated by initials: D, Droitwich; P, Pershore; W, Worcester.

THE DISTRIBUTION OF PROSPERITY AND POPULATION

Some idea of the nature of the information in the Domesday folios for Worcestershire, and of the form in which it was presented, may be obtained from the entry relating to the village of Hanbury (174), five miles to the east of Droitwich. This village was held entirely by the Church of Worcester, and so it is described in a single entry:

This same Church [of Worcester] holds Hanbury. There 14 hides pay geld. In demesne are 2 ploughs, and 16 villeins and 18 bordars and a priest and a reeve.

Between all (*Inter omnes*) they have 24 ploughs. There [are] 4 serfs, and 1 bond-woman, and 20 acres of meadow. Wood 1 league in length and half a league in width, but it is in the King's forest. From the salt-pans in Droitwich 105 mitts of salt. It was worth £7, now £6. Of this land 2 hides are waste.

This entry does not include all the kinds of information that are given for some other places; there is no mention, for example, of mills, churches and fisheries. But it is representative enough, and it does contain the recurring standard items that are found for most villages. These are four in number: (1) hides; (2) plough-teams; (3) population; (4) values. The bearing of these four items upon regional variations in prosperity must now be considered. First, however, the almost complete absence of any reference to plough-lands must be noted. In this Worcestershire resembles the adjoining counties of Hereford and Gloucester. Many entries, however, do tell us something about the potential, as opposed to the existing, arable, and these are discussed under the heading of plough-lands below.

(1) *Hides*

The Worcestershire assessment is stated in terms of hides and virgates; no acres are mentioned. There were four virgates to the hide; although this is nowhere directly stated, the equation can be readily deduced, sometimes from rather neat additions such as the following:

Fo. 172. To this manor [of Bromsgrove] belonged and belong Grafton, where are 3½ hides, and Cooksey, where are 2½ hides, and Willingwick where are 2 hides and 3 virgates, and Chadwick where are 3 hides; in all, 12 hides less 1 virgate.

Occasionally there was an awkward fraction, as at Feckenham (180b), where a radman held 'half a hide and two-thirds of a half a hide', i.e. five-sixths of a hide in all. We also find departures from the straightforward statement of the assessment. Sometimes, for example, the hides are said to have been there in 1066 (*T.R.E. fuerunt*), as in a number of successive entries following that for Badsey on fo. 175 b.

There are several references to 'free' or 'non-gelding' hides, i.e. hides that did not pay geld. At Offenham (175 b) there is mention of a *hida libera*. At Phepson (174) there were 6 hides, one of which was exempt from geld (*Una ex his non geldat*); at North Piddle one hide out of nine was free (175 *bis*); and at Inkberrow 5½ hides were free (174). Mention must also be made of the entry '*n* hides in wood and field' (*inter silvam et planum*), which appears in three entries—for Defford (174b), for Beoley (175) and for Ombersley (175b).

There are many examples in the Worcestershire folios of the five-hide unit; thus Croome D'Abitot was held for 5 hides (173), Naunton Beauchamp for 10 hides (175), Severn Stoke for 15 hides (175), and Longdon for 30 hides (174b). The number of villages assessed at five hides or a multiple of five, amounts to about one-sixth of the total. If this number seems small, it must be remembered that it is possible that some villages were grouped in blocks for the purpose of assessment. There are, moreover, also several entries of 2½-hide holdings. Round, in fact, seems to imply a widespread basis of the five-hide unit: 'By a purely artificial arrangement, counties, hundreds, and "vills" (or to speak loosely, villages), were respectively assessed in lump sums, based, it is essential to remember, on a "five-hide unit".'[1] In any case, the absence of information about the blocks or groups makes it impossible to be definite about the full extent of the five-hide unit in Worcestershire.

Whatever the incidence of the five-hide unit, it is clear that the assessment was largely artificial in character. Maitland emphasised that the Worcestershire hidage afforded 'strong evidence of a neat arrangement of a whole county'.[2] He pointed out that the total hidage of the Domesday shire was 1,204,[3] which was near what he called 'the beautiful figure' of 1,200, that is, the total hidage of the twelve Domesday hundreds, if the relationship of a hundred to a hundred hides be accepted.

Consider Oswaldslow, owned by the Church of Worcester, which was indeed 'a very artificial aggregate of land', and yet formed a great 'triple hundred'. It was made up as shown below:

	Hides		Hides
*Blockley	38	Kempsey	24
Bredon	35	Northwick	25
Cropthorne	50	Overbury	6
Droitwich	15	Ripple	25
Fladbury	40	Sedgeberrow	4
Grimley	3	*Shipston on Stour	2
Hallow in Grimley	7	*Tredington	23
Harvington	3		
		Total	300

* Outside the present county.

[1] *V.C.H. Worcestershire*, I, p. 236.   [2] F. W. Maitland, *op. cit.* pp. 451–5.
[3] This should be 1,203, see below, p. 231.

Again, when we add the hidages of the manors within the hundred of Fishborough, we find that the total was only sixty-five. But the Domesday Book tells us that this hundred was perfected by the addition of 20 hides from Doddingtree and 15 from the city of Worcester (175 b). The final count by Maitland of hides in Worcestershire resulted in the following:

	Hides
Oswaldslow	300
Westminster	199
Pershore	100
Fishborough	65
Five territorial hundreds	539
Total	1,203[1]

As Maitland wrote, 'The hides in the vill are imposed from above, not built up from below,'[2] but it is not often that the system of geld assessment can be related so closely to the Domesday hundreds.

In view of its obviously conventional character, it is unlikely that the hidage reflected the agricultural realities of the time. The figures certainly bear no relation to those of plough-teams or population, as the variation among a selection of five-hide vills shows:

	Teams	Population
Holt (172 b)	12	36
Kingston (176 b)	4	19
Abbots Lench (173)	8	13
Rushock (177 b)	8	19
Suckley (180 b)	29	58

Thus the composite entry for Cropthorne and Netherton (174) gives 14 hides and 17 plough-teams, while, on the other hand, Abberley (176), assessed at only 2½ hides, had 19 plough-teams. There are entries, it is true, where hides and teams are either equal, or nearly so. One thing is certain: it would be rash to rely upon the hidages to give us any idea of the relative prosperity of different parts of the county.

The total assessment (including non-gelding hides) amounted to

[1] F. W. Maitland, op. cit. pp. 452–5. The total of 1,203 does not agree with Maitland's total of 1,204, although the latter was derived from the table above.
[2] F. W. Maitland, op. cit. p. 451.

1,367½, but it must be remembered that this refers to the area included in the modern county. The fact is that the nature of some of the composite entries makes exact calculation exceedingly difficult. All the figures can do is indicate the order of magnitude involved.

(2) *Plough-lands*

The Domesday folios for Worcestershire do not record the number of plough-lands, as distinct from that of plough-teams, on each holding. There are, however, two exceptional entries on fo. 177 which seem to have crept, as if by accident, into the final text for Worcestershire. These must be quoted in full:

> (1) The same Drogo holds Stildon.... There half a hide pays geld. There is land for 2 ploughs. It was worth 5s. It is now waste (*Modo est Wasta*).
>
> (2) The same William holds Bellington in the jurisdiction of his castle (*in castellaria sua*). Ælfric and Holand held [it] as 2 manors. There [are] 5 hides. There is land for 5 ploughs. It was and is waste (*Wasta fuit et est*). There 4 furlongs of wood, but it is in the King's forest. The meadows of this manor are worth 4d.

It is noteworthy that both holdings were waste in 1086.

In addition, there are other references to plough-lands at places which were in Domesday Warwickshire or Staffordshire, counties where plough-lands were regularly recorded. These places comprise Ipsley and Bickmarsh from Warwickshire, and Upper Arley and *Alia Ernlege* from Staffordshire; the Ipsley entry (244) may be quoted as an example:

> The same Hugh holds 3 hides in Ipsley. There is land for 7 ploughs. In demesne is 1 plough and 2 serfs, and 7 villeins with a priest and 13 bordars with 4 ploughs.

This entry seems to imply that two more ploughs could be employed.

Although the Worcestershire folios do not normally mention plough-lands, they sometimes seem to give an indication of potential as opposed to actual arable. Thus at Wychbold (176b) we are told that there was one team on the demesne and 18 teams with the peasantry, but we are also told that 2 ploughs more could be employed (*ii carucae plus possent esse*). This is the most common formula but two variants are used. At Bockleton (174) we are told *ibi possunt esse plus iiii carucae*; this phrase is used for about half a dozen places. The entry for Tardebigge (172b) tells us *in dominio est i caruca et alia potest fieri*; and that for Eardiston and Knighton runs: *et adhuc iii carucae possent fieri* (174).

The deficiencies may indicate changes in prosperity between 1066 and 1086, for it is sometimes difficult to escape the conclusion that the Commissioners must have based their figures on a comparison with the numbers employed in 1066. Thus at Elmbridge (176b), 10 ploughs were employed in 1086 and 10 more could be employed, while the value of the holding showed a proportionate decline from 100*s.* in 1066 to 50*s.* in 1086. But it must be confessed that this exact relationship only occasionally holds good.

Deficiencies are encountered in about 13% of the entries that record teams in the Worcestershire folios. These deficiencies are plotted on Fig. 77; most of them were in places situated on or near the uplands in the northern part of the county. There are few entries of deficiencies for the south-centre and south of the county, that is, in the Plain of Worcester and the Vale of Evesham. In fact, the Commissioners were at pains in one entry, in the account of the holdings of the Church of Worcester, to state specifically '*In omnibus his maneriis non possunt esse plus carucae quam dictum est*' (174). It may well be that the total deficiency of 146½ teams, as compared with the 2,008¾ present in 1086, was as low as it was because the Church was such a powerful landowner in the county.

(3) *Plough-teams*

The Worcestershire entries, like those of other counties, draw a distinction between the teams held by the lord of a manor in demesne and those held by the peasantry. Fractions of plough-teams are sometimes given; thus at Clent (172b) there were 1½ teams in demesne and 9½ held by the peasantry, while at Halesowen (176) there was a total of 51½ teams. There are occasional variations from the normal formula. A few entries make reference to oxen as well as to plough-teams; thus at Leopard there were '1 plough and 6 oxen' (174). A most unusual piece of information occurs in the composite entry for Offenham, Littleton and Bretforton (175 b), where there were 'oxen for 1 plough, but they draw stone to the church' (*Ibi sunt boves ad i carucam sed petram trahunt ad ecclesiam*). Additional details are sometimes given, and on fo. 174b there are three consecutive entries which mention the operations of ploughing and sowing:

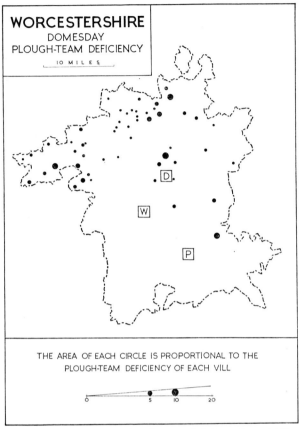

Fig. 77. Worcestershire: Domesday plough-team deficiency in 1086 (by settlements). Domesday boroughs are indicated by initials: D, Droitwich; P, Pershore; W, Worcester. *Halac*, *Hantune* and *Hatete* are unidentified.

Bricklehampton: *Ibi x villani et x bordarii cum vi carucis et arant et seminant vi acras de proprio semine.*

Defford: *Ibi viii villani et x bordarii cum vi carucis et arant iiii acras et seminant de suo semine.*

Eckington: *Ibi vi coliberti reddunt per annum xi solidos et ii denarios et arant et seminant de proprio semine xii acras.*

The total number of plough-teams amounted to 2,008¾, but it must be remembered that this refers to the area included in the modern county. Maitland estimated the number for the Domesday county at 1,889.[1]

[1] F. W. Maitland, *op. cit.* p. 401.

(4) *Population*

The main bulk of the population was comprised in the three main categories of bordars, villeins and serfs. In addition to these were the burgesses, together with a very varied miscellaneous group. The details of these categories are summarised on p. 238. Several estimates have been published of the population of the Domesday county. Sir Henry Ellis's[1] original figures appeared in 1833, while other estimates were made by J. H. Round in 1901[2] and by R. C. Gaut in 1939,[3] but they are comparable neither with one another nor with the present estimate, which has been made in terms of the modern county. Definitive accuracy rarely belongs to a count of Domesday population, and all that can be claimed for the precise figure of 4,409 that it indicates the order of magnitude involved. This figure is of recorded population, and must be multiplied by some factor, say 4 or 5, in order to obtain the actual population; but this does not affect the relative density as between one area and another.[4] That is all that a map, such as Fig. 80, can roughly indicate.

Moreover, a few of the entries are not specific: thus at Besford we find that William the priest, 'with his men' (*cum suis hominibus*) had 1½ ploughs (174b). There are also some obscure references to individuals, who might or might not be tenants, such as 'one man' at Pull Court (180b) who has not been included in our total. Some entries combine different categories of population in one total. Thus at Alvechurch (174) there were '7 serfs male and female' (*Inter servos et ancillas sunt vii*), and at Tenbury (176b) there were '14 villeins and bordars' (*xiiii inter villanos et bordarios*). Such entries do not necessarily affect the total figure, but in estimating individual categories the only course possible is to divide them arbitrarily.[5] In all, for example, there are combined returns of sixty-two male and female serfs; thirty-one have therefore been included in the male population figures. Finally, it must be remembered that a few entries seem to be

---

[1] Sir H. Ellis, *A General Introduction to Domesday Book* (London, 1833), II, pp. 504–6. Ellis's figures were quoted by J. W. Willis-Bund, 'Worcestershire Doomsday', in *Associated Architectural Societies' Reports and Papers* (Worcester Diocesan Architectural and Archaeological Society, Lincoln, 1894), vol. XXII, part II, pp. 88–108.

[2] J. H. Round, *V.C.H. Worcestershire*, I, pp. 273–9.

[3] R. C. Gaut, *A History of Worcestershire Agriculture and Rural Evolution* (Worcester, 1939), p. 22.

[4] But see p. 430 below for the complication of serfs.

[5] At Powick (174b) there was an unspecified number of bordars and serfs (*plures bordarii et servi*).

defective and contain no reference to population, but it is impossible to be certain about the significance of these omissions. Several holdings were waste, e.g. at Hollin (177) and Hillhampton (178), and supported no population.

Bordars constituted the most important element in the population, and amounted to 39% of the total. Two unusual entries seem to refer to poverty among the bordars. In the unidentified *Hatete* (177b) we read of '1 bordar who has nothing', presumably no stock, and at Suckley (180b) there were '10 other bordars [who are] poor' (*Ibi alii x bordarii pauperes*). Villeins made up 38% of the total, but they present no special feature of interest for our purpose. The serfs formed some 16% of the total population, a proportion which is broadly similar to that in other west midland counties.[1] As stated above, where a combined number of male and female serfs is given, such figures have been divided equally among the sexes for purposes of computation; *ancillae* have not been included in the total population.

The remaining categories of population are small in number, but are of considerable interest. As Round said, '...we are confronted by a hierarchy of classes bewildering enough in its variety. Indeed, it would be difficult in any county to find a greater variety.'[2] One of the most striking features about the composition of the Worcestershire population is that there were only seven freemen (*liberi homines*)—three at Upper Arley, two at Astley and one each at Moor (in Rock) and Wolverley. There is no mention of sokemen. To this group of seven freemen must be added thirty-three radmen (*radmanni* or *radchenistres*), a category which is almost restricted to the western counties along or near the Welsh border —in Gloucestershire, Worcestershire, Herefordshire, Shropshire and Hampshire. The significance of their occurrence thus is discussed by Maitland[3] and Round;[4] the conclusion seems to be that these men were freemen, specially distinguished by the fact that they owed 'riding-duty', which in Round's opinion was escort duty but not military service.[5]

---

[1] J. W. Willis-Bund (*op. cit.* pp. 102–7) devoted much of his paper to '...the position of the monasteries as slave-owners...', and produced some interesting theories concerning the high proportion of serfs in the Welsh marchlands.

[2] *V.C.H. Worcestershire*, I, p. 273.

[3] F. W. Maitland, *op. cit.* pp. 57, 66, 305–8.

[4] *V.C.H. Worcestershire*, I, pp. 250–1.

[5] There were T.R.E. 10 radmen, 9 freemen and 13 thegns. For the identity of radmen and freemen in Gloucestershire, see p. 17 above.

Priests totalled fifty-five; those priests who were obviously sub-tenants have not been counted here, but the decision is not always an easy one. Furthermore we cannot include the monks living in religious houses, to whom only occasional and unspecific references are made.[1] It is interesting to note that no less than twenty-one of the total of fifty-four cottars were found in Upton Snodsbury.[2] There were only nine coliberts —six at Eckington and three at Powick, where the word *coliberti* is inter-lined above *buri* (174b). Another category is that of the oxmen (*bovarii*), of whom there were seventy-seven.[3] These men were confined, with a few exceptions, to the West Midland counties. There were twenty-five *francigenae*, including two *francigenae servientes* at Upton Snodsbury. This category in all probability comprised Frenchmen by birth. Round[4] considered that they were serjeants '...whose services were rewarded by land...'. Then there was the group of village officials: reeves (11), beadles (5), smiths (7) and millers (3). There is also mention of seven swineherds (six of whom were at Hanley Castle where they held four ploughs), of four salt-workers, of two foresters, of one cowman (*vaccarius*), of one huntsman (*venator*), of one keeper of the bees (*custos apium*) and of one 'man'. These various categories are far from complete; the total of three millers compares strangely with the total of over 100 mills, and the entries of village officials and other countrymen are equally deficient. Obviously the other millers and the like were entered under the category of villeins.

Finally, mention must be made of the female population. This con-sisted for the most part of bondwomen (*ancillae*). These are not recorded for many counties, and only infrequently for others. Worcestershire, with 131, had one of the largest totals for any county; this figure includes those entered under the formula *inter servos et ancillas*. Three other women are mentioned, two dairymaids (*daia*) at Bushley (180b) and at Queenhill (180b), and a widow (*una vidua*); the last, at Badsey (175b), is most unusual.

---

[1] J. W. Willis-Bund (*op. cit.* p. 91) says '...we know from other sources that the Monastery of Evesham alone had probably over 100 monks belonging to it at that date, and it is not unlikely that Pershore and Worcester had nearly as large a number...'.

[2] The Worcestershire folios make a distinction between the *cotarii* and *cotmanni*. Maitland (*op. cit.* pp. 39–40) believes that there is some evidence that the latter had heavier rents and services, and therefore had larger holdings.

[3] See pp. 127–8 above.

[4] *V.C.H. Worcestershire*, I, p. 273.

Recorded Population of Worcestershire in 1086

A. Rural Population

Bordars	1,732
Villeins	1,653
Serfs	721
Miscellaneous	303
Total	4,409

There were also 131 bondwomen (*ancillae*), two dairymaids and one widow who have not been included in the above total.

Details of Miscellaneous Rural Population

Oxmen . . .	77	Beadles . . .		5
Priests . . .	55	Salt-workers . .		4
Cottars . . .	54	Millers . . .		3
Radmen . . .	33	Foresters . . .		2
Francigenae . . .	23	*Francigenae servientes* .		2
Reeves (*Prepositi*) .	11	Cowman . . .		1
Coliberts . . .	9	Huntsman . . .		1
Freemen . . .	7	Bee-keeper . . .		1
Swineherds . . .	7	A man (*Homo*) . .		1
Smiths . . .	7			
		Total . . .		303

B. Urban Population

Villeins and salt-workers are also included in the table above.

WORCESTER   7 burgesses; 125 *domus*; 28 *masurae*; 5 *masurae vastae*.
DROITWICH   116 burgesses; 35 *domus*; 7 villeins; 4 salt-workers.
PERSHORE   28 burgesses—see pp. 265–6.

(5) Values

The valuation of estates in Worcestershire appears to have been carried out with care. The value of an estate is normally given in pounds and shillings, although occasionally pence are used; Besford (174b), for example, was worth 16d. The majority of entries give values for only two dates—1066 and 1086; but there are about thirty entries which add intermediate valuations. The amounts had sometimes risen, sometimes fallen, or had occasionally remained the same. There are a few entries in which only one value is given, either for 1066 or for 1086; thus Queenhill (173b) was worth 40s. in 1066, and Pensham (174b) was worth £3 in 1086. There are a few composite entries; thus Mucknell, Stoulton and Wolverton (172b) were together worth 100s., the value of Feckenham is given with that for Hollow Court (180b), and a combined value is also given for six Worcestershire manors entered among the Herefordshire folios (180b). The number of these composite entries is greatly increased if the berewicks are included; thus Bromsgrove and its eighteen berewicks were valued together at £24 (172). No values appear for any date in a few entries, e.g. in that for *Cochehi* (177b).

The values are usually entered as plain statements of money, but there are entries in which the values were reckoned by weight. The account of Worcester itself (172) refers to amounts by weight (*ad pensum*) and by tale (*ad numerum*). The value of the whole manor of Kidderminster (172) was £10. 4s. by weight (*ad peisam*). Some places were valued in terms of the *ora*; thus the value of Martley (180b) was £24 at 20d. to the ounce (*xxiiii librae denariorum de xx in ora*). Occasionally the value was expressed in kind; holdings at Droitwich, for example, returned salt as well as money; the Droitwich holding attached to Tardebigge paid 20s. and 100 mitts of salt (172b).

Generally speaking, the greater the number of plough-teams and men on an estate, the higher its value, but it is impossible to discern any constant relationships, as the following figures for five manors, each yielding 50s. in 1086, show:

	Teams	Population	Other resources
Elmbridge (176b)	10	35	Wood, meadow, salt-pan
Kington (176b)	4	14	Wood, 'hay'
Pedmore (177)	6½	17	Underwood
Shelsley (176)	10	17	Fishery, meadow, wood
Upton Warren (177b)	7	25	Wood, mill, salt-pan

It is occasionally possible to correlate changes in value with economic changes as between 1066 and 1086. Where there was a decline in value, we often find the formula '*n* ploughs more can be employed', as at Elmbridge (176b) and at Glasshampton (177).

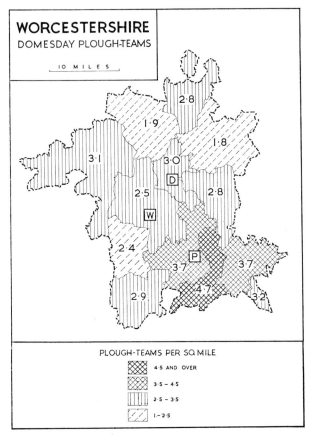

Fig. 78. Worcestershire: Domesday plough-teams in 1086 (by densities).
Domesday boroughs are indicated by initials: D, Droitwich; P, Pershore;
W, Worcester.

Conclusion

The varied size of the Domesday hundreds and their dispersed nature,
particularly of the Church hundreds, do not make them convenient
units for calculating densities. The straggling Pershore-Westminster
hundreds have been divided into two, and the widely separated fragments
of Oswaldslow grouped into four units. The five territorial hundreds
have been broadly adopted, but have been simplified and rounded off.
The result of these modifications has been to produce thirteen units as
a basis for calculating densities.

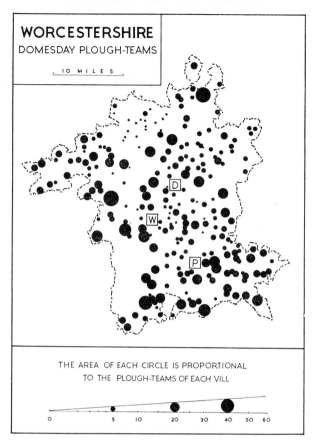

Fig. 79. Worcestershire: Domesday plough-teams in 1086 (by settlements).
Domesday boroughs are indicated by initials: D, Droitwich; P, Pershore;
W, Worcester.

Of the four standard formulae, those relating to plough-teams and
population are most likely to reflect something of the distribution of
wealth and prosperity throughout the county (Figs. 78 and 80). Taken
together, they supplement one another, and, when they are compared,
certain common features stand out. The most important of these features
is the contrast between the south-eastern part of the county and the
remainder. Over the Avon valley and the southern part of the Plain of
Worcester the plough-teams vary between 3·7 and 4·7 per square mile.
The corresponding densities over the rest of the county are much lower,

only between 2·4 and 3·1 per square mile, except in two areas. These two are the Triassic sandstone area in the north and the uplands in the north-east, where the densities are below 2. The density of population shows a like contrast; in the south-east it is about 9 or 10; over the rest of the

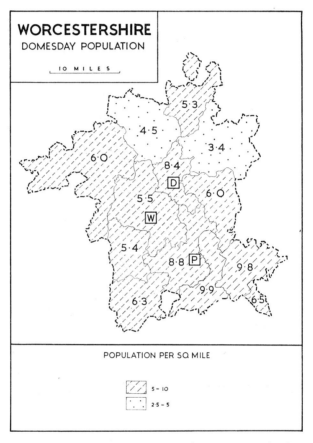

Fig. 80. Worcestershire: Domesday population in 1086 (by densities).
Domesday boroughs are indicated by initials: D, Droitwich; P, Pershore;
W, Worcester.

county it varies between 5·4 and 8·4, except in the north, where it falls as low as 3·4. Clearly, then as now, the Vale of Evesham and the southern part of the Plain of Worcester, gently undulating, covered either with heavy loams or with alluvial soils on the floodplains, was the most attractive part of the county for agriculture. The thin soils of the marginal uplands and the hungry sandy soils of the Triassic sandstone area of the north were very much less favourable.

Figs. 79 and 81 are supplementary to the density maps, but it is necessary to make one reservation about them. As we have seen on

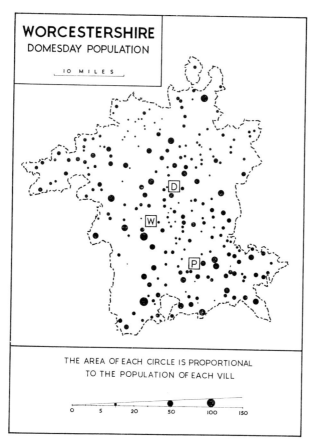

Fig. 81. Worcestershire: Domesday population in 1086 (by settlements). Domesday boroughs are indicated by initials: D, Droitwich; P, Pershore; W, Worcester.

p. 224, it is possible that some Domesday names may have covered two or more settlements. A few of the symbols should therefore appear as two or more smaller symbols. The absence of the unidentified places must also be remembered, but these hardly affect the pattern. The maps reveal certain obvious blank areas on the one hand, certain groupings and alignments on the other, and therefore help to amplify the density maps.

WOODLAND AND FOREST

Types of entries

The usual term for woodland is *silva*, although *nemus* is occasionally used, and *silvula*, in a single place at Pedmore (177), may denote underwood. The extent of woodland on each holding is recorded normally by giving its length and breadth in terms of leagues and furlongs. The league is usually thought to have contained twelve furlongs, i.e. $1\frac{1}{2}$ miles, but Round maintained that a league might well have comprised only four furlongs (i.e. half a mile) because, he wrote, the Worcestershire folios never mention a figure higher than three furlongs below the league.[1] This is not so because three Worcestershire entries each mention four furlongs— for Croome D'Abitot (173), Grafton (175) and Bellington (177).[2] In any case, the exact significance of these dimensions is far from clear, and we cannot hope to convert them into modern acreages.[3] All we can do is to plot them diagrammatically as on Fig. 82. The largest individual entries are of 7 leagues by 4 at Bromsgrove (172), 6 leagues by 3 at Beoley (175), and 3 leagues by 2 at Alton in Rock (176). The first of these illustrates some of the problems involved in reconstructing an impression of the extent of the woodland; the entry refers to Bromsgrove with its eighteen berewicks. But we cannot tell whether the figure represents a continuously wooded area, or whether it implies some process of addition whereby the dimensions of separate pieces of woodland within the lands of these nineteen different places were consolidated into one sum. On Fig. 82 these and similar dimensions have been plotted for the caput of the manor or for the first-named village in a composite entry. The associated villages in all such entries have also been indicated.

There are two entries, for Suckley and Hanley Castle respectively, in which the linear formula is peculiar:

Suckley (180b): *Silva habet v leuuas inter longitudinem et latitudinem.*
Hanley Castle (180b): *Silva v leugarum inter longitudinem et latitudinem.*

---

[1] *V.C.H. Worcestershire*, I, pp. 271–2.

[2] The *V.C.H.* translation omits any mention of wood (and meadow) from the Croome entry, and omits the Grafton entry completely.

[3] See p. 437 below.

About thirty other entries give merely one dimension. Thus at Kidder-minster there was *Silva de iiii leuuis* (172) and at Chaddesley Corbett there was 'wood of 2 leagues and other wood (*alia silva*) of 1 league' (178). It would seem that both peculiar types of measurement are variants

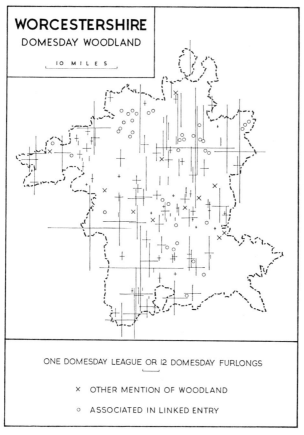

Fig. 82. Worcestershire: Domesday woodland in 1086.
Where the wood of a village is entered partly in linear dimensions and partly in some other way, only the dimensions are shown.

used when length and breadth are the same.[1] The wood for Suckley and for Hanley Castle has been plotted on this assumption on Fig. 82, but the single dimensions have been shown as single lines to indicate their frequency.

[1] See p. 437 below.

While leagues and furlongs form the usual units of measurement, there are some exceptional entries that add further detail or that refer to wood in other ways:

Atch Lench (175b): 6 acres of wood (*vi acrae silvae*).

Crowle (176b): Half a league of wood for 100 swine (*Silva dimidia leuua ad c porcos*).

Inkberrow (174): Wood 2 leagues long and 1 league wide. For pannage it renders 100 swine (*de pasnagio reddit c porcos*).

Hanley Castle (180b): And 6 swineherds render 60 swine and they have 4 plough-teams. There [is] a mill yielding (*de*) 2s. Wood 5 leagues between length and breadth. This is placed outside the manor. There is a hawk's eyry. The forester holds half a virgate of land (*et vi porcarii reddunt lx porcos et habent iiii carucas. Ibi molinum de ii solidis. Silva v leugarum inter longitudinem et latitudinem. Haec missa est foris manerium. Ibi est airea Accipitris. Forestarius tenet dimidiam virgatam terrae*).

Bredon (173): Wood 2 leagues long and 1½ leagues wide. The bishop has from it 10s. and all its proceeds in honey and hunting and all else.

Fladbury (172b): Wood 2 leagues long and a half wide. The bishop has all its proceeds in hunting and honey and fuel (*lignis*) for the salt-pans of Droitwich, and 4s.

Malvern (173): Wood half a league long and 3 furlongs wide in Malvern. From this the bishop used to have the honey and the hunting and all profits, and 10s. over and above. It is now in the forest, but the bishop has its pannage and [wood for] firing and the repair of houses (*Pasnagium vero et ignem et domorum emendationem inde accipit episcopus*).

Huddington (173b): Wood yielding 3s., and it is in the king's forest (*Silva de iii solidos et est in foresta regis*).

Sheriff's Lench (176): There, wood rendering 2s. (*Ibi silva reddens ii solidos*).

Beoley (175): There, a wood 6 leagues long and 3 leagues wide, and it renders 40d.

Astley (176): There is a wood. It renders nothing (*Silva est. Nil reddit*). (A marginal note adds: *Silva i leuua longa et dimidia leuua lata.*)

Rochford (186b): The wood renders nothing (*Silva nil reddit*).

Horton (177b): There, a small wood (*Ibi parva silva*).

Moor in Rock (176): There, a small wood (*Ibi parva silva*).

Whittington and *Rodeleah* (173b): Wood sufficient for firing (*Silva ad ignem tanta*).

Pedmore (177): One league of underwood (*una leuua silvulae*).

Laughern (172b): 12 oaks (*xii querci*).

Martley (180b): The villeins and bordars render 12s. in lieu of fish and fuel (*pro pisce et pro lignis*).

The references to swine in three of the above entries are of particular interest. In some counties the wood is indicated almost exclusively in terms of the number of swine for which it could afford pannage. In the entries for Crowle and Inkberrow, the swine are mentioned side by side with linear dimensions, but it is obvious that the two units cannot be correlated, and that no formula for the conversion of one unit into the other is possible.

Mention must be made of a phrase, unusual in the Domesday Book, which does little more than hint at the presence of wood—'*n* hides in wood and field' (*inter silvam et planum*). This form appears three times in the Worcestershire folios—for Defford (174b), Beoley (175) and Ombersley (175b). There is no further reference to wood in the entry for Defford, but for the two latter places the dimensions of the woodland are also given. There is also a number of other references to wood in the entries relating to the king's forest discussed below.

Distribution of woodland

A general idea of the distribution of the Domesday woodland of the county is given by Fig. 82. It must be remembered that the distribution shown here is plotted under the settlements for which the wood was entered. Where settlements are scanty, there are therefore blanks on the woodland map. Apart from these blank areas, the map does suggest that the Domesday woodland was widespread over the Lias clay lowland, but that the Vale of Evesham in the south-east was much less wooded. It is a striking fact that wood is recorded for only three places in the Vale of Evesham. The correlation of these scanty woodland entries in the Vale with the high plough-team and population densities is perhaps too obvious to labour; the distribution of the woodland in a broad way was complementary to that of the plough-teams.

Forests

There were 22 places at which woodland had been placed in the forest. The formula usually runs *in foresta* or *in foresta regis*, but sometimes the wording is more complicated. Thus for Feckenham (180b) we read: *Silva hujus manerii foris est missa ad silvam regis*. At Alvechurch (174) the king took only one half the wood: *Silva iiii leuuedes. Inde rex tulit medietatem in suam silvam.* In addition to these, afforestation must also have taken place at 5 other places for the wood of which some such phrase as 'outside the manor' is used:

Eldersfield (180b): *Silva ii leuuis longa et tantundem lata. Hoc extra manerium est missa.*

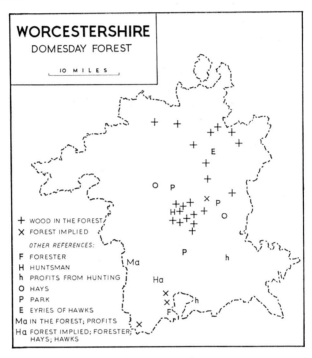

Fig. 83. Worcestershire: Domesday Forest in 1086.
Ha, Hanley Castle; Ma, Malvern.

Hanley Castle (180b): *Silva v leuugae inter longitudinem et latitudinem. Hoc missa est foris manerium.*
Queenhill (180b): *Silva est foris manerium missa.*
Queenhill (173): *Nunc est in manu regis* (For the whole holding consisting of meadow and wood. No T.R.W. value).
Hollow Court (180b): *Ibi est parcus ferarum sed missus est extra manerium cum tota silva.*
Shell (176b): *Silva missa est in defenso.*

Afforestation is therefore indicated at 27 places. There are also a number of other entries which may imply the presence of forests in 1086. There was a forester at Bushley (180b) and another at Hanley Castle (180b); an entry for Pull Court (180b) says that Earl William put both these foresters outside his manors 'for the keeping of the woods' (*propter silvas custodiendas*). There was a huntsman at Leopard (174); there were profits from hunting and honey at Bredon (173), Fladbury (172b) and Malvern (173); and there were eyries of hawks (*airea accipitris*) at Bromsgrove (172) and Hanley Castle (180b). There were three 'hays' (*haiae*), or enclosures into which deer were driven, at Holt (172b), Kington (176b) and Hanley Castle (163b). There were three parks, at Salwarpe (174, 176), Wadborough (175) and Hollow Court (180b); the last-named, as we have seen, was said to have been placed outside the manor. Finally, Barley (173) was in the king's hands (*Nunc est in manu regis*); it had no T.R.W. value, and on the analogy of Queenhill, it, too, may well have been in the forest, but unlike Queenhill no wood is entered for it.

Fig. 83 shows that most of the entries referred to places in the central part of the county, but the outlying ones for the south-west indicate the presence of the great Malvern forest.

MEADOW

Types of entries

The entries for meadow are for the most part comparatively straight-forward. Almost all the entries which refer to meadow repeat the same monotonous formula—'*n* acres of meadow' (*n acrae prati*). The amount of meadow in each vill varied from 4 acres at Ombersley (175b) to 160 acres at Pershore and its berewicks (174b, 175). The usual amount seems to have been between 10 and 50 acres. As in the case of other counties, no attempt has been made to translate these figures into modern acreages. The Domesday acres have been treated merely as conventional units of measurement, and Fig. 84 has been plotted on that assumption.

While measurement in acres is normal, there are two entries that refer to meadow in other ways. At Bellington the amount of meadow was unspecified, but we are told that it was worth 4d. (177); at Inkberrow we are simply told that there was meadow for the oxen (173). Furthermore, in the entry for Abbots Lench the words 'and meadow' are followed by a blank in the manuscript, where, apparently, the amount was intended to be filled in (173). Occasionally, a little additional detail is given us about meadow. For six places—Eckington, Dormston, Grafton, Longdon, Powick, and Upton Snodsbury—there are references to services connected with mowing in the meadows. They are all described on fos. 174b–175, and the relevant phrase in each refers to conditions in 1066. At Eckington, for example, we are told that the English tenants 'used to mow in the meadows of their lord for a day as a customary service' (*secabant in pratis domini sui pro consuetudinem* [*sic*] *unam diem*). It seems as if the scribe, at this point in his digest, departed from his usual terse statement about meadow.

Distribution of meadowland

Fig. 84 shows with remarkable clarity the contrast between the Plain of Worcester and the Vale of Evesham on the one hand and the western and northern parts of the county on the other. In all the northern third of the county, meadow is entered for one place only, and that is for Bellington (177), where it was worth only 4*d*. But elsewhere the meadow seems to have been distributed fairly widely and uniformly. It was found both on the heavy clay-lands of the Lower Lias, with its slow and imperfect drainage, and along the valleys of the Severn and the Teme in the west and of the Avon in the south. Along the Severn valley there were amounts of 40 acres at Kempsey (172b) and 20 acres at Severn Stoke (175), while near the banks of the Teme, Eastham (176) had 60 acres, and Shelsley (176) and Leigh (175b) had 30 acres each. Elmbridge (176b), near a small tributary of the Severn, had 50 acres. Meadow occurred especially near the junctions of the Avon and its various right-bank tributaries, such as the Bow and Piddle, where the main river wanders over its flood-plain in a series of sweeping meanders. Here are the biggest entries of meadow: over 100 acres at Pershore (174b, 175), 98 acres at Longdon (174b), 80 acres at Bredon (173) and 50 acres at Fladbury (172b).

FISHERIES

Fisheries (*piscariae*) are specifically mentioned in connection with at least 16 places in Worcestershire. It is difficult to be sure of the exact number because of composite entries. One place, *Broc* (176b) with half a fishery, is at present unidentified. Some entries record merely the presence of a fishery. Other entries add a render in eels or in money. The number of eels is usually expressed in *stiches* or sticks of twenty-five; thus at Orleton (177) two fisheries rendered forty *stiches*. At Ombersley (175 b) one and a half fisheries rendered 2,000 eels; the other half may have been that rendering an undefined number of *stiches* at Grimley

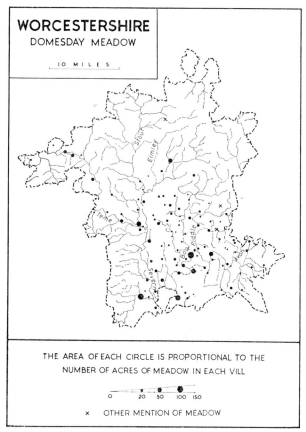

Fig. 84. Worcestershire: Domesday meadow in 1086.
Where the meadow of a village is entered partly in acres and partly in some other way, only the acres are shown.

nearby (173 b). When expressed in money, the render also varied considerably. A fishery at Hamcastle (176 b) rendered 2*s.*; one at Holt (172 b) 5*s.*; and one at Kidderminster (172) rendered 100*d.* A fishery on one holding (176 b) at Shelsley rendered 16 *stiches*, and on another holding (176) a second fishery rendered 2*s.*

In addition to the 16 places for which *piscariae* are specifically mentioned, there were also at least four other places with mills that rendered eels (some places again in composite entries); thus at Fladbury (172b) a mill rendered 10s. and 20 *stiches*. Finally the entry for Martley (180b) records not only a mill rendering 8s. but also two weirs (*gurgites*) which rendered 2,500 eels and 5 *stiches*; and the villeins and bordars of the manor rendered 12s. in lieu of fish and fuel. Altogether there were at least 21 places for which fisheries were recorded or implied. In view of the reference to weirs, it is interesting to note that Heming's Cartulary, produced at Worcester early in the twelfth century, refers to the construction of brushwood hedges across the Severn to form fish traps.[1] The distribution of places with fisheries is shown on Fig. 85. They were situated along the Severn and its tributaries, particularly the Teme and the Avon; most of those along the latter streams were fisheries associated with mills.

SALT

Domesday salt-pans (*salinae*) were for the most part located on the coast, where salt was obtained by the evaporation of sea-water. There were, however, two very important inland groups of saltworks, in Cheshire and in Worcestershire. Among the Keuper Marl beds of the Triassic system there is a stratum of rock-salt, from whose brine-springs salt has been extracted for many centuries.

A total of at least 305 salt-pans was recorded for Worcestershire in the Domesday Book (Fig. 86). No less than 160 were entered for Droitwich itself, or *Wich*, as it was then known. It was a borough, and as such is described on pp. 263–5 below, but its life clearly revolved around the salt industry. The main entry (172) relating to the salt-pans at Droitwich must be quoted in full:

In Droitwich King Edward had 11 houses, and in 5 brine-pits (*puteis*) King Edward had his share. In 1 brine-pit—Upwich—54 salt-pans (*salinae*) and 2 *hocci* pay 6s. 8d. In another brine-pit—Helpridge—17 salt-pans. In a third brine-pit—Middlewich—12 salt-pans and two-thirds of a *hoccus* pay 6s. 8d. In 5 other brine-pits, 15 salt-pans. From all these King Edward had a rent of £52. In these brine-pits Earl Edwin had 51½ salt-pans and from the *hocci* he had 6s. 8d. All this paid a rent of £24. Now King William has in demesne both what King Edward and Earl Edwin had.

[1] *V.C.H. Worcestershire*, I, p. 272.

This single entry refers therefore to a total of 149½ salt-pans owned in 1086 by the king. In addition to these, 7 were held by Herald son of Earl Ralf (177), 1½ by Roger de Laci (176b), 1 by the Church of Westminster (174b), and 1 by the Church of St Denis (174).

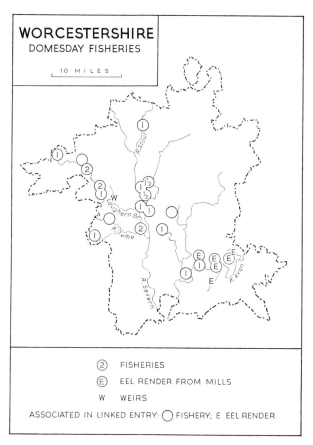

Fig. 85. Worcestershire: Domesday fisheries in 1086.
The fishery entered for Kidderminster and its sixteen berewicks has been plotted at Kidderminster itself. *Broc*, with half a fishery, is unidentified.

A further 76½ salt-pans at least[1] were stated to be in Droitwich, but belonged to outside manors under which they are entered. The entry for

[1] Allowing two each at Fladbury (172b), Hanbury (174), Martin Hussingtree (174b) and Phepson (174), because of the unspecific but plural mention of salt-pans for each place.

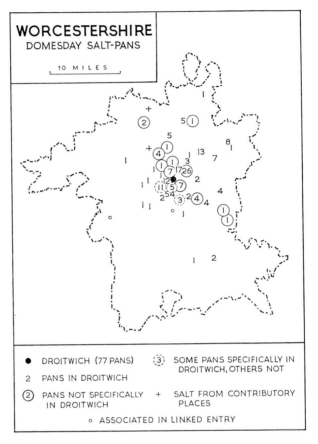

Fig. 86. Worcestershire: Domesday salt-pans in 1086.
Crowle (174, 176b), Salwarpe (174, 176) and Witton (177b *bis*) had some salt-pans
entered for Droitwich, and also others not so specifically attributed. The differences
may be merely scribal, but the three places have been differentiated on the map.
Hantune (177b), with one salt-pan, is unidentified.

Bromsgrove (172) is typical: 'To this manor belong 13 salt-pans in
Droitwich.' Many of these entries refer only to a single salt-pan, which
obviously supplied the needs of its parent village, but apart from the
13 at Bromsgrove, 8 belonged to Alvechurch (174), 7 to Tardebigge
(172b), 6 to Salwarpe (174), 5 to Chaddesley Corbett (178), and there
were many other manors with fewer salt-pans.

In addition to these 236½ salt-pans which were definitely in Droitwich, the Worcestershire folios also mention another 68½ salt-pans;[1] these are entered among the details for various villages. Thus at Inkberrow (174) there was 1 salt-pan which rendered 15 mitts of salt, and at Wychbold (176b) 26 salt-pans rendered £4. 12s. But it is not clear where these and the rest of this category of 68½ salt-pans were situated, and J. H. Round's warning must be borne in mind: 'Some misapprehension has arisen from the entry of *salinae* under other places, without the explanation that they were situated in Droitwich; the existence of local salt-works has been wrongly deduced from these entries.'[2] All we can say is that the villages near Droitwich probably had their own salt-pans, but equally probably those away to the north and to the south-east had not.[3] On Fig. 86 a distinction is made between those places owning salt-pans which are specifically stated to be situated in Droitwich, and the others.

With one exception, that at Ombersley (175b), the value of each salt-pan is given, sometimes in money, sometimes in terms of the amount of salt which is rendered. The four examples that follow illustrate the types of entry that are encountered:

Wychbold (176b): 26 salt-pans rendering £4. 12s.
Bell Hall (177): a salt-pan worth (*de*) 2 ounces of silver.
Horton (177b): a salt-pan worth 40d.
Alvechurch (174): In *Wich* 8 salt-pans. One of these renders 50 mitts of salt, the other seven render 70 mitts of salt.

The last entry, incidentally, gives some indication of the very variable size and output of the different salt-pans. The measure used here, the *mitta* or *metta*, is somewhat obscure; it has been suggested that it was equivalent to eight bushels.[4] Probably it was merely used as a rough measure for a horse-load. The word *summa*, however, was used more commonly than *mitta* in the Worcestershire folios, and probably also indicated a horse-load.

[1] This figure includes part of a salt-pan (*partem de una salina*) worth 10 mitts of salt, which is counted as a half at Witton (177b).   [2] *V.C.H. Worcestershire*, I, p. 269.
[3] Some of the places with contributory houses and burgesses in Droitwich also rendered *mittas* of salt—see p. 265. Whether these imply the existence of yet other *salinae* is not clear; but they have been marked on Fig. 86.
[4] E. B. Pillans, *V.C.H. Worcestershire* (London, 1906), II, p. 257. T. Habington, *A Survey of Worcestershire* (printed by the Worcester Historical Society, 1895–6), II, p. 297, stated: 'Foure of these barowes, conteygninge about towe bushells of Salt are named a Mit'. Round, *op. cit.* p. 270, quotes from Hale's *Registrum* (34a): *Invenient singulis annis equos diebus Dominicis ad portandum sal de Wich apud Wigorniam… quilibet equus portabit unam mittam.*

Although the Domesday Book gives full details of the numbers, values and dues of the salt-pans, there are few allusions to the actual process of salt-making. It is true that several technical terms are used, but many of these are obscure. The word *salina* itself, for example, may have referred to the actual salt-pan, or to the place at which the brine was boiled, or possibly it even represented shares of brine measured out from wells to those who had the rights of salt-making. The brine-pits themselves were called *putei*. There are several references to the unexplained *hocci*, as seen in the Droitwich entry quoted on p. 252. In this entry the term 'two-thirds of a *hoccus*' adds to the mystery; this would seem to indicate that it was a measure of some kind.[1] Leaden pans or vats (*plumbi*) are mentioned at Bromsgrove (six) and Tardebigge (two). At Northwick there was a *fabrica plumbi*, which might refer to a lead-works for making the vats. At Droitwich there were four furnaces (*furni*), which rendered 60s. and 100 mitts of salt in 1066. But these references are isolated and casual; no leaden vats, for example, are mentioned at Droitwich itself. There is only one reference in the Worcestershire folios to the workers (*salinarii*) employed in the salt-industry—three belonging to the manor of Bromsgrove who worked in Droitwich.

One final point of interest concerns the fuel supply for the furnaces which evaporated the brine. This destruction of English wood by industry is a familiar theme. There are, however, only four direct references to fuel for the furnaces.[2] It is curious to note in the first two references that the numbers of cartloads of wood and mitts of salt coincide, but it can hardly be more than coincidence.

> Bromsgrove (172): To this manor belong 13 salt-pans in Droitwich and 3 salt-workers who render, from these salt-pans, 300 mitts of salt, for which they used to be given 300 cartloads (*caretedes*) of wood by the keepers of the wood in the time of King Edward.
>
> Northwick and Tibberton (173b): And in Droitwich one salt-pan renders 100 mitts of salt for 100 cartloads of wood (plotted at Northwick).
>
> Fladbury (172b): Wood 2 leagues long and a half wide. The bishop has

---

[1] There seems to be no agreement about the meaning of *hoccus*, as may be seen from the following three definitions: (i) 'A small measure of salt water', R. Kelham, *Domesday Book Illustrated* (London, 1788), p. 232; (ii) 'a salt pit', C. T. Martin, *The Record Interpreter* (London, 1892), p. 211; (iii) '(possibly) shed for drying salt', F. H Baxter and C. Johnson, *Medieval Latin Word-List* (Oxford, 1934), p. 205.

[2] For two other references linking wood and salt (for Leominster and Marcle) in the Herefordshire folios, see p. 93 above.

all its proceeds in hunting and honey and fuel (*lignis*) for the salt-pans of Droitwich (*ad salinas de Wich*), and 4*s*.

Martin Hussingtree (174b): There, 11 villeins have 4 plough-teams and they render annually 100 cartloads of wood for the salt-pans of Droitwich.

It is clear that, despite the absence of detailed information concerning salt-making, the Droitwich area was of very great importance. For salt was an indispensable commodity in medieval times, and Droitwich must have supplied the needs of a considerable part of the west Midlands and perhaps even of a wider area.

We have seen that 305 salt-pans definitely lay within the boundary of the modern county, but the Domesday folios for Worcestershire do not exhaust the list. Many places in adjoining counties possessed either salt-pans in Droitwich or rights to salt there (Fig. 87). In the Herefordshire folios there are eleven references to Droitwich salt-pans; six of these state that a particular manor actually owned a salt-pan, or part of one, in Droitwich, while the other five record that various manors had rights to a certain number of *mittae* or *summae* of salt at *Wich*.[1] The Gloucester-shire folios mention salt in connection with ten places. At six of these, the holdings are specifically stated to be at or near Droitwich, although at three others there is no mention of the location of the salt-pans. In addition, the entry for Chedworth (164) refers to the toll of salt (*theloneum salis*) brought to the hall at that place.[2] Apart from these entries in the Gloucestershire folios, the Worcestershire folios themselves speak of holdings in Droitwich belonging to the King's hall at Gloucester and to the Church of St Peter at Gloucester (172b, 174), and it is not unlikely that these interests were concerned with the salt industry. Furthermore, the Warwickshire folios mention salt in connection with six places (of which two specifically mention Droitwich);[3] those for Shropshire, three;[4] those for Oxfordshire, two; and, surprisingly enough, those for Bucking-hamshire mention one—Princes Risborough. The Oxfordshire (154b and 160b) and Buckinghamshire (143b) entries both specifically name Droitwich. The entry for Princes Risborough (143b) mentions that a salt-worker was employed at Droitwich.

The salt was carried by pack-horse from Droitwich to these various manors, and the roads or tracks used were known as 'saltways' or 'salt-roads'. Names such as Saltway, Salter's Corner, Salter's Well and Salford

[1] See p. 93 above.   [2] See p. 98 above.
[3] See pp. 301–2 below.   [4] See p. 154 above.

are preserved on the One-Inch Ordnance Survey maps of today. Probable saltways from Droitwich have been traced by linking up place-names which include the element 'salt'.[1]

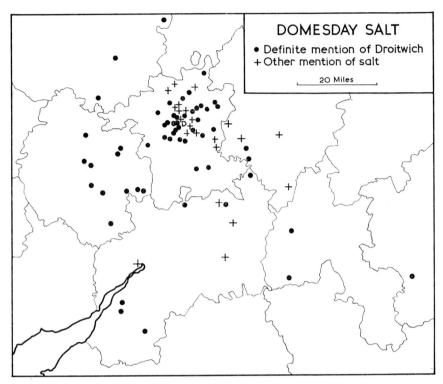

Fig. 87. The Midland Counties: Domesday salt.
D indicates Droitwich.

WASTE

Waste was relatively unimportant in Worcestershire. Only four places were wholly waste and only eight partly waste in 1086. Of these places, Bellington and Hollin were also wholly waste at some earlier date, and Besford was partly waste. Himbleton, too, had been waste but was tilled by 1086. The manor of Kidderminster once, apparently, completely waste, seems to have recovered. It had 19 teams in 1086 when its value

[1] A. Mawer and F. M. Stenton, *The Place-Names of Worcestershire* (Cambridge, 1927), pp. 4–9; A. H. Smith, *The Place-Names of Gloucestershire* (Cambridge, 1964), Part I, pp. 19–20.

was £10. 4s. as compared with £14 in 1066. It is possible that the manor
of 1066 had been wasted at about, say, 1068 and had not fully recovered
by 1086. Here are the relevant entries:

Bellington (177): 5 hides. *Wasta fuit et est. Ibi iiii quarentenae silvae sed est in
foresta regis. Pratum hujus manerii valet iiii denarios.*
Besford (174b): 1 hide. *Wasta est et fuit et tamen valebat et valet xvi denarios.*
Cleeve Prior with Atch Lench (174): 10½ hides. *De hac terra sunt wastae ii
hidae una virgata minus.*
Cooksey (177b): 2 hides. *Haec terra ex multa parte est wasta.*
Cropthorne with Netherton (174): 50 hides. *Ibi sunt wastae v hidae.*
Hanbury (174): 14 hides. *De hac terra sunt ii hidae wastae.*
Hillhampton (178): *Ibi i virgata terrae et est wasta. T.R.E. valebat xii
denarios.*
Himbleton (173b): *fuit wasta. Ibi sunt modo ii villani et ii bordarii cum i
caruca et dimidia.*
Hollin in Rock (177): 1 hide. *Wasta est et wasta fuit. T.R.E. valebat v solidos.*
Kempsey (172b): 24 hides. *De his hidis sunt v hidae wastae.*
Kidderminster (172): 20 hides. *Hoc manerium fuit totum wastum... Totum
manerium T.R.E. reddebat xiiii libras de firma. Modo reddit x libras et iiii
solidos ad peis.*
Stildon (177): ½ hide. *Valuit v solidos. Modo est wasta.*

There is also a cursory reference to the borough of Worcester (175b)
which mentions five waste houses.

That some vills were only partly waste, we may judge from two types
of evidence. In the first place, there were those vills comprising two or
more holdings of which at least one was waste; Besford alone in Worcestershire belonged to this category. In the second place, some single holdings
were, apparently, themselves partly waste; this was so at Cleeve Prior
with Atch Lench, Cooksey, Cropthorne with Netherton, Hanbury and
Kempsey.

That a holding was 'waste' did not necessarily imply that it was
entirely without profit, presumably from wood or meadow, as at Bellington, which has been counted as entirely waste in the sense that it had no
teams or population. The waste holding at Besford also returned a small
value.

MILLS

Mills are mentioned in connection with at least 70 of the 260 Domesday
settlements in Worcestershire. It is difficult to be sure of the exact total

because mills sometimes appear in composite entries covering a number of places. Thus a mill of 10s. is entered for Eardiston and Knighton (174), and three mills of 13s. 4d. for Bromsgrove and its 18 berewicks. For each composite entry the mill or mills have been plotted for the first-named place.[1] The associated villages in all such entries have also been indicated on Fig. 88.

In each entry the number of mills is given and (with a few exceptions) their annual value is stated, usually in pounds, shillings and pence but occasionally in ounces of silver, or in loads of grain, or in sticks of eels. At Wychbold (176b), for example, there were 5 mills worth £4. 8s., while at the other end of the scale, at Bishampton (172b), the mill was worth only 12d. At Kidderminster (172) one mill was worth 5 ounces of silver, and at Stoke Prior (174) 2 mills rendered 7 ounces. A few entries refer to renders of loads of grain (summae annonae), although in one of these the mills had a money value as well. At Comberton (175) the mill rendered 30 loads. At Kyre Magna (176b) it is interesting to read that the render was 10 loads of wheat (frumenti), as distinct from the more usual reference to grain. A few entries specifically connect renders of eels with the mill-pond.[2] Finally, there was one unusual render at Cleeve Prior (174), where one mill rendered a sester of honey. Three entries of mills contain no mention of value. At Cofton Hackett (177b), there was a mill working for the hall of one of the owners (serviens aulae unius eorum); at Grimley (173b), there was another 'without profit' (sine censu); and at Powick (174b), there was simply one mill 'for the use of the hall' (serviens aulae). The only recorded fraction of a mill is the half mill (dimidia molina) at Wyre Piddle (175), but there is no clue to the other half.

Domesday Mills in Worcestershire in 1086

Under 1 mill	1 settlement
1 mill	47 settlements
2 mills	12 settlements
3 mills	7 settlements
4 mills	1 settlement
5 mills	2 settlements

[1] The figure of 70 results from counting as one group the mills of a composite entry. When more than one mill is recorded in such an entry, and it is assumed that each was at a separate place, the total of places with mills is 81. If all the places named in the relevant composite entries are counted, the total becomes 114, a very artificial figure.   [2] See p. 252 above.

The group of four mills was at Astley (176). The groups of five were at Wychbold (176b) and at Pershore (174b, 175 *bis*), but two of the latter group were entered for Pershore itself and seven berewicks.

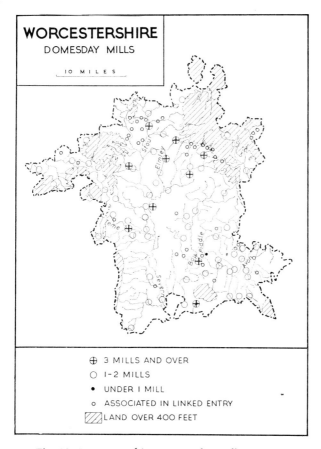

Fig. 88. Worcestershire: Domesday mills in 1086.

Fig. 88 shows that the mills were widely distributed over the county. Many occurred along the Teme and the Avon. Others were situated on the numerous small left-bank tributaries of the Severn and the right-bank tributaries of the Avon, which flow from the north-eastern and eastern uplands on the Plain of Worcester.

CHURCHES

Churches (*ecclesiae*) are mentioned in connection with only ten places in Worcestershire, apart from the cathedral at Worcester and the abbey churches elsewhere. Five places had both churches and priests—Astley (176), Bellbroughton (177b), Doverdale (177b), Feckenham (180) and Halesowen (176), the last-named with two priests. In addition to its abbey, Pershore had a church rendering 16s., the only example of a render. A church is mentioned in a remarkable entry for Offenham, Littleton and Bretforton (175b): *Ibi sunt boves ad i carucam sed petram trahunt ad ecclesiam*; this has been counted as one place in our reckoning. The list of places with churches is completed by Ashton under Hill (164), Beckford (164) and Martley (180b); each of the entries refers to tithes. Church-scot (*circset*) appears for Leopard (174) and in connection with the lands of Pershore Abbey (175b). It is also entered for Bishampton in 1066 (173).

It is worth repeating that a total of fifty-five priests was mentioned at fifty-two places, but even if a priest implies the presence of a church, the total number of settlements with churches recorded or implied thus amounts to only fifty-seven. There must certainly have been many more.

URBAN LIFE

Three places in Worcestershire seem to have been regarded as boroughs— Worcester itself, Droitwich and Pershore. The information about all three places is very unsatisfactory. Beyond a few scattered references to houses, burgesses and values, and to the salt industry at Droitwich, we are told practically nothing; there is, for example, only one very incidental reference to a market at Worcester. This slender evidence is set out below.

Worcester

The description of Worcestershire starts with an account of the city (*civitas*) of Worcester on fo. 172. This entry refers almost entirely to financial and fiscal matters. There is little that reflects any urban activity beyond a reference to a mint; when the coinage was changed, we are told that each moneyer had to give 20s. at London on receiving the dies for the money (*monetarius dabat xx solidos ad lundoniam pro cuneis monetae accipiendis*). There is no mention of burgesses, houses, parish churches and the like in this entry. The cathedral church itself appears elsewhere as a great land-holder in the county.

What other information we can deduce about the city has to be garnered from scattered entries. The existence of a market at Worcester is indicated by the words 'in the market-place (*in foro*) of Worcester', which occur in the entry for Northwick (173b), a settlement some three miles to the north of the city. One interesting feature that the city shares with a number of Domesday boroughs is that some of its burgesses are recorded as belonging to rural manors. These are entered not in any account of Worcester but in those of the respective manors which are widely scattered through the county (Fig. 89). For the most part, these 'contributory burgesses' and houses rendered money, but one house returned plough-shares, and some nothing at all. The full list is as follows:

Astley (176): 2 burgesses of (*de*) 2*s.*
Bushley (180b): 1 house (*domus*) rendering 1 silver mark.
Chaddesley Corbett (178): 2 burgesses rendering 12*d.*
Coddington (182): 3 houses (*masurae*) rendering 30*d.*
Evesham, church of (175b): 23 houses (*masurae*) rendering 20*s.*; 5 waste houses.
Halesowen (176): 1 house (*domus*) of 12*d.*
Hollow Court (180b): 1 house (*domus*) rendering 2 plough-shares, 2 others nothing.
Kidderminster (172): 1 house (*domus*) rendering 5*d.*
Martley (180b): 3 houses (*domus*) rendering 12*d.*
Northwick and Tibberton (173b): 90 houses (*domus*); 25 *in foro*, rendering 100*s.* a year.
Osmerley (177b): 1 house (*domus*) of 16*d.*
Pedmore (177): 2 houses (*masurae*) rendering 2*s.*
Suckley (180b): 1 burgess but he renders nothing (*sed nil reddit*).
Upton Warren (177b): 1 burgess of 2*s.*
Witton (177b): 1 burgess of 2*s.*

There were, therefore, 7 burgesses and 158 houses (of which 5 were waste), counting the 25 *in foro* as additional to the 90 of the same entry. It would seem then, that there were at least 160 burgesses in the Domesday city. But this figure may well be only a fraction of the total number.

Droitwich

The description of Droitwich, scattered through many folios, is dominated by the salt industry which, as we have seen, was as important

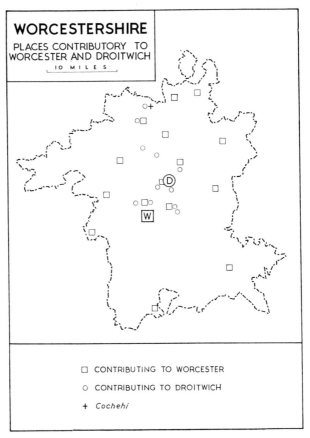

Fig. 89. Worcestershire: Places contributory to Worcester (W) and Droitwich (D). The cross indicates *Cochehi*, a village of uncertain identification in Doddingtree hundred, but J. H. Round located it in the extreme north of the county. Its solitary burgess is not ascribed specifically to Droitwich, but as his render consisted in part of salt, it is fairly safe to assign him there. Coddington, included in the list on p. 263, does not appear on this map as it was, and is, in Herefordshire.

and had such a wide influence extending into many counties.[1] Six entries tell us something of the urban population of this industrial centre. There were 18 burgesses rendering 4*s*. 6*d*. (174); 31 rendering 15*s*. 8*d*. (174b); 9 rendering 30*s*. (176); 11 rendering 32 mitts of salt (176b); 20 rendering 50 mitts of salt (177); and 11 houses (172). But there were also others. Droitwich, like Worcester, had contributory burgesses and houses

[1] See pp. 257–8 above.

recorded in connection with various rural manors around (Fig. 89). The full list is as follows:

Cochehi (177b): 1 burgess of (*de*) 16*d.* and 4 mitts of salt.

Crowle (176b): 1 burgess of 2*s.*

Elmley Lovett (176): 5 houses (*domus*) of 20*d.*, and there 7 villeins render 3*s.*

Hallow in Grimley (173b): 10 houses (*domus*) of 5*s.*

Hartlebury (174): 5 houses (*domus*) rendering 5 mitts of salt.

Kidderminster (172): 1 house (*domus*) rendering 5*d.*

Northwick and Tibberton (173b): 3 houses (*domus*) rendering 3 mitts of salt and 2*s.* from the lead-works.

Abbots Morton (176b): 1 burgess rendering 10*s.*

Salwarpe (174): 4 burgesses.

Witton (177b): 7 burgesses.

Wychbold (176b): 13 burgesses had to mow for two days in August and March and do service at the lord's court.

The burgesses at *Cochehi*, Crowle and Abbots Morton are not specifically assigned to Droitwich, but as they are in each case associated with salt or salt-pans, we may safely assign them there.

It seems therefore that there were at least 35 houses and 116 burgesses, representing 151 burgesses in all. The 7 villeins may have been salt-workers. At any rate, an entry for Bromsgrove mentions 3 salt-workers at Droitwich (172), and an entry for Princes Risborough, in Buckinghamshire, mentions yet another salt-worker there (143b).

Alongside this industrial activity there was also some agriculture. The mowing duties of the burgesses entered under Wychbold hint at this other side of the economic activity of the settlement. The grand total of population, as entered in the Domesday Book, is thus seen to consist of 151 burgesses, 4 salt-workers and 7 villeins; these amount to 162. It is very unlikely that this figure represents the full number of male adults in the boroughs in 1086. But meagre and unsatisfactory as the information is, we can at any rate glimpse something of the great industrial activity that was superimposed upon the more normal life of an eleventh-century settlement.

Pershore

Pershore is described in two entries on fos. 174b and 175. The first of these entries gives an indication of its borough status, for we are told that 28 burgesses rendered 30*s.* and that the toll (*theloneum*) yielded 12*s.*; this last item may indicate the presence of a market. The entry also mentions villeins, a *francigena*, serfs and a bondwoman, and also plough-teams, wood, meadow, mills and a church. Unfortunately, in the second entry the account of Pershore is combined with that of seven berewicks in a complicated fashion, and it is impossible to disentangle the resources of Pershore itself. It is, however, clear, that Pershore was largely an agricultural settlement whose general character must have resembled that of the villages around.

MISCELLANEOUS INFORMATION

Livestock

Like the other counties recorded in the Great Domesday, livestock were not usually entered for Worcestershire. There are three references to swine in the woodland entries.[1] But the only other reference to live-stock is the somewhat mysterious entry for Lower Sapey (176b), where there was 'nothing but 9 beasts (*animalia*) in demesne', which we have counted as plough-oxen. At Offenham, the oxen were used for drawing stone to the church.[2]

Beehives

There are eight references either to hives of bees (*vasculi apium*) or to returns of honey (*mellis*), the latter usually in conjunction with holdings of wood. At Suckley (180b) there was a keeper of the bees (*custos apium*) with 12 hives. At seven other places, we read of honey-rents, and in some of these, the honey was one of the several items comprising the woodland revenue. Thus at Powick (174b) 3 coliberts rendered 3 sesters (*sextarii*) of honey; at Wolverley (174) a freeman rendered 2 sesters; at Witley (172b) the priest rendered one sester to the Church; and at Cleeve Prior (174) the mill rendered one sester.

Vineyards

The only reference to a vineyard in Worcestershire was for Hampton by Evesham (175b), where there was a newly planted vineyard (*et vinea novella ibi*).

Castles

Only one castle is entered for the county in 1086—that of William Fitz Ansculf at Dudley (177)—*ibi est castellum eius*. On the same folio we

[1] See p. 246 above.   [2] See p. 233 above.

are told that Bellington was in the jurisdiction of William's castle (*in castellaria sua*); Bellington is nine miles to the south-west of Dudley.

Other references

At Wadborough (175) there was a hide of land on which was the monks' dairy farm (*vaccaria*) in 1066, a reference to the estates of the Abbey of Pershore. At Feckenham (180b) there is the laconic mention of a croft (*crofta*) held by a radman. The entry for Northwick and Tibberton (173b) mentions the existence of a market at Worcester, and of a render of 2s. from pasture (*de pascuis*), both unique items in the Worcestershire folios. The same entry also refers to a *fabrica plumbi*, possibly a leadworks making vats for use in the salt industry.

REGIONAL SUMMARY

Worcestershire consists essentially of a gently undulating central plain, crossed by the Severn and its tributaries, and bordered on the west, north and east by a discontinuous upland rim. In the extreme south-east there is a small area of the Cotswold country. Seven regions may be distinguished (Fig. 90).

(1) *The West Worcester Hills*

These hills extend for about twenty miles along the western edge of the county in a narrow interrupted 'chain', rarely more than two miles in width. They are composed for the most part of old resistant rocks. The various groups of hills form prominent features in the landscape; the Malverns rise almost to 1,400 ft. above sea-level, presenting a steep face to the Plain of Worcester on the east, while the hills further north attain 900 ft. in places.

Domesday settlements were scarce over the region. This is to be expected, especially as much of these bleak upland areas lay in the royal forest. The blank spaces on Figs. 79 and 81 are indicative of the emptiness of these uplands in 1086.

(2) *The North-west Uplands*

In the north-west of the county lie rolling uplands, consisting for the most part of Old Red Sandstone rocks. The river Teme flows across this upland region, with many sudden changes of direction, in a steep-sided

valley, more or less across the 'grain' of the country. To the south of the
Teme valley, in the Kyre Uplands, the hills reach 800 ft. above sea-level.
North-east of the Teme lies the much dissected upland of Wyre Forest,
with an average elevation of some 600 ft. Beyond the Severn is a third

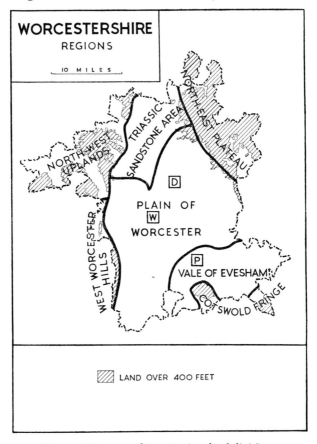

Fig. 90. Worcestershire: Regional subdivisions.
Domesday boroughs are indicated by initials: D, Droitwich; P, Pershore;
W, Worcester.

hilly section—the small Trimpley Upland, of somewhat more subdued
relief and averaging only about 450–500 ft. above sea-level. The soils
on these hills are generally coarse-grained and sandy, although there are
occasional patches of stiff clays.

The uplands carried many more Domesday settlements than might have been expected. Some, like Tenbury and Lower Sapey, lay in deeply cut valleys; others, like Hanley and Bayton, were on the uplands. The densities of population (6·0) and of teams (3·1) were surprisingly high. There is no mention of meadow, but woodland was widespread. Numerous fisheries and mills were to be found along the course of the Teme.

(3) The North-east Plateau

In the north-east of Worcestershire there is an area of hilly country, extending from the broad rolling Clent Hills to the Lickey Hills. These uplands form, in fact, the most southerly part of the South Staffordshire plateau, and consist of ancient rocks, covered in part by Upper Coal Measures and by various Triassic rocks. The Clent Hills exceed 1,000 ft. above sea-level, and the Lickeys attain 850 ft. Coarse sandy or gritty soils, gravels, and in places patches of Boulder Clay cover the surface.

There were few Domesday settlements here, except along the margins of the uplands and in the valley of the Arrow. Entries of woodland are scarce, no meadows or fisheries were recorded, and only three places had mills.

(4) The Triassic Sandstone Area of North Worcestershire

Between the uplands of the north-west and the north-east, is a lower area of Triassic rocks. The relief is varied, due in large measure to differences in the Triassic rocks—sandstones, marls and pebble beds. In the west, near Ombersley, the sandstone forms a low plateau that lies for the most part between 300–500 ft., and drains by numerous streams to the Severn and to the Stour. The valleys of these tributaries contrast with the areas of undulating heathland between them. Much of the region has shallow hungry soils, but these improve southwards.

Domesday settlements were surprisingly numerous, but most of them were small, and their arable lands were of limited extent. The population density (4·5) over the region as a whole was low, as was that of teams (1·9), but on the better soils of the southern part, population and plough-team figures were higher. Meadow was recorded only once, at Bellington. There were considerable areas of woodland.

(5) The Plain of Worcester

The western and northern parts of the Plain of Worcester consist of a rolling lowland of reddish Keuper Marl, sloping south-westwards from over 400 ft. to less than 100 ft. above sea-level. The south-eastern part of the plain, mostly under 200 ft., is floored with heavy grey or blue Lower Lias clays. There are extensive sheets of glacial material. Both the marls and clays develop heavy ill-drained soils, often waterlogged in winter, and both have a poor water-supply. But locally patches of sand and gravel provide more favourable sites for settlement.

The river Severn flows southwards across the plain. Above Worcester its floodplain is narrow, for the river is incised in a shallow trench. Below the city the river winds over a more extensive flood-plain, although numerous small hills diversify the valley floor. Just below Worcester the Severn is joined by the Teme.

Domesday settlements were distributed fairly uniformly over the region, varying little in size, except for the two boroughs of Worcester and Droitwich. The salt deposits of the Keuper Marl in the north of the plain were responsible for the growth of Droitwich and neighbouring settlements. The densities of population and of teams were higher than in the north, but lower than in the Vale of Evesham. The most striking distribution is that of meadow; nearly every settlement had from 10 to 20 acres. There was much woodland on the heavy clays, although the axe and the plough were already making steady inroads. Many Domesday fisheries and mills were situated along the Severn and Teme or their tributaries.

(6) The Cotswold Fringe

The small area of the Cotswolds that lies within Worcestershire consists of Broadway Hill and the outlying mass of Bredon Hill. The former rises to over 1,000 ft. above sea-level and the latter to 961 ft. From the Domesday record we find that the Oolite escarpment of Broadway Hill and the dip-slope beyond, as well as Bredon Hill, were devoid of settlements, ploughs and meadows. But at the foot of the escarpment, near the spring-line, lies Broadway village itself, with 25 teams and 51 recorded people in 1086. On the plain to the south of Bredon Hill was Beckford, another large manor with 38 teams and 75 people.

(7) The Vale of Evesham

The Avon flows across a broad area of Lower Lias clay, covered for the most part by drift deposits, ranging from river-terrace sands and gravels to brick-earths. The river in sweeping meanders wanders across this gently undulating lowland to join the Severn near Tewkesbury. Its fall is slight, and a considerable tract suffers periodic inundation, although there has been some modern regularisation. Below Pershore the meanders arc incised some fifty feet below the general level of the Vale, and several extensive terraces may be traced, on which lie light well-drained soils which are very suitable for agriculture. The present floodplain is floored with extensive arcas of alluvium, while away from the river are the heavy loams derived mainly from glacial drifts.

The Vale of Evesham has long been an attractive area for settlement. Its gentle relief, its extensive tracts of easily worked soils, its early spring and generally attractive climate, its settlement sites on the terraces above flood level—these must have compared favourably with conditions in the rest of the county at the time of the Domesday survey as today. A glance at the various distribution maps confirms this. The densities per square mile of both population (up to 9·9) and plough-teams (up to 4·7) were higher than over the rest of the county; extensive tracts of meadow occurred along the Avon; and there were numerous mills and fisheries. There are few references to woodland, indicating perhaps the progress of clearance for agriculture.

BIBLIOGRAPHICAL NOTE

(1) It is interesting to note that there was a facsimile copy of the Domesday folios for the county, together with observations on the text, in T. R. Nash's *Collections for the history and antiquities of Worcestershire* (London, 1781–99), 2 vols. and supplement.

There was also a nineteenth-century extension and translation of the text: W. B. Sanders, *A literal extension of the Latin text; and an English translation of Domesday Book in relation to the county of Worcester* (Worcester, 1864).

(2) A translation of the Latin text, with an Introduction, by J. H. Round, is in the *Victoria County History: Worcestershire* (London, 1901), I, pp. 235–340.

(3) The following deal with various aspects:

J. W. WILLIS-BUND, 'Worcestershire Doomsday', *Associated Architectural Societies' Reports and Papers* (Lincoln, 1894), XXII, pt. II, pp. 88–108. This is mainly concerned with the religious houses and their servile populations.

R. C. GAUT, *A History of Worcestershire Agriculture and Rural Evolution* (Worcester, 1939). Chapter III is concerned with Domesday statistics.

G. B. GRUNDY, 'The Saxon Settlement in Worcestershire', *Trans. Birm. Archaeol. Soc.* (Birmingham, 1931), LIII, pp. 1–17, introduces some reference to Domesday information.

REGINALD LENNARD, 'The hidation of "demesne" in some Domesday entries', *Econ. Hist. Rev.* (Cambridge, 1954), 2nd Ser. VII, pp. 67–70.

(4) Some maps of Domesday plough-lands, meadow and wood appeared in K. M. Buchanan, *Worcestershire* (London, 1944), p. 488, being Part 68 of *The Land of Britain* ed. L. Dudley Stamp.

(5) The subsidiary sources of information are translated and discussed by J. H. Round in *V.C.H. Worcestershire* (London, 1901), I, pp. 324–31. See also (i) J. H. Round, 'The Worcestershire Survey' in *Feudal England* (London, 1895), pp. 169–80; and (ii) P. H. Sawyer, 'Evesham A, a Domesday Text', *Miscellany I, Worcs. Hist. Soc.* (Worcester, 1960), pp. 3–36.

(6) Before the volume of the English Place-Name Society dealing with Worcestershire was published in 1927, there were two studies of the place-names of the county. A short paper, entitled 'Worcestershire Doomsday Place-Names', was read by H. Kingsford before the Worcester Diocesan Architectural and Archaeological Society in 1894, and printed in the *Associated Architectural Societies' Reports and Papers* (Lincoln, 1894), XXII, pt. II, pp. 108–108*f*. Eleven years later there appeared W. H. Duignan's *Worcestershire Place Names* (London, 1905). The authoritative study of the place-names of the county is, of course, *The Place-Names of Worcestershire*, by A. Mawer and F. M. Stenton, in collaboration with F. T. S. Houghton (Cambridge, 1927).

CHAPTER VI

WARWICKSHIRE

BY R. H. KINVIG, M.A.

'The Warwickshire portion of the Great Survey,' wrote J. H. Round, 'is interesting and fairly full.'[1] But despite this, its information occasionally bears witness to the human element in its assembling. Thus, as will be seen later, the units of measurement for meadow are missing in one entry, while in another we are told nothing of the people who worked the four plough-teams recorded. Moreover, there are four successive entries on fo. 238 where the number of plough-lands was never filled in. But these were exceptions, and the Warwickshire text, as a whole, does not contain many such omissions or apparent omissions.

The present-day county of Warwickshire, in terms of which this study is written, is far from being the same as the Domesday county. The main differences were on the western boundary with Worcestershire and Gloucestershire, where the three counties were intermingled in a strange manner that reflected the disposition of the scattered holdings of the bishop of Worcester (Fig. 2). These complicated arrangements proved remarkably stable until the nineteenth century. Successive boundary Acts in the nineteenth and twentieth centuries have, however, obliterated the detached portions of Worcestershire, and so given territorial continuity to each of the three counties. The Domesday holdings thus transferred from Northamptonshire and Worcestershire are set out on p. 385 and p. 219 respectively. Those transferred from Gloucestershire are as follows:

Admington (165 b)	Little Compton (166)	Sutton under Brailes (166)
Clifford Chambers (163 b)	Long Marston (166)	Welford on Avon (166)
Clopton (167)	Meon (163)	Weston on Avon
Dorsington (168)	Preston on Stour (166)	(166, 169)
Lark Stoke (166)	Quinton (169 *bis*)	Willicote (169)
		Wincot (163 b, 167)

Finally from Staffordshire have come Handsworth (250), Harborne (247) and Perry Bar (250).

[1] *V.C.H. Warwickshire* (London, 1904), I, p. 269.

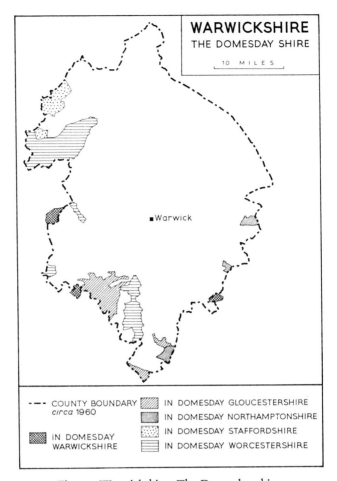

Fig. 91. Warwickshire: The Domesday shire.

Conversely, transferences of holdings from Warwickshire, on the other hand, have been few in number. Bickmarsh and Ipsley, surveyed in the Warwickshire folios, are now in Worcestershire; Mollington and Spelsbury are now in Oxfordshire, and Tamworth is now wholly in Staffordshire. Before 1890, part of Tamworth, including the castle, was in Warwickshire; the remainder, including the church, was in Staffordshire. These various changes are shown on Fig. 91.

Within the Domesday county there were ten hundreds, some of which were not compact territorial units, but included a number of detached

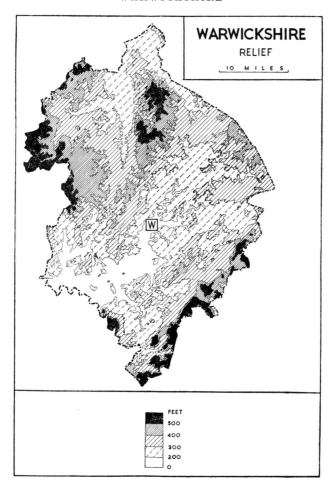

Fig. 92. Warwickshire: Relief.
W indicates the Domesday borough of Warwick.

portions.[1] Any attempt at reconstructing these hundreds raises many difficulties, partly because the Domesday rubrication is unsatisfactory and seems to be incomplete, and partly because the boundaries and names

[1] For Benjamin Walker's reconstructions, see the papers listed on p. 311. The footnotes to the *V.C.H.* translation indicate many of the complications. There is also a map of Domesday hundreds in R. H. Kinvig, 'The Birmingham District in Domesday Times', being pp. 113–34 of J. M. Wise (ed.), *Birmingham and its Regional Setting* (British Association, Birmingham, 1950).

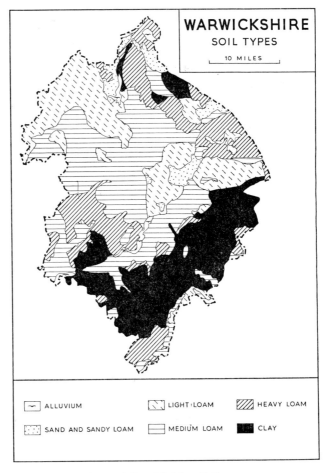

Fig. 93. Warwickshire: Soil types.
Based on a map prepared by the Warwickshire Office of the National Agricultural Service, 1959. For help over this map we are indebted to Professor Harry Thorpe.

of the hundreds changed greatly in later times until only four hundreds were left. Moreover, the names of some parishes do not appear in the Domesday text, which further complicates the problem.

One difficulty in handling the material from a geographical point of view is the fact that the information about two or more places is sometimes combined into one statement. There are about a dozen such composite entries in the Warwickshire folios, and, although the hidages of their respective vills are occasionally set out separately, their other resources are described collectively, so that we have no means of apportioning exactly the details of population, plough-teams, mills, meadow and wood among the constituent vills. For the purpose of constructing distribution maps the details of these combined entries have been divided equally among all the villages concerned. The woodland measurements, however, cannot be so divided and where information is given about wood, the details have been plotted for the first village in each group. Mills have been treated likewise.[1]

Finally, there is the minor problem of duplicate entries. The identical entries for Flecknoe (238 b and 244 b) are almost certainly so, and only one has been counted here. Those for Middleton (242 and 244 b), on the other hand, may well refer to separate halves of what was once a single estate, and both have been counted. The same may be true of Shuttington (240 *bis*) which has been treated likewise. The entries for Barston (241 and 242 b) have also been taken to refer to separate holdings.[2]

SETTLEMENTS AND THEIR DISTRIBUTION

The total number of separate places mentioned in the Domesday Book for the area now included in the modern county of Warwick seems to be 281, including Warwick, the only place for which burgesses are recorded. This figure, however, may not be quite accurate because there are some instances of two or more adjoining villages bearing the same surname today, and it is not always clear whether more than one unit existed in the eleventh century. There is no indication that the Over and Nether Eatington of today existed as separate villages; the Domesday information about them is entered under only one name, though there may well have been separate settlements in the eleventh century. The same applies, for example, to the two Pillertons and to the two Whitacres. The distinction between the respective units of these groups appears late in time. Thus the names Lower Shuckburgh and Upper Shuckburgh date from the thirteenth century. Only where there is some evidence of a second village has it been included in the total of 281. Thus the existence of Bishop's Itchington and Long Itchington is quite clear—they were in different hundreds; the former was in Stoneleigh, the latter in Marton. Leamington Priors (now Leamington Spa) and Leamington Hastings can

[1] See pp. 294 and 303. [2] *V.C.H. Warwickshire*, I, p. 296.

be similarly distinguished. In the case of Bickenhill the Domesday Book records two places as *Bichehelle* and *alia Bichehelle* (240b); the first of these became Church Bickenhill and the other was divided into Hill and Middle Bickenhill about the time of Henry III, but neither of these grew into a separate parish. Similarly, Compton Scorpion is distinguished from the other Warwickshire Comptons and is entered as *parva Contone* (242b).

Not all the Domesday names appear on the present-day map of Warwickshire villages. Some are represented by hamlets, by individual houses, or by the names of topographical features.[1] Thus *Brome* is now the hamlet of Broom in Bedford parish, and *Bernhangre* is that of Barnacle in the parish of Bulkington. *Cintone*[2] was a small settlement in Solihull, until it developed into the modern housing estate of Kineton Green, and *Holehale* is the hamlet of Ullenhall in Wootton Wawen. *Langedone* is represented by Langdon Hall in the parish of Solihull, *Soulege* by Sole End Farm in Astley, and *Winchicelle* has become Wiggins Hill Farm in Sutton Coldfield. Again the Domesday *Smitham* survives as Smite Brook and Smeeton Lane in the parish of Combe Fields. Disappearance has been even more complete in the case of *Offeworde* in Wootton Wawen, which in Dugdale's time was marked only by a mill; the site was near Pennyford, but the name itself has been lost.[3] Similarly, the Domesday *Smerecote* in the parish of Bedworth appears on Morden's map (*c.* 1680), but there is no indication of the place on modern maps. These are but some of the changes in the Warwickshire villages. To them must be added a number of unidentified names: *Donnelie* (240),[4] *Optone* (238), *Surland* (238b) and *Werlavescote* (240). Whether they will yet be located, or whether the places they represent have completely disappeared leaving no record or trace behind, we cannot say.

On the other hand, some villages on the modern map are not mentioned in the Domesday Book. Their names do not appear until the twelfth and

[1] For the deserted villages see M. W. Beresford, 'The deserted villages of Warwickshire', *Trans. Birm. Archaeol. Soc.* (Birmingham, 1950), LXVI, pp. 49–106.

[2] On fo. 243 and distinct from the *Cintone* on fo. 240 which was Kington.

[3] J. E. B. Gover, A. Mawer and F. M. Stenton, *The Place-Names of Warwickshire* (Cambridge, 1936), p. 244.

[4] Dugdale's identification of *Donnelie* with Beaudesert (adjoining Henley in Arden) cannot reasonably be accepted, and the most likely solution is that it referred to a part of Wedgnock Park in the parish of Hatton. Indeed Dugdale (*Antiquities of Warwickshire* (1730), I, p. 272) speaks of that park being enlarged by 'certain woods called Wegenok Donele', but the name 'Donele' now seems to be lost.

thirteenth centuries, and presumably, if they existed in 1086, they are accounted for under the statistics of neighbouring settlements. Thus, as far as record goes, Allesley was first mentioned in 1176, Packwood in 1194, and Willenhall in 1195. Other examples of units that came into being after 1086 are those of Bushwood, first mentioned in 1197, and

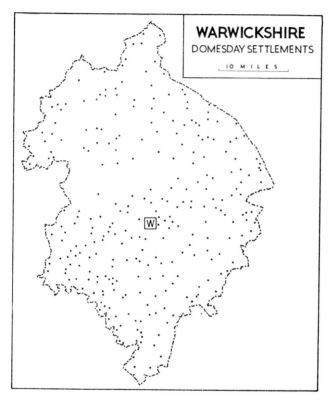

WARWICKSHIRE
DOMESDAY SETTLEMENTS
10 MILES

Fig. 94. Warwickshire: Domesday place-names.
W indicates the Domesday borough of Warwick.

Solihull in 1242. As might be expected these new places are more frequent in the Arden region than in Feldon. Some of the places not mentioned in the Domesday Book must have existed, or at any rate been named in Domesday times, because they appear in pre-Domesday documents. Nuthurst, for example, was mentioned as a woodland given with Shottery to the Church of Worcester by Offa *c.* A.D. 705, and the name appears again in the thirteenth century; but it is not mentioned in the Domesday text. Similarly, Kings Newnham is mentioned in 1043, but does not appear again until the twelfth century. Sometimes, although a modern parish is not named in the Domesday Book, other names within the parish are mentioned. An interesting example of this is the Domesday *Ailespede* in the modern parish of Meriden; the latter does not appear in the Domesday Book, but *Ailespede* survives as the name of Alspath Hall. Another missing parish is Withybrook, and it contains a Domesday name (*Apleford*), represented by Hopsford Hall.[1] From this account it is clear that there have been many changes in the village geography of the county, and that the list of Domesday names differed considerably from a list of present-day parishes.

Domesday names are to be found in most parts of the county, but are more numerous in the south-eastern half—from the Avon to the Edge Hill Fringe—and more thinly distributed towards the north-west (Fig. 94). This distinction emphasises the contrast between the relatively open Feldon, and the wooded area of Arden, and is a fundamental feature of the geography of Warwickshire. There were very few settlements on the watershed between the Blythe basin in the north and the Alne, draining southwards to the Avon. Settlements were also sporadic in the East Warwickshire Plateau.

THE DISTRIBUTION OF PROSPERITY AND POPULATION

Some idea of the nature of the information in the Domesday folios for Warwickshire and of the form in which it was presented, may be obtained from the entries relating to the villages of Tysoe in the Feldon area, and to Wootton Wawen in Arden. Each village was held by a single landowner, and so is described in one entry:

> Tysoe (242b): The same Robert [de Statford] holds Tysoe. There are 23 hides. There is land for 32 ploughs. In demesne are 11, and 9 serfs, and 53 villeins, with a priest and 28 bordars, have 23 ploughs. There, 16 acres of meadow; and in Warwick 3 houses pay 18*d*. It was worth £20. Now £30. Waga held it freely.
>
> Wootton Wawen (242b): The same Robert [de Statford] holds 7 hides in Wootton Wawen. There is land for 9 ploughs. There are 23 villeins with

[1] The dates in this paragraph are from J. E. B. Gover, A. Mawer and F. M. Stenton, *The Place-Names of Warwickshire* (Cambridge, 1936).

a priest and 22 bordars having 6 ploughs. There 2 mills of (*de*) 11*s*. and 8 sticks of eels. Wood 2 leagues long and 1 league broad. It is worth £4. Waga held it freely.

These entries do not include all the items of information that appear elsewhere in the folios for the county. They do not mention, for example, the categories of population known as freemen, knights and *ancillae*. Nor do they mention waste or salt-pans. But although not comprehensive, they are fairly representative and straightforward entries, and they do set out the recurring standard items that are found for most villages. These are five in number: (1) hides; (2) plough-lands; (3) plough-teams; (4) population; and (5) values. The bearing of these five items of information upon regional variations in the prosperity of the county must now be considered.

(1) *Hides*

The Warwickshire assessment is stated in terms of hides and virgates; the formula employed is simply 'there are *n* hides and *m* virgates'. There are, however, a few exceptions; Oldberrow (175 b), for example, a place which has been transferred from Worcestershire, was assessed at '12 acres of land' (*xii acrae terrae*). Another variation that occurs infrequently is recorded at Weston under Wetherley (241 b) where there were 'one and a half virgates of land' (*virgatae terrae*). In two entries mention is made of 'inland'. At Offord (242 b) in Wootton Wawen, in addition to 5 hides, there was 'one carucate of inland' (*i carucata de inland*); similarly at Lighthorne (243) there were '5 hides beside inland' (*Ibi sunt v hidae praeter inland*).[1]

A glance through the Warwickshire folios reveals many examples of the five-hide unit. Just over one quarter of the Warwickshire villages were assessed at five hides or a multiple of five. In general there were many more instances of the five-hide unit in the south-east of the county than in the north-west. Arlescote with five hides, Oxhill and Butlers Marston with ten each, and Priors Hardwick with fifteen, are all in the south, and many other examples can be found. When a village was divided amongst a number of owners the same feature can sometimes be demonstrated; thus Frankton (239, 240) consisted of two holdings, one of

---

[1] 'Inland' was land free from contributing to the geld (*V.C.H. Warwickshire*, I, p. 270).

4 hides less one virgate, the other of one hide plus one virgate. Similarly, the three holdings at Napton together accounted for 5 hides:

Countess Godiva (240): 3 hides and 3 virgates.
Turchil of Warwick (241): 3 virgates.
Turchil of Warwick (241): half a hide.

These combinations of assessments also suggest, incidentally, that there were four virgates to the hide. It is possible that some villages were grouped in blocks for the purpose of assessment, as in Leicestershire.[1] In any case the absence of information about these groups makes it difficult to be definite about the full extent of the five-hide unit in Warwickshire.

Whether the five-hide unit can be demonstrated or not, it is clear that the assessment was largely artificial in character, and bore no constant ratio to the agricultural resources of a vill. A striking instance of low hidation is that of Rowington. It answered for only three hides, yet it had nine teams, a total recorded population of fifty-two and was valued at £5 in 1086 (242). The variation among a representative selection of ten-hide vills speaks for itself:

	Teams	Population
Butlers Marston (242)	10	43
Dunchurch (244)	6	27
Hampton in Arden (243 b)	15	69
Hodnell (240, 241, 243)	9½	32
Preston Bagot (240, 240 b)	3½	16

The total assessment amounted to 1,502 hides, $\frac{5}{6}$ virgate, 12 acres and 1 carucate, but it must be remembered that this refers to the area included in the modern county. Maitland estimated the number of hides in the Domesday county at 1,338.[2]

(2) Plough-lands

Plough-lands are systematically entered in the Warwickshire folios and the normal formula is simply 'there is land for n ploughs' (terra est n carucis). There are occasional omissions; thus, on fo. 238, there are four successive entries in which 'terra est' is followed by a blank where

[1] See p. 325 below.
[2] F. W. Maitland, *Domesday Book and Beyond* (Cambridge, 1897), p. 400.

a figure was apparently intended to be inserted. As plough-lands are not recorded in the Domesday folios for Gloucestershire or Worcestershire, it is impossible to state a total of plough-lands for modern Warwickshire. The total for that part of the county with plough-lands, or potential teams,[1] amounts to 2,154½. If we assume that plough-lands equalled teams in those blank 'terra est' entries, the total then becomes 2,265½; the corresponding total for teams is 2,060¾. As Fig. 99 shows, these totals refer to only part of the modern county. Generally speaking the figures for plough-lands and plough-teams are very similar in the south of the county. It is in the northern half that there is some deficiency, and one possible implication is that settlement and economic development had not proceeded as far in the Arden region as it had in much of the Avon basin and the other parts of the south-east.

Of the entries which record both plough-lands and teams, there was a deficiency of teams in 48% and an excess in 24%. The existence of excess teams is occasionally emphasised in the text, and it may be of interest to note that, with one exception, the entries concerned are all on fo. 241 and its reverse. They are set out below:

Wolfhamcote (241): There is land for 2 ploughs and yet there are 3 ploughs (*Terra est ii carucis et tamen sunt ibi iii carucae*).

Elmdon (241): There is land for half a plough. There is however 1 plough in demense (*Terra est dimidia caruca. Ibi tamen est in dominio i caruca*).

Newton (241): There is land for half a plough. There is however 1 plough (*Terra est dimidia caruca. Ibi est tamen i caruca*).

Walcote, Willoughby and Caldecote (241): There is land for 1 plough. In demesne however there is 1 plough and 2 serfs and 4 villeins and 6 bordars with 1½ ploughs (*Terra est i caruca. In dominio tamen est una caruca et ii servi et iiii villani et vi bordarii cum i caruca et dimidia*).

Ladbroke (241): There is land for half a plough. In demesne however there is 1 plough (*Terra est dimidia caruca. In dominio tamen est i caruca*).

Holme (241): There is land for half a plough. There is however 1 plough (*Terra est dimidia caruca. Ibi est tamen i caruca*).

Lillington (241 b): There is land for half a plough. There is however one (*Terra est dimidia caruca. Una tamen ibi est*).

Norton Lindsey (242 b): There is land for 2 ploughs. In demesne however

---

[1] Of the places in Domesday Worcestershire but in modern Warwickshire, only one has potential teams recorded for it: Northfield (177) had 14 teams and there could be 5 more (*v carucae possent fieri*).

there are 2 and 4 serfs and 5 villeins and 2 bordars with 2 ploughs (*Terra est ii carucis. In dominio tamen sunt ii et iiii servi et v villani et ii bordarii cum ii carucis*).

The precise significance of these excess teams is obscure. It may be that the holdings with excess teams were overstocked, or that the figures for plough-lands may be relics of some older assessment. But it must be pointed out that, as far as we can see at present, no conventional element is to be discerned among the Warwickshire figures, and this is in contrast with, say, Leicestershire and Northamptonshire, where the plough-land is clearly artificial.[1] The following table illustrates some of the varying relationships between hides, plough-lands and plough-teams in the county:

	Hides	Plough-lands	Plough-teams
Deficient teams:			
Coton End (238)	1	20	4
Shrewley (242)	3	12	3½
Equal teams:			
Caldecote (238b)	2	6	6
Burton Hastings (242)	4	8	8
Excess teams:			
Wormleighton (240b)	1½	5	9
Monks Kirby (243b)	15	20	35

Fig. 99 showing the density of plough-lands also indicates those parts of the modern county for which no Domesday plough-lands are stated or implied.

(3) *Plough-teams*

The Warwickshire entries, like those of other counties, draw a distinction between the teams of the demesne and those held by the peasantry. Occasionally there is no mention of teams apart from the demesne. In one of the holdings at Cubbington (240b) there was only one team in demesne, with 3 bordars; similarly, there was only half a team with 2 bordars in demesne on one holding at Shuckburgh (241). Another interesting entry is that for Aston near Birmingham (243) which reads: 'there is land for 20 ploughs. In demesne is land for 6 ploughs, but the ploughs are not there' (*sed carucae ibi non sunt*). The peasants, however,

---

[1] See pp. 329 and 395 below.

had 18 teams among them. There is a solitary reference to the use of oxen at Whitacre (242) where there was one villein ploughing with 2 oxen (*Ibi est i villanus cum ii bobus arans*). The record of teams is usually complete and straightforward.

The total number of plough-teams amounted to 2,275¾, but it must be remembered that this refers to the area included in the modern county. Maitland estimated the number for the Domesday county at 2,003.[1]

(4) *Population*

The main bulk of the population was comprised of the three categories of villeins, bordars and serfs. In addition to these were the burgesses, together with a large miscellaneous group that included knights, freemen, *francigenae*, Flemings and others. The details of the groups are summarised on p. 290. Ellis's estimate of total population (excluding burgesses) is 6,542,[2] but this figure refers to the Domesday county and includes tenants in chief, under-tenants and *ancillae* (bondwomen). The present estimate has been made in terms of the modern county, and includes the population of many places transferred from neighbouring counties,[3] but excludes tenants, burgesses and *ancillae*. The totals do, at any rate, indicate the order of magnitude involved, and the relative size of the different groups of the population. The figures are those of recorded population, and they must be multiplied by some factor, say 4 or 5, in order to obtain the actual population; but this does not affect the relative density as between one area and another.[4] That is all that a map, such as Fig. 97, can roughly indicate.

One entry is defective, that for Weethley (239), where there were 4 plough-teams, but no recorded population. Two other entries do not specify the precise numbers in each of the categories mentioned. These entries refer to *Optone* (238), an unidentified vill, where there were 10 villeins and bordars together (*x inter villanos et bordarios*), and to Haselor (244) where there were 5 serfs and bondwomen (*v inter servos et ancillas*). In estimating individual categories the only course possible is to divide them arbitrarily.

[1] F. W. Maitland, *op. cit.* p. 401.
[2] Sir Henry Ellis, *A General Introduction to Domesday Book* (London, 1833), II, pp. 499–500.
[3] See p. 273.
[4] But see p. 430 below for the complication of serfs.

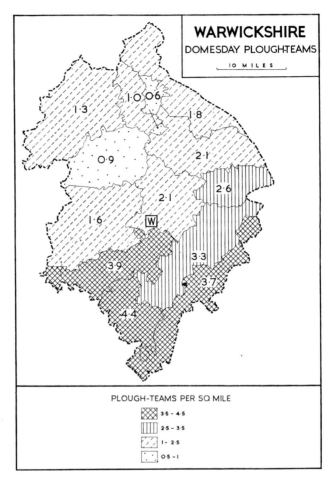

Fig. 95. Warwickshire: Domesday plough-teams in 1086 (by densities).
W indicates the Domesday borough of Warwick.

Villeins constituted the most important element in the population, and
amounted to 56% of the total; the bordars took second place with 29%,
and the serfs came far below with only 13%. The three groups thus
together accounted for 98% of the rural population, which suggests that
the structure of the population was very different from that in the adjacent
counties of Leicester and Northampton, where there was a substantial
free element in the population.

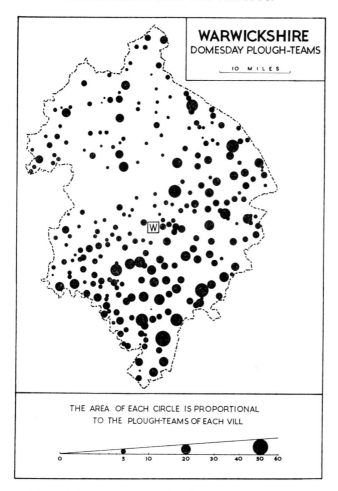

Fig. 96. Warwickshire: Domesday plough-teams in 1086 (by settlements).
W indicates the Domesday borough of Warwick.

The miscellaneous category is a very varied one; as Round says, 'the variety of classes, and even of nationalities named in the Warwickshire survey, is exceptionally large'.[1] Seventy-one priests are mentioned at sixty-five places. Twenty-two knights (*milites*) are recorded and five of these are described as English knights (*milites angli*). At Aston Cantlow (244) there were 9 Flemings (*flandrenses*) and at Birdingbury (241)

[1] *V.C.H. Warwickshire*, I, p. 284.

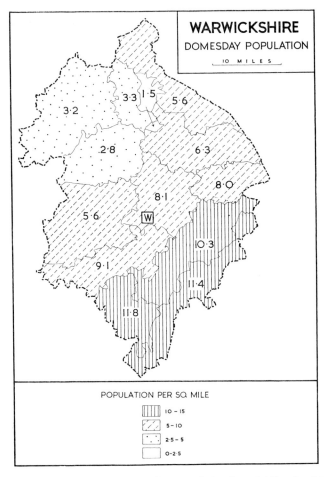

Fig. 97. Warwickshire: Domesday population in 1086 (by densities).
W indicates the Domesday borough of Warwick.

3 franklins (*francones homines*); furthermore, there was a total of 12
Frenchmen (*francigenae*). Only 19 freemen are specifically mentioned and
no sokemen. Of the 13 men (*homines*), 8 at Myton (241 b) rendered 32*d.*
Another interesting group is that of 6 coliberts at Nuneaton (239b) who
were in some way intermediate between serfs and villeins. Among the
smaller groups were the oxmen (*bovarii*) at Selly Oak (177) and Chester-
ton (242), and the smiths (*fabri*) recorded at Wilnecote (240), where

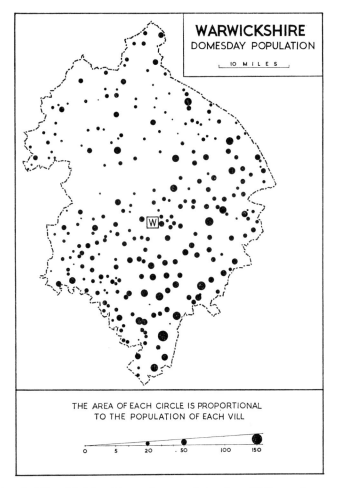

Fig. 98. Warwickshire: Domesday population in 1086 (by settlements).
W indicates the Domesday borough of Warwick.

there was an ironworks (*ferraria*) worth 5*s*. In addition to these, thegns
are mentioned at Pillerton (242) and Eatington (242), and there were
2 swineherds (*rustici porcarii*) at Oldberrow (175 b). The 6 cottars were
at Northfield (177). Two radmen were at Alderminster (175) and one
each at Clopton (167) and at Tredington (173).

Recorded Population of Warwickshire in 1086

A. Rural Population

Villeins	3,807
Bordars	1,965
Serfs	893
Miscellaneous	177
Total	6,842

There were also 54 bondwomen (*ancillae*) who have not been included in the above total.

Details of Miscellaneous Rural Population

Priests	71	English knights . .	5
Freemen	19	Thegns . . .	5
Knights	17	Radmen . . .	4
Men (*Homines*) . .	13	*Francones homines* .	3
Francigenae . . .	12	Oxmen . . .	3
Flemings . . .	9	Smiths . . .	2
Coliberts . . .	6	Swineherds . .	2
Cottars	6		
		Total	177

B. Urban Population

The bordars are also included in the table above.

WARWICK   19 burgesses; 225 (*domus* or *masurae*); 4 *masurae vastae*; 100 bordars *extra burgum*.

(5) Values

The value of an estate is normally given in a round number of pounds for two dates, 1066 and 1086. In about one-sixth of the entries a value for an intermediate period (i.e. *c.* 1070) is also given. This again is usually a round figure. There are also a few entries in which only one value is given; for Bramcote (239b) we are told the *valuit*, but for Wilnecote (240) the *valet*. The royal manors of Kineton and Wellesbourne, Bidford, Stoneleigh, Coleshill and Coton End were valued collectively for three dates—1066, *c.* 1070 and 1086 (238). An unusual defect occurs in fo. 242b in the entry for Wolvey: there is a smudge of ink in the original and it is not clear whether the 1066 value was £2 or £3.

The figures are rarely so detailed as to give pence. It is true that

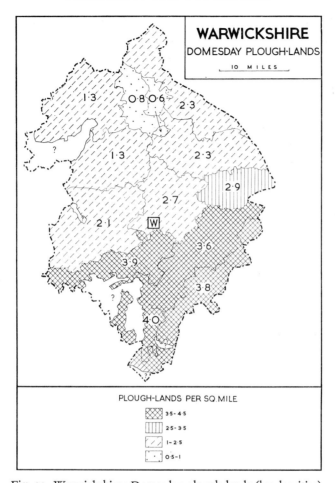

Fig. 99. Warwickshire: Domesday plough-lands (by densities).

W indicates the Domesday borough of Warwick. The blank areas are the parts of modern Warwickshire in Domesday Worcestershire and Gloucestershire, and so are without plough-lands.

amounts of 10s. 8d. and 5s. 4d. occasionally appear, e.g. 10s. 8d. for Lawford (241b), but these are a mark and half a mark respectively on the Scandinavian reckoning of 16d. to the *ora*. They may therefore be but round figures like 20s. and 10s. About 54% of the Warwickshire vills had increased in value over the twenty-year period; a further 29% had maintained their 1066 level, and the remainder showed a decline. When

10-2

three values were recorded, the intermediate value was usually the lowest, e.g. at Arrow (238b): '*T.R.E. valebat lx solidos et post xl solidos. Modo iiii libras.*' But there were some exceptions to this generalisation, and at Long Itchington, for example, the intermediate value was the highest: '*Valuit xii libras. Modo xx libras. Quando rex dedit Cristinae reddebat xxxvi libras*' (244). The greatest increases were generally recorded in the southern half of the county, including the Avon valley; thus the value of Monks Kirby rose from £5 to £10 (243b). But even in this well-developed area there were some reductions; Luddington, for example, fell in value from £8 to £6 (240b).

The 'values' are usually entered as plain statements of money, but in view of the circulation of debased coins, some values were reckoned by weight. Thus at Kingsbury (239b) and Coventry (239b) the 1086 value was expressed in terms of weighed money (*librae ad pondus*). A variation not found elsewhere in the county is recorded at Brailes (238), which was 'worth £55 and 20 loads of salt' (*summas salis*) in 1086. One holding at Weston under Wetherley (241b), yielding 6s. in 1066, was waste in 1086 and rendered nothing (*modo nil reddit*).

Generally speaking, the greater the number of plough-teams and men on an estate, the higher its value, but it is impossible to discern any constant relationships as the following figures for six holdings each yielding £2 in 1086 show:

	Teams	Population	Other resources
Ashow (241b)	4	22	Mills, meadow, wood
Bedworth (240b)	3	10	Meadow, wood
Bericote (241b)	3	9	Mill, meadow
Burton Hastings (242)	8	21	Mills
Grendon (242)	8	40	Mill, meadow, wood
Rugby (241)	6	18	Mill, meadow

It is clear that the values were not directly related to the amount of arable, or to the recorded population. Grendon was roughly three times the size of Bedworth in terms of plough-teams and men, but they were both valued at £2; even when the other resources are taken into account the figures are not easy to explain.

Conclusion

The varied size of the Domesday hundreds and the dispersed nature of some of them do not make them convenient units for the purpose of calculating densities.[1] Some thirteen units of a more geographical nature have therefore been adopted, and in forming them, local considerations of relief, surface deposits and drainage have been borne in mind. As far as possible, they have been devised to emphasise the contrast between the north-western and south-eastern parts of the county.

Of the five standard formulae, those relating to population and plough-teams are most likely to reflect something of the distribution of wealth and prosperity throughout the county in the eleventh century. Taken together they supplement one another to provide a general picture (Figs. 95 and 97). The county may usefully be divided into three areas: (1) the prosperous well-developed Feldon area of the south-east; (2) the relatively poor Tame-Blythe region of the north-west including the small East Warwickshire Plateau; and (3) the valley of the middle and upper Avon, including the basins of the Alne-Arrow and Anker, together with the small upland of Dunsmore. The degree of development in this third division was intermediate between that of the south-east and that of the north-west. This threefold division can be illustrated statistically as follows (see Fig. 105):

Region	Plough-teams per sq. mile	Population per sq. mile
The Feldon	3·3–4·4	10·3–11·8
Avon Valley	1·6–3·9	5·6–9·1
Tame-Blythe Basin	0·6–1·3	1·5–3·3

Figs. 96 and 98 are supplementary to the density maps, but it is necessary to make one reservation about them. As we have seen on p. 277, it is possible that some Domesday names may have covered two or more settlements, e.g. the present-day villages of Pillerton Hersey and Pillerton Priors are represented in the Domesday Book by only one name. A few of the symbols should therefore appear as two or more smaller symbols, but this limitation does not affect the main pattern of the maps. Generally speaking, they confirm and amplify the information of the density maps. In view of the doubtful nature of the plough-land entries, the implications of Fig. 99 are uncertain, but the map has been included for comparison with Fig. 95.

[1] See p. 274 above.

WOODLAND

Types of entries

The amount of woodland on a holding in Warwickshire was normally recorded by giving its length and breadth in terms of leagues and furlongs, and very occasionally of perches. The following entries are typical:

Sambourn (239): *Silva i leuua longa et dimidia leuua lata.*
Coundon (238b): *Silva iii quarentenis et xxx perticis longa et iii quarentenis lata.*

When the quantities are in furlongs, the number is always less than the twelve commonly supposed to make a Domesday league, except in the entry for Middle Bickenhill (240b) which tells of wood 12 furlongs by 6. The entry for Edgbaston (243) is unusual in that it puts breadth before length.

The exact significance of these measurements is far from clear, and we cannot hope to convert them into modern acreages.[1] All we can do is to plot them diagrammatically, as on Fig. 100. Occasionally one set of measurements is found in a combined entry covering two or more villages. This seems to imply, although not necessarily, some process of addition whereby the dimensions of separate tracts of woodland were consolidated into one sum. On Fig. 100 these dimensions have been plotted for the first-named village in a composite entry, e.g. for Kineton in the entry for Kineton and Wellesbourne (238). The associated villages in all such entries have also been indicated. The wood at Wilnecote was valued as well as measured—*Silva i leuua longa et dimidia lata. Valet v solidos* (240). Eleven entries, in addition to giving dimensions, also state that the wood was 'worth *n* shillings when it bears mast' (*cum oneratur*).[2] Additional information is also given for Stoneleigh (238) where the wood was 4 leagues long and 2 leagues broad, and this we are told provided 'feed for 2,000 swine' (*Pastura ad ii milia porcorum*). Similarly, at Coughton (241b), the wood which measured 6 by 4 furlongs yielded 'feed for 50 swine' (*Pascua l porcis*). The linear formula in the entry for Nuneaton

[1] See p. 437 below.
[2] The places were as follows: Sutton Coldfield (238); Fillongley (238b); Roundshill Farm (239b); Claverdon (240); Preston Bagot (240); Smercote and Sole End Farm (240b); Bedworth (240b); Packington (241); Ulverley Green (244); Arley (244); Astley (240b). The phrase *cum oneratur* refers 'to the mast it bore' (*V.C.H. Warwickshire*, I, p. 292). There was a twelfth entry with the phrase *cum oneratur* but without dimensions—that for Fillongley (244b); see p. 295 below.

(241b) is peculiar: *Silva i leuua in longitudine et latitudine*. Six other entries give merely one dimension. Here are two examples:

Claverdon (240): *una leuua silvae; cum oneratur valet x solidos.*
Fillongley (238b): *Ibi silvae quarta pars leuua; cum oneratur valet x solidos.*

It would seem that both peculiar types of measurements are used when length and breadth are the same.[1] The wood for Nuneaton has been plotted on this assumption on Fig. 100, but the single dimensions have been shown as single lines to indicate their frequency.

While leagues and furlongs form the usual units of measurement, there are a number of exceptional entries that refer to wood in other ways. At Bevington (175b) the wood was recorded in terms of acres, but this entry is from the Worcestershire folios. At Lighthorne (243) there was a grove (*Grava*) 2 furlongs long by 20 perches broad, and at Weston under Wetherley (240b) we are told that there was a spinney (*Spinetum*) 2 furlongs long and one furlong broad. The entry for Oldberrow (175b) is interesting: there were only 12 acres of land with one league of wood and 2 swineherds (*rustici porcarii*), and it appears to have been a small woodland settlement which was yet to grow into a village with cultivation.

A few entries seem to be defective. Thus at Kington, and again on a holding in Fillongley, we are told that the wood was worth 10*s*., but no measurements are entered. The two entries read as follows:

Kington (240): *Silva valet per annum x solidos. Tantundem valuit T.R.E.*
Fillongley (244b): *Silva de x solidis cum oneratur.*

It is curious too that in the entry for *Donnelie* there is no mention of wood, though there was a hay (*haia*) there 'half a league long and as much broad' (240). There is no reference to forests in Domesday Warwickshire, but they are implied in two entries that refer to wood set apart for, or in the hands of, the king:

Erdington (243): *Silva i leuua longa et dimidia lata, sed in defenso regis est.*
Southam (238b): *Silva i leuua longa et dimidia leuua lata. Haec silva est in manu regis.*

Distribution of woodland

The main fact about the distribution of wood in Domesday Warwickshire is the contrast between the wooded Arden region of the north and

[1] See p. 437 below.

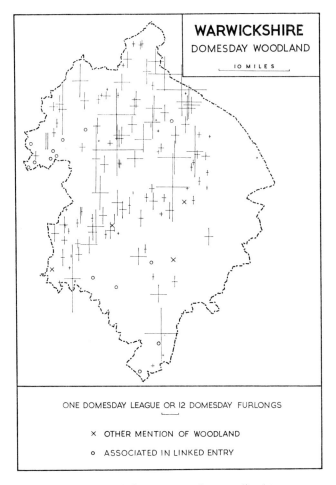

Fig. 100. Warwickshire: Domesday woodland in 1086.

west, and the open region of the Feldon in the south and east (Fig. 100).
The boundary between the two areas was very roughly a line joining
Welford on Avon with Rugby. Generally speaking, the basis of the
twofold division of the county lies in the distinction between the outcrop
of the Lower Lias which forms the essential Feldon of the present day,
and the Keuper Marl and Coal Measures which is the solid foundation
of much of the historic Arden of central Warwickshire.

The Forest of Arden is not mentioned in the Domesday Book, but later evidence shows that it extended as far south as Henley in Arden, as far east as Weston in Arden, and in the north to the line of the river Tame, thus including most of the East Warwickshire Plateau. The varying loams to the north of the Avon carried considerable amounts of wood as Fig. 100 shows. Substantial quantities were entered for several villages, e.g. wood 3 leagues by 3 for Hampton in Arden (243b), the same amount for Caldecote (238b), and wood 4 leagues by 2 for Stoneleigh (238b). This was, however, by no means a continuously wooded district and there were some cleared areas. It should be noted, too, that below Leamington the boundary of the wooded area lay to the north of the river; the Avon terrace deposits in this district must have been well drained and brought into cultivation at an early date. The picture presented for 1086 is that of a number of settlements steadily extending their tracts of cleared land from the south and south-east particularly into the Alne–Arrow and upper Avon areas. From the north, too, settlements were spreading up the Blythe basin, e.g. at Hampton in Arden.

In the Feldon area of the south-east there were less than a dozen wood entries, and later evidence shows that the largest of these entries, that for Brailes, refers in part at least to Tanworth in Arden, a place which belonged to Brailes and which is not mentioned in the Domesday Book.[1] It is surprising that the heavy intractable clays of the Lower Lias, a suitable habitat for dense oakwoods, were so bare of woodland in the eleventh century; the original wood must have been cleared at a relatively early date.

It is interesting to compare the contrast between the open Feldon and the wooded Arden with the distribution of place-names indicating the former presence of wood.[2] The name Ansley, for example, indicates a lonely dwelling in the wood, and it no doubt reflects the conditions that must have prevailed in a good deal of the East Warwickshire Plateau. The distinction between Arden and Feldon was often recognised by English topographers, e.g. by William Dugdale in his *Antiquities of Warwickshire* (1656). The former was a land of hamlets and isolated farms, while Feldon was a land of compact villages.

[1] *V.C.H. Warwickshire*, I, p. 326.
[2] See J. E. B. Gover, A. Mawer and F. M. Stenton, *The Place-Names of Warwickshire* (Cambridge, 1936). Map in folder at end of volume.

MEADOW

Types of entries

The entries for meadow in the Warwickshire folios are for the most part straightforward. For holding after holding the same phrase is repeated monotonously—'*n* acres of meadow' (*n acrae prati*). The amount of meadow in each vill varied from one acre at Ufton to 164 acres at Bidford, but generally speaking amounts of more than 50 acres for individual vills are rare. As for other counties, no attempt has been made to translate these figures into modern acreages.

While measurement in acres is normal, there are eleven entries that refer to meadow in terms of linear dimensions. The following entries are typical:

> Stratford on Avon (238b): meadow 5 furlongs long and 2 furlongs broad.
> Morton Bagot (242b): meadow 3 furlongs long and 6 perches broad.

For Salford, and again for Wolverton there are two meadow entries, one in acres and one in furlongs:

> Salford (239): *Pratum vi quarentenis et dimidia longa, et i quarentena et dimidia lata.*
> Salford (244): *xii acrae prati.*
> Wolverton (242b): *i quarentena prati.*[1]
> Wolverton (243): *Ibi xx acrae prati.*

For these two places only the acres are shown on Fig. 101. An entry for Fenny Compton (241b) is defective in that it reads *Ibi xxxiiii prati*, and the word *acrae* has presumably been omitted. Money renders are stated in three entries. At Coton End (238), with 80 acres of meadow, the meadow and the pasture were valued together (see p. 300 below). For Long Marston (166) and for Meon (163) no measurements are given and for each we are merely told: *Pratum de x solidis*.

Distribution of meadowland

The distribution of meadow on Fig. 101 reflects fairly faithfully the main river pattern, but it also reveals the importance of other factors which had tended to retard or encourage settlement in various places. Thus in the north-west the paucity of meadow is striking, although on

---

[1] The only single dimension—see p. 437.

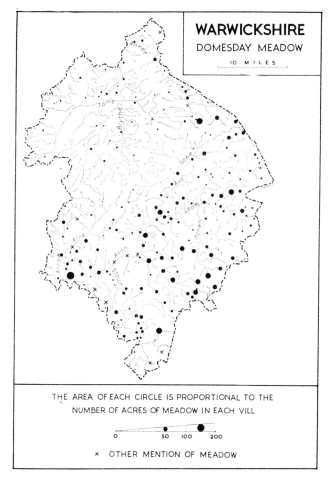

Fig. 101. Warwickshire: Domesday meadow in 1086.
Where the meadow of a village is entered partly in acres and partly
in some other way, only the acres are shown.

physiographic grounds a greater amount might have been expected; the
amounts recorded do not exceed 16 acres for individual places. Some
higher totals were to be found in the Anker basin in the north-east,
e.g. 100 acres at Bulkington and 50 acres at Wolvey. Most of the meadow
of the county was located in the Liassic clay and Keuper Marl country of
the south, and some well-marked linear distributions are evident. These

were associated with the river Avon and its tributaries—the Sowe, Leam, Itchen, Stour, Arrow and Alne. Some high individual amounts were recorded—90 acres for Fenny Compton, 69 for Warmington, and 60 for Farnborough, and there can be little doubt that along this Edge Hill Fringe the existence of water-supplies in the form of springs must have been an important factor in this abundance.

PASTURE

Pasture is mentioned in connection with only six places in Warwickshire; for one place there are two entries. At four places the measurements were in terms of linear dimensions; at one place in terms of acres; and at the remaining place in terms of a joint render also covering meadow.

> Long Itchington (244): *Pastura ii quarentenis longa et una quarentena lata.*
> Stretton on Fosse (243b): *pasturae xl perticis longa et tantundem lata.*
> Thurlaston (240): *ii quarentenae pasturae.*
> Thurlaston (242): *una quarentena pasturae.*
> Wolvey (242b): *Pasturae dimidia leuua in longitudine et latitudine.*
> Napton (240): *Ibi x acrae prati et totidem pasturae.*
> Coton End (238): *De pratis et pascuis iiii librae.*

This meagre list stands in great contrast to the regular entries for pasture that occur in the folios relating to the adjoining county of Oxford. It can only mean that the record of pasture was deliberately excluded for Warwickshire, but that these few references have crept in by chance. For the formulae for Thurlaston and Wolvey, see p. 437.

FISHERIES

No fisheries are mentioned in the Warwickshire folios, but reference is made to eels in connection with mills at eleven places in the county. The location of these places on the lower Avon, Alne, Arrow and Stour is shown on Fig. 102. The complete list of references is as follows:

> Alveston (238b): *Ibi iii molini de xl solidis et xii stichis anguillarum et mille.*
> Aston Cantlow (244): *Ibi molinum de viii solidis et v stichis anguillarum.*
> Atherstone on Stour (238b): *Ibi molinum de x solidis et x stichis anguillarum.*
> Barford (244): *Ibi molinum de ii solidis et xiii stichis anguillarum.*
> Binton (243): *De parte molini iiii summas annonae et viii stichas anguillarum.*
> Salford (239): *Ibi molinum de x solidis et xx stichis anguillarum.*
> Spernall (243b): *Ibi molinum de iiii solidis et vii stichis anguillarum.*

Stratford on Avon (238b): *Ibi molinum de x solidis et mille anguillis.*
Wasperton (239): *Ibi molinum de xx solidis et iiii summis salis et mille anguillis.*
Wixford (239): *Ibi molinum de x solidis et xx stichis anguillarum.*
Wootton Wawen (242b): *Ibi ii molini de xi solidis et viii stichis anguillarum.*

There surely must have been many other fisheries along the Warwickshire rivers, but they are unrecorded in the Domesday Book.

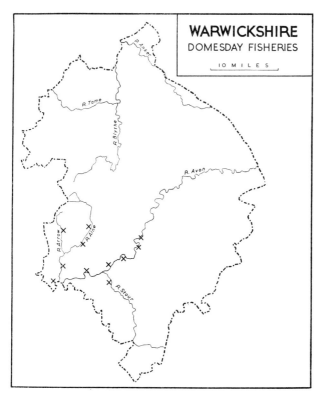

Fig. 102. Warwickshire: Domesday fisheries in 1086.

SALT

The Warwickshire folios mention salt in connection with six places (Fig. 87), and the entries are as follows:

Binton (243): *de Wich iii summas salis.*
Brailes (238): *T.R.E. reddebat xvii libras et x solidos. Modo valet lv libras et xx summas salis.*

Haselor (244): *salina reddit iiii solidos et ii summas salis.*
Hillborough (243 b): *salina in Wich reddit iii solidos.*
Studley (243): *Salina reddit xix summas salis.*
Wasperton (239): *Ibi molinum de xx solidis et iiii summis salis et mille anguillis.*

Two of the places it will be noted are specifically linked with *Wich* (i.e. Droitwich). The location of all six places in the south and west of the county is perhaps significant; so, too, is the occurrence of the Domesday name *Salford* in the south western extremity of the county; 'the name must have referred to a ford where an ancient salt-way passed. There was probably a salt-way from Abbots Morton which passed through the Arrow at Salford and proceeded to Hillborough and Binton'.[1] The entries suggest that the brine springs of Worcestershire formed an important element in the economy of some at least of the Warwickshire villages.[2]

WASTE

Waste was relatively unimportant in Warwickshire. Only three places were wholly waste and only six partly waste in 1086. No mention is made of waste at an earlier date. The reference to wasting 'by the army of the king' at Harbury is unusually explicit. Here is a list of the relevant entries:

Eatington (244 b): 1 hide. *Haec vasta est.*
Harbury (239): 1¼ hides. *Vasta est per exercitum regis. Ibi ii acrae prati. Valuit x solidos. Modo ii solidos.*
Kington (240): 1½ hides. *Vasta est. Valet v solidos. Silva valet per annum x solidos. Tantundem valuit T.R.E.*
Marston juxta Wolston (241 b): 1 hide. *Vasta est. Ibi iii acrae prati. Valuit x solidos. Modo xvi denarios.*
Roundshill (239 b): 1 hide. *Vasta est. Silva ibi dimidia leuua longa et ii quarentenis lata. Cum oneratur valet x solidos.*
Tredington and Tidmington (173): 23 hides. *Una ex his est Wasta.*
Weston under Wetherley (241 b): 1½ virgates. *Vasta est. Ibi iiii acrae prati. Valuit vi solidos, modo nil reddit.*
Whitacre (224): *Idem comes tenet in Witacre dimidiam hidam vastam et valet xii denarios.*

A cursory reference in the account of the borough of Warwick (238)

---

[1] J. E. B. Gover, A. Mawer and F. M. Stenton, *op. cit.* p. 220.
[2] See p. 257 above.

should also be noted; we are told that four houses had been laid waste to make room for the castle.

That some vills were only partly waste, we may judge from two types of evidence. In the first place, there were those vills comprising two or more holdings, of which at least one was waste; this was true of Harbury, Eatington, Weston and Whitacre. In the second place, some single holdings were, apparently, themselves partly waste; this was so at Tredington and Tidmington.

That a holding was 'waste' did not necessarily imply that it was entirely without profit, presumably from wood or meadow, and sometimes specifically said to be so, e.g. at Roundshill. Such holdings have been regarded not as partly waste but as entirely waste in the sense that all their arable seems to have been devastated and that they had no population. They do not necessarily appear as 'wholly waste' on Fig. 159 because some vills had fully stocked holdings in addition to their devastated land.

MILLS

Mills are mentioned in connection with at least 94 of the 281 Domesday settlements within the area covered by modern Warwickshire. It is difficult to be sure of the exact total because mills sometimes appear in composite entries covering a number of places. Thus three mills of 32s. 6d. are entered for Tredington and Tidmington (173). For each composite entry the mill or mills have been plotted for the first named place.[1] The associated villages in all such entries have also been indicated on Fig. 103.

In each entry the number of mills is given, and also their annual value ranging from one mill worth only 16d. at Oxhill (242) to others worth much larger sums; the mill at Milverton rendered 50s. (239b), and that at Eatington 18s. (242). Part of one of the mills at Binton rendered 4 loads of grain and 8 sticks of eels (243), while the mill at Wasperton yielded 20s., 4 loads of salt and 1,000 eels (239). Mills at nine other places returned money and eels; the three mills at Alveston (238b), for example, rendered 40s. and 12 sticks of eels and 1,000 eels. Fractional parts of mills were rare. At

[1] The figure of 94 results from counting as one group the mills of a composite entry. When more than one mill is recorded in such an entry, and it is assumed that each was at a separate place, the total of places with mills is 96. If all the places named in the relevant composite entries are counted, the total becomes 114, a very artificial figure.

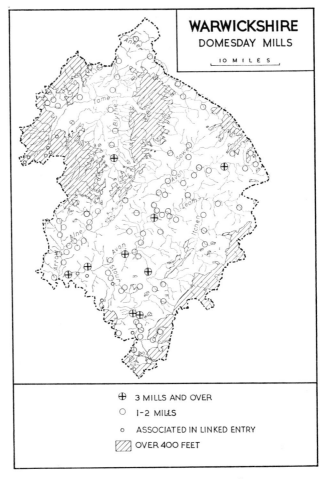

Fig. 103. Warwickshire: Domesday mills in 1086.

Shuttington, two holdings each had half a mill (240 *bis*). A part of a mill is also entered for one holding at Binton (243), but the relation of this to the mill entered in the composite entry for Binton and Hillborough (244) is not clear.

Domesday Mills in Warwickshire in 1086

1 mill	67 settlements
2 mills	18 settlements
3 mills	6 settlements
4 mills	3 settlements

The groups of four mills were at Bidford and at Myton, both on the Avon, and at Honington on the Stour. The most important alignments of mills were along the rivers Avon, Stour, Alne and Arrow and almost all the mill groups were on these streams. Linear distributions are also evident in the Tame, Blythe and Anker basins, but the total number of mills in these areas was small. The only parts of the county devoid of mills were the Edge Hill Fringe in the south-east, and the upland area of the north and west.

URBAN LIFE

The Domesday description of the county begins with an account of the borough (*burgus*) of Warwick (238). Apart from an incidental reference to Tamworth, this is the only place for which burgesses are mentioned.[1] The information for Warwick is very unsatisfactory, and it provides us with hardly any idea of its life and activities. There can be little doubt, however, that in 1086 the town was of some considerable size. The king had 113 houses and his barons another 112 between them.[2] In addition, there were 19 burgesses with 19 messuages (*masurae*). The entry also says that the 112 properties of the barons belonged to, and were valued with, the lands they held outside the borough (*Hae masurae pertinent ad terras quas ibi ipsi barones tenent extra burgum et ibi appreciatae sunt*). Some entries relating to baronial holdings in nearby villages do record the existence of properties appurtenant to Warwick, and a full list is set out on p. 306. Thus the main entry attributes two properties to William Fitz Corbucion, and one each appears for Bearley and Wolverton in the account of his fief on fo. 243. But not all the properties of the main entry can be accounted for in this way. Of Ralf de Limesi's nine properties, for example, only seven appear in the account of his fief—in the entry for Budbrooke (243). The total of the properties thus entered for surrounding manors amounts only to 26. We are left to surmise whether the remainder have been omitted from the accounts of their respective fiefs or whether the 26, after all, are additional to the properties of the main entry. We have assumed that the remainder have been so omitted and that the 26 are included in the figures of the main entry. For one nearby village, Coton

[1] Under Coleshill (238) we are told that there were 10 burgesses in Tamworth. This is the only reference to Tamworth in the Warwickshire folios, but it seems to have lain partly in Warwickshire and partly in Staffordshire, and so continued until 1890 when the whole borough was included in Staffordshire. See p. 208 above.

[2] Fo. 238 gives a detailed list of these houses belonging to the barons.

End (238), there is an unusual statement: *Extra burgum c bordarii cum hortulis suis reddebant l solidos.* It is difficult to know whether or not we should associate these bordars with the borough. On the assumption that they can be associated, the grand total for Warwick is 225 houses, 19 burgesses and 100 bordars. This may imply a population of up to 1,500. We are given no clue to the activities—agricultural or commercial—of these inhabitants.

Fig. 104. Warwickshire: Places contributory to Warwick.

Alveston (238b) *iiii domus*   Edstone (242b) *i domus*
Bearley (243) *i domus*   Hampton Lucy (238b) *iii domus*
Billesley (242) *i domus*   Haselor (244) *unus burgensis**
Budbrooke (243) *vii domus*   Pillerton (242) *i masura*
Butlers Marston (242) *ii burgenses* Tysoe (242b) *iii domus*
Coughton (241b) *i domus*   Wolverton (243) *i domus*

* Warwick is not specifically named, but must be implied.

The description of Warwick makes no reference to churches, but that there was at least one we may judge from an entry relating to Myton which says 'the church of St Mary of Warwick holds 1 hide in Myton' (241b). That there was a castle we know because 4 houses had been destroyed to make room for it (*iiii sunt vastae propter situm castelli*). Finally, honey renders are mentioned, together with a variety of other dues. This is a meagre collection of details, and provides merely a few glimpses of what was evidently an important centre.

Smiths

Two smiths (*fabri*) were recorded at Wilnecote (240), where we are told there was an ironworks (*ferraria*) worth 5*s*. This is the only mention of smiths in the Warwickshire folios.

Churches

Apart from a cursory reference on fo. 241 b to the Church of St Mary at Warwick, no mention is made of churches in the Warwickshire folios, but one appears for Clifford Chambers in the Gloucestershire folios (163 b). There were, however, some 71 priests recorded at 65 places, and it may well be that the mention of priests implies the existence of churches, as is probable in Northamptonshire. The Domesday names *Chircheberie* (Monks Kirby) and *Donecerce* (Dunchurch) can leave little doubt that there were churches at these places. Two priests were recorded at Monks Kirby (243 b), and one at Dunchurch (244). At Alveston (238 b), although neither church nor priest is mentioned, there is a reference to churchscot (*cerset*) a payment in grain made at Martinmas. These details, taken together, suggest that there were certainly churches in the county, although they were unrecorded.

REGIONAL SUMMARY

Warwickshire is traditionally divided into two distinct parts by the Avon valley—the Feldon or 'open country' to the south-east, and the Weldon or 'wooded country' of the Forest of Arden to the north-west. The contrast stands out well, for example, on the map of Domesday woodland (Fig. 100). In the light of other Domesday distributions, e.g. of plough-teams and population, another division is more convenient for our purpose. The county falls quite readily into three portions, each trending from south-west to north-east, and the broad impression is that of a general diminution in economic activity and population density, when passing from the south-east of the county towards the north-west (Fig. 105). The first division is largely that of the historic Feldon, and excludes the land lying immediately to the south of the Avon. The second division is that of the Avon basin, and its extension northwards to include the Anker valley. The third division is that of the Tame-Blythe basin together

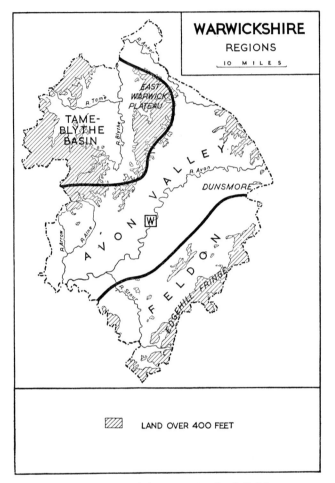

Fig. 105. Warwickshire: Regional subdivisions.

with the East Warwickshire Pleatau. Each of these three areas is far from uniform and some of the more important variations are noted below. But the main threefold division provides a convenient basis for summarising the geography of the county in the eleventh century.

(1) *The Feldon*

The Feldon, or open area of the south-east, is physically a platform, for the most part between 300 and 400 ft. above sea-level. It is based on the

Lower Lias outcrop which is here fairly free from drift. The boundary towards the Avon is marked by a well-defined if small escarpment at the junction of the Lower Lias and the Keuper Marl. In the extreme south-east, where the Middle and Upper Lias are exposed, the country rises to some 700 ft. in Edge Hill and Burton Hill. The sloping plateau surface is broken by a series of streams which flow north-west to join the Avon, the Leam, the Itchen, and most important of all, the Stour. The soils are mainly heavy clays derived from the Lower Lias, but more mixed and better soils are characteristic of the Edge Hill Fringe.

The general prosperity of the Feldon, as indicated by the densities of plough-teams and population, shows that the area was the most intensively settled and cultivated part of the county. The density of plough-teams ranged from 3 to over 4 per square mile; that of population between 10 and 12. The valleys were marked by numerous villages, each with some meadow. Many of these villages, particularly those of the Stour valley, also had mills, some in clusters of two or three. The area as a whole was largely devoid of wood, and it is evident that by Domesday times the Feldon was already the 'open country', so well described by Speed and later topographers.

(2) *The Avon Valley*

The valley of the Avon below Warwick is marked by rich easily worked soils developed on the numerous river terraces as well as on drift over-lying the Keuper Marl, and on the numerous river terraces. These favourable conditions are reflected in the high plough-team density, which was nearly 4 per square mile, and in the relatively high popu-lation density of about 9. Settlements were numerous in the valley, and were marked by a good deal of meadow and numerous mills. The woodland that once covered the region had been for the most part cleared by 1086.

Above Warwick the valley lies at about 200–400 ft. above sea-level, and is marked by a diversity of soils. There are clays derived from the Liassic and Keuper formations, some lighter soils from glacial drifts, and also sands and shales derived from the Upper Coal Measures. The densi-ties of plough-teams and population are very much less than those of the valley below Warwick. The plough-team density was only about 2, and that of population about 8. These low figures may in part be explained by the fact that there was a substantial amount of woodland still remaining.

Despite the relatively low prosperity of the area, its villages carried a fair amount of meadow and some mills.

With the upper Avon valley may be grouped the valley of the Anker to the north. This area also lies at between 200 and 400 ft. above sea-level. Its soils are derived for the most part from drift overlying the Keuper Marl. Its densities of plough-teams and population resembled those of the upper Avon valley, but were just a little lower. There was a good deal of wood, some meadow and about half a dozen mills.

The Alne-Arrow basin is closely connected with the valley of the Avon below Warwick. Its general level ranges between 200 and 400 ft. above sea-level. Its mixed soils are derived mainly from the Lower Lias and from the drift that covers the Keuper Marl. The area also contains some narrow outcrops of Arden Sandstone, which has influenced the siting of various settlements because of its water supplies. One of the outstanding features of the area was its woodland. As might be expected therefore, the densities of plough-teams and population were low, and comparable with those of the upper Avon and Anker valleys. There were small amounts of meadow and a few mills.

Finally, the Avon valley includes the small oval-shaped area of Duns-more. This consists of a low plateau rising to 390 ft. bounded on the north by the Avon and on the south by the Leam. Its soils are for the most part derived from glacial sands and gravels which overlie the Lower Lias and Keuper clays. The lighter deposits are characteristic of Dunsmore Heath, which in the eleventh century was almost devoid of settlements. Large settlements on the margins of the heath bring up the density of teams to 2·6, and that of people to 8 per square mile, and these figures indicate considerable development. No wood was recorded in the district, but there were some small amounts of meadow, and a few mills on the Avon to the north.

(3) *The Tame-Blythe Basin*

The upland country of north-west Warwickshire, drained by the Tame-Blythe system of streams, lies mainly between 200 and 400 ft., but rises to well over 500 ft. in the East Warwickshire Plateau and on the Stafford-shire border. The soils are very mixed in character and quality, and are derived from a wide variety of drift deposits overlying the Keuper Marl. The density of settlement was everywhere low, and many villages were small. The general poverty and lack of development is reflected in the low densities for plough-teams (1·3 and below) and population (1·5 to 3·3). A good deal of wood was recorded in the area, but only very small amounts of meadow, and very few mills. The East Warwickshire Plateau seems to have been the most backward region in the county.

BIBLIOGRAPHICAL NOTE

(1) It is interesting to note that Sir William Dugdale, in his famous *Antiquities of Warwickshire* (London, 1656), made great use of the Domesday Book. Portions of it were quoted by him, and his identifications of places provide an excellent basis for subsequent study, even if they cannot all be now accepted. A second edition was revised and augmented in two volumes by William Thomas in the following century (London, 1703).

The first complete text of the Warwickshire folios was that prepared with a translation by William Reader—*Domesday Book for the County of Warwick* (Coventry, 1835). Only a hundred copies of this were printed, but a second edition containing a useful introduction and further notes by E. P. Shirley appeared later in the century (Warwick, 1879).

(2) The standard text now available is that prepared by W. F. Carter in *V.C.H. Warwickshire* (London, 1904), I, pp. 299–344. This contains a valuable series of footnotes and identifications and is preceded by one of the famous Introductions by J. H. Round (pp. 269–97).

(3) The following articles deal with various aspects of Domesday study in so far as it affects the county:

C. TWAMLEY, 'Notes on the Domesday of Warwickshire', *Archaeol. Journ.* (London, 1864), XXI, pp. 375–6.

B. WALKER, 'Some notes on Domesday Book', *Trans. Birm. Archaeol. Soc.* (Birmingham, 1901), XXVI, pp. 33–80.

B. WALKER, 'The hundreds of Warwickshire', *ibid.* (Birmingham, 1906), XXXI, pp. 22–46.

B. WALKER, 'The hundreds of Warwickshire at the time of the Domesday Survey', *Antiquary* (London, 1903), XXXIX, pp. 146–51, 179–84. This paper contains an analysis of each hundred (with map).

(4) Domesday maps appear in the following:

P. N. Nicklin, 'The early historical geography of the Forest of Arden', *Trans. Birm. Archaeol. Soc.* (Birmingham, 1934), LVI, pp. 71–6. Includes a Domesday population map of the Forest of Arden.

R. H. Kinvig, 'The Birmingham District in Domesday Times', being pp. 113–34 of M. J. Wise (ed.) *Birmingham and its Regional Setting* (British

Association, Birmingham, 1950). Includes maps of Domesday hundreds, wood, plough-teams and population.

(5) The pioneer study of the place-names of the county is W. H. Duignan, *Warwickshire Place-names* (Oxford, 1912). This has been superseded by J. E. B. Gover, A. Mawer and F. M. Stenton's *The Place-Names of Warwickshire* (Cambridge, 1936), which is a very valuable aid to the study of the county in Domesday times.

CHAPTER VII

LEICESTERSHIRE

BY D. HOLLY, M.A.

The Leicestershire section of the Domesday Book gains greatly in interest from the existence of the so-called Leicestershire Survey which was discovered by J. H. Round, and which can be assigned to the years 1129–30.[1] Unfortunately, the Survey covers mainly the northern and eastern parts of the county (Fig. 106) and is concerned largely with assessment. In spite of these limitations, it is of paramount importance in any Domesday study of the county. Its information is arranged geographically and not tenurially, that is in terms of the vills themselves and not of their land holders. This helps in the identification of a number of holdings, and it also throws light on the method of assessment. The survey, moreover, contains a number of village names not found in the Domesday folios, and helps in the attempt to obtain a realistic picture of Domesday conditions.

The Leicestershire Survey is not the only source that provides an indication of the imperfections of the Domesday Book; the Domesday folios themselves sometimes bear witness to the human element in their assembling. Thus, seven entries on fo. 234b each state 'there is land', followed by a blank; the number of teams for which there was land in each of these holdings was never filled in. There are also blanks in the manuscript on fos. 230b and 237; and on the same fo. 237, there is a duplication of holdings described on fo. 235. But these are exceptions; the text of the Leicestershire folios is, in general, full and complete.

The present-day county of Leicester, in terms of which this study is written, differs somewhat from the Domesday county. The main difference is in the west where Leicestershire and Derbyshire meet. The present-day Derbyshire villages of Nether Seal and Over Seal are surveyed in the Leicestershire folios. On the other hand, Chilcote, Oakthorpe, Measham and Willesley, now in Leicestershire, are surveyed in the Derbyshire folios. Moreover, the folios for both Leicestershire and Derbyshire contain entries relating to the present-day Leicestershire

[1] J. H. Round, *Feudal England* (London, 1895), pp. 196–214. A later edition incorporating new material is C. F. Slade's *The Leicestershire Survey* (Leicester, 1956).

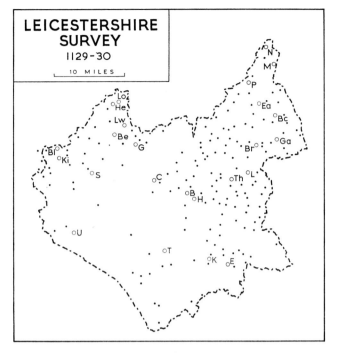

Fig. 106. The Leicestershire Survey, 1129–30.

All identified vills mentioned in the Leicestershire Survey are indicated. Those mentioned in the Leicestershire Survey, but not specifically mentioned in the Domesday Book, are as follows: B, Barkby Thorpe; Be, Belton; Bc, Bescaby; Bl, Blackfordby; Br, Brentingby; C, Cropston; E, East Langton; Ea, Eaton; G, Garendon; Ga, Garthorpe; H, Hamilton; He, Hemington; K, Kibworth Beauchamp; Ki, Kilwardby; L, Little Dalby; Lo, Lockington; Lw, Long Whatton; M, Muston; N, Normanton; P, Plunger; S, Snibston; T, Thorpe by Cosby; Th, Thorpe Satchville; U, Upton. In addition to these are two unidentified names not mentioned in the Domesday Book—*Chilteston* and *Dailescroft*. Two other unidentified names are *Gillethorpe* and *Widesers*, but these do appear in the Domesday Book as *Godtorp* (235b) and *Windesers* (233b). The three places beyond the western boundary are not in the modern county (Linton, Nether Seal and Over Seal).

villages of Appleby, Ravenstone, Stretton en le field and Donisthorpe,[1] and to the Derbyshire vill of Linton. In the south-east corner of the county lies Little Bowden which appears in the Domesday account of

[1] The Derbyshire Domesday vill of *Trangesbi* has been identified as the Leicestershire village of Thringstone. See F. T. Wainwright, 'Early Scandinavian Settlement in Derbyshire', *Journ. Derbyshire Archaeol. and Nat. Hist. Soc.* (Derby, 1947), New Series, xx, p. 108.

Northamptonshire and which was transferred to Leicestershire as late as 1888.

One difficulty in handling the Leicestershire material from a geographical point of view is the fact that the information about two or more places is sometimes combined in one statement. This is particularly true of the four groups of holdings associated with the manors of Rothley

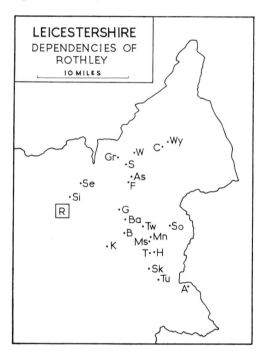

Fig. 107. Leicestershire: Dependencies of Rothley (R).

A, Allexton; As, Asfordby; B, Baggrave; Ba, Barsby; C, Chadwell; F, Frisby on the Wreake; G, Gaddesby; Gr, Grimston; H, Halstead; K, Keyham; Mn, North Marefield; Ms, South Marefield; S, Saxelby; Se, Seagrave; Si, Sileby; Sk, Skeffington; So, Somerby; T, Tilton; Tu, Tugby; Tw, Twyford; W, Waitnaby; Wy, Wycomb.

(230b), Melton Mowbray (235b), Great Bowden (230b), and Barrow on Soar (237), although it is not confined to them (Figs. 107–109). Thus, the manor of Rothley consisted of Rothley itself, together with holdings in some twenty-two vills which are scattered over a wide area. Their respective assessments, meadow, woodland and mills are entered separately, but their population, plough-teams and renders are given collectively, so that we have no means of apportioning these latter resources among the twenty-two. For the purpose of constructing distribution maps, the details of this and other combined entries have been divided exactly among all the vills concerned. There are entries, however, in which some information cannot be so divided. Thus, in the account of Holwell and Ab Kettleby (234b), the meadow measured three furlongs by half a

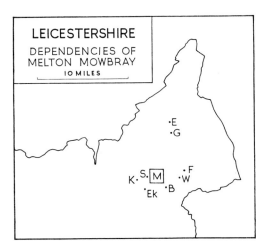

Fig. 108. Leicestershire: Dependencies of Melton Mowbray (M).
B, Burton Lazars; E, Eastwell; Ek, Eye Kettleby; F, Freeby; G, Goadby Marwood; K, Kirby Bellars; S, Sysonby; W, Wyfordby.

furlong, and it has been allocated to Ab Kettleby because there is a separate entry of 10 acres for Holwell (231). Woodland measuring one and a half furlongs by one furlong, and recorded for Hathern and Dishley (237), has been allocated to Hathern, the first-named village. The mills of two other composite entries have likewise been plotted for the first-named place in each entry.[1]

SETTLEMENTS AND THEIR DISTRIBUTION

The total number of separate places mentioned in the Domesday Book for the area now included in the modern county of Leicestershire seems to be 296, including the borough of Leicester itself. This figure, however,

[1] See pp. 342 and 348 below.

may not be quite accurate because there are some instances of two or more adjoining villages bearing the same surname today, and it is not always clear whether more than one unit existed in the eleventh century. Only where there is specific mention of a second village in the Domesday Book

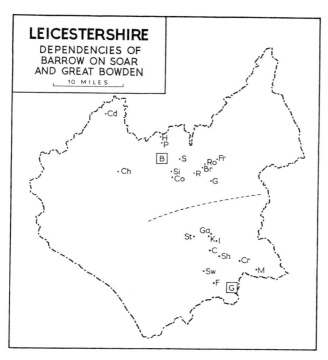

Fig. 109. Leicestershire: Dependencies of Barrow on Soar (B) and Great Bowden (G).

(a) *Barrow on Soar*: Br, Brooksby; Cd, Castle Donington; Ch, Charley; Co, Cossing-ton; Fr, Frisby on the Wreake; G, Gaddesby; H, Hoton; P, Prestwold; R, Rearsby; Ro, Rotherby; S, Seagrave; Si, Sileby. (b) *Great Bowden*: C, Carlton Curlieu; Cr, Cranoe; F, Foxton; Ga, Galby; I, Illston; K, King's Norton; M, Medbourne; Sh, Shangton; St, Stretton; Sw, Smeeton Westerby.

itself has it been included in the total of 296. Thus, the existence of Peatling Magna and Peatling Parva in the eleventh century is indicated by the mention of *Petlinge* and *Alia Petlinge*, and of Ashby Magna and Ashby Parva by *Essebi* and *Parva Essebi*. There is no indication, on the other hand, that, say, the Great Dalby and Little Dalby of today existed as separate villages; the Domesday information about them is entered under one name, though there may well have been separate settlements in the eleventh century.[1] The same is true also of, say, Claybrooke Magna and Claybrooke Parva which are described under the name of *Claibroc*.

The total of 296 includes some places about which very little information is given. *Legham* is mentioned by name under Leicester City (230)

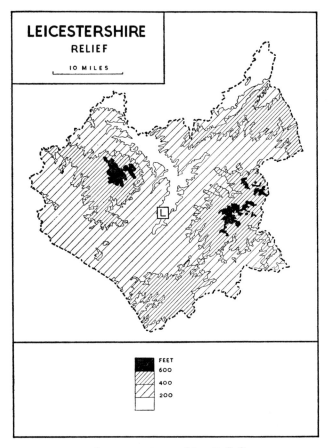

Fig. 110. Leicestershire: Relief.
L indicates the Domesday borough of Leicester.

but does not occur again. Villages recorded as waste have, as might be expected, very brief entries. Apart from these entries, the resources of most Leicestershire villages are set out with the usual detail.

[1] It seems likely because the Domesday *Dalbi* appears in the Leicestershire Survey as *Dalbia* and *Magna Dalbia*.

Not all the Domesday names appear on the present-day map of Leicestershire villages. Some are represented by hamlets, by individual houses and farms, or by the names of topographical features. Thus, *Tunge* is now the hamlet of Tonge in Breedon parish, and *Bortrod* is that

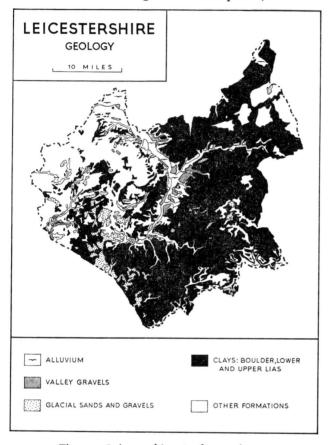

LEICESTERSHIRE
GEOLOGY

10 MILES

◠ ALLUVIUM

▓ VALLEY GRAVELS

░ GLACIAL SANDS AND GRAVELS

▓ CLAYS: BOULDER,LOWER AND UPPER LIAS

☐ OTHER FORMATIONS

Fig. 111. Leicestershire: Surface geology.
Based on Geological Survey One-Inch Sheets (New Series) 126,
141–3, 155–7 and 169–71.

of Boothorpe in Ashby de la Zouch. *Queneberie* has become Quenby Hall in the parish of Hungerton; and *Chitebie* has become Eye Kettleby Farm in the parish of Melton Mowbray. Other settlements are now represented merely by traces left on the ground; that is all that remains, for example, of the Danish settlement of Ingarsby, which contained thirty-one recorded inhabitants in Domesday times; it was abandoned in 1469, 'the line of the main street is now marked by ancient, twisted thorn trees'. Others have left no trace of either their name or site; such appears to have been the fate of *Burgo*, which probably stood somewhere in the modern parish of Launde. W. G. Hoskins has brought together the evidence relating to the deserted villages of Leicestershire.[1] As many as thirty-nine of the villages he discusses had appeared in the Domesday Book.[2] These villages, most of them flourishing settlements in the eleventh century, decayed largely as a result of enclosure for pasture between 1450 and 1600. Only eight of the thirty-nine sites are in the western half of the county; the richer pastures of the east and south took a heavier toll of the ancient settlements. These are but some of the changes in the Leicestershire villages. To them must be added a number of unidentified names, viz. *Elvelege* (236), *Godtorp* (235 b), *Legham* (230), *Lilinge* (235 b), *Netone* (236b), *Plotelai* (237) and *Windesers* (233b). Whether these will yet be located, or whether the places they represent have completely disappeared leaving no record or trace behind, we cannot say.

On the other hand, some villages on the modern map are not mentioned in the Domesday Book. Their names do not appear until the twelfth or thirteenth centuries, and presumably, if they existed in 1086, they are accounted for under the statistics of neighbouring settlements. Tonge, described *cum omnibus appendiciis* (233), is the one example in Leicestershire where such inclusion is apparent from the Domesday entry itself. By the time of the Lay subsidy of 1327, Tonge, together with Wilson and Anderchurch, were hamlets of Breedon. It is thus possible that in Domesday times, Breedon, Wilson and Anderchurch, none of which is mentioned by name in the Domesday Book, were the 'appendages' of Tonge. It seems almost certain that many, if not all, of the new names which appeared in the Leicestershire Survey were those of villages surveyed with others in the Domesday Book. Thus, the Domesday Book speaks

¹ W. G. Hoskins, 'The Deserted Villages of Leicestershire', *Trans. Leics. Arch. Soc.* (Leicester, 1945), XXII, pp. 242–64. See also W. G. Hoskins, *Essays in Leicestershire History* (Liverpool, 1950), pp. 67–107.

² The list of these deserted villages which appeared in the Domesday Book is as follows: Alton Grange, Brascote, Dishley, Gopsall, Lubbesthorpe, Potters Marston, Woodcote and Weston in the west; in the east, Baggrave, Bittesby, Brooksby, Cotes-de-Val, Eye Kettleby, Foston, Frisby, Goldsmith Grange, Great Stretton, Holyoak, Ingarsby, Keythorpe, Knaptoft, Leesthorpe, Lowesby, Misterton, Newbold Folville, Noseley, North Marefield, Othorpe, Prestgrave, Poultney, Quenby, Shoby, Storms-worth, Sysonby, Welby, Whatborough, Willowes, Wistow, Withcote.

of *Langetone*, but the Leicestershire Survey distinguishes between *Langeton* and *Alia Langeton*. Or again, the Domesday Book describes the assessment and resources of Shepshed, while in the Leicestershire Survey, Hemington, Lockington and Long Whatton share the assessment of that vill, and so on. These new names which appeared in the Leicestershire Survey numbered as many as twenty-six (Fig. 106). The names of other modern villages appear even later than the date of the Leicestershire Survey. Thus, as far as record goes, Launde was first mentioned in 1163, Bushby in 1175, Isley Walton in 1220. Some of these new names occur in the area not covered by the Leicestershire Survey and there is occasional evidence to show that they existed in 1086. One example is that of Bringhurst. The name first appeared in 1188, and in 1086 the village was 'undoubtedly silently included in the small soke of Great Easton'.[1] But it is certain that, at any rate, some of these were names of post-Domesday settlements. Thus, Market Harborough was first mentioned in 1177 and we know that in 1086 it was 'merely an outlying part of the fields of the royal manor of Great Bowden'.[2] Then, too, some of the coal-mining villages, e.g. Coalville and Ellistown, are of very late date.

The essential feature of the distribution of Domesday settlements is the contrast between the west and the east of the county (Fig. 112). In the west, the Charnwood region, and the area corresponding roughly with the modern parishes of Hinckley, Higham on the Hill and Sutton Cheney, are almost devoid of settlement; elsewhere in the west, settlement is relatively sparse. The valley of the Soar, separating west from east, is marked by numerous villages fairly closely placed on the patches of valley gravel. To the east of the river, most of the county is quite densely settled but there are variations. The numerous villages of the Wreak valley contrast with the rather less dense settlement to the north and south and with the open strip in the extreme south-east.

One important factor in this distribution, especially in the contrast between east and west, is water-supply. The county is not well served with water from underground sources, and village sites are usually related to pockets of water-bearing sand and gravel in the Boulder Clay which covers much of the county (Fig. 111). In Anglian and Scandinavian times, writes Dr Hoskins, 'dependence on the sands and gravels [within

[1] W. G. Hoskins, 'The Origin and Rise of Market Harborough', *Trans. Leics. Arch. Soc.* (Leicester, 1949), xxv, p. 56.
[2] For a full account of this town see W. G. Hoskins, *ibid.* p. 56.

the Boulder-Clay area] is revealed in almost every village site to the east of Leicester, whether of the heathen or of the Christian period'.[1] The frequency with which gravel patches occur in the eastern Boulder Clay as compared with that in the west, is thus an important element in the pattern of Domesday settlement.

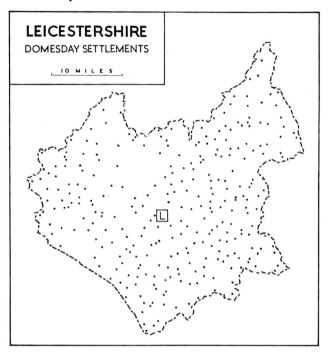

Fig. 112. Leicestershire: Domesday place-names.
L indicates the Domesday borough of Leicester.

THE DISTRIBUTION OF PROSPERITY AND POPULATION

Some idea of the information in the Domesday folios for Leicestershire, and of the form in which it was presented, may be obtained from the entries relating to Belgrave. It is situated on the river Soar, immediately to the north of Leicester and now lies within the boundaries of that city.

Fo. 232. The same [Hugh] holds 7 carucates of land in Belgrave. There is land for 6 ploughs. In demesne there are 3 ploughs and 3 serfs; and 8 villeins

[1] W. G. Hoskins, 'The Anglian and Scandinavian Settlement of Leicestershire', *Trans. Leics. Arch. Soc.* (Leicester, 1935), XVIII, p. 125.

with 5 bordars and 7 sokemen have 4 ploughs. There is a mill rendering (*de*) 12*s*., and 24 acres of meadow. Wood 5 furlongs in length and 3 furlongs in breadth. It was worth 60*s*. Now 100*s*.

Fo. 236b. Adeliz, Hugh's wife, holds of the king one carucate of land in Belgrave. There is land for half a plough; 3 villeins have this there. It was worth 5*s*. Now 4*s*.

It is interesting to note that no pasture or fisheries are recorded for Leicestershire. The basic items are five in number: (1) carucates; (2) plough-lands; (3) plough-teams; (4) population; and (5) values. The bearing of these five items of information upon regional variations in prosperity must now be considered.

(1) *Carucates*

The Leicestershire assessment is normally stated in terms of carucates and bovates. The size of the holdings varied greatly. There were some with only 2 bovates, e.g. at Illston (230b), and others with up to 28 carucates, e.g. Stockerston (232).[1] No holdings in Leicestershire were specifically exempted from assessment but very infrequently the assessment of a holding appears to have been omitted, e.g. at Theddingworth (235b). Frequently, the assessment is stated in a complicated fashion; thus at Enderby there were '6 carucates of land less 3 bovates' (232b); and at Stathern there were '4½ carucates of land and 3 bovates' (234). Occasionally, two or more holdings were assessed collectively. Thus, Pickwell and Leesthorpe were jointly assessed at 14 carucates (235b); similarly, 27½ carucates was the joint assessment of Edmondthorpe and Wymondham (233). On the other hand, there are examples of the assessment of a single holding being split between the demesne and the land not in demesne. Of the 5 carucates at which Rothley was assessed, 2 carucates were in demesne (230); of the 14 carucates at Stapleford, 4 carucates were in demesne (233); and similar entries occur for Edmondthorpe and Wymondham (233) and for Tonge (233) and Wigston Magna (232).

A glance through the Leicestershire folios reveals many examples of the six-carucate unit. At times, it is very obvious; Cossington is assessed at 6 carucates (237), Loddington at 12 carucates (234b), Croxton Kerrial

---

[1] The assessment of Melton Mowbray (235b) at 7 hides and 1 carucate and 1 bovate seems open to question—see *V.C.H. Leicestershire* (London, 1907), I, p. 295 and p. 326 below.

at 24 carucates (230); and there are many other equally clear examples.
These are most easily recognised when the vill was in the hands of one
landowner, although in one entry the Domesday Book simplifies the
situation when more than one landowner was involved. Thus, at Bottesford (234) the holdings of ten sub-tenants are given with a concluding
sentence—'Altogether [there are] 12 carucates of land'. But the same
feature can also be seen when the various holdings in some villages are
assembled from the different pages of the Domesday Book. Here, for
example, are the different holdings in Peatling Magna and Peatling Parva
respectively:

Peatling Magna

Land-holder	Folio	Carucates	Bovates
(1) Godwin the Priest	231	0	4
(2) Robert de Buci	234	3	0
(3) Countess Judith	236	4	0
(4) Robert	237	4	4
	Total	12	0

Peatling Parva

Land-holder	Folio	Carucates	Bovates
(1) Huard	232b	3	4
(2) Levric	236b	8	4
	Total	12	0

Frequently the sum of the holdings in a village does not come to such
a convenient total, but the duodecimal system is readily apparent in
nearly one-third of the Leicestershire villages.

Further evidence of the duodecimal assessment of the county comes
from the Leicestershire Survey and this was discussed by J. H. Round.

But my strong evidence is found in an invaluable survey of Leicestershire,
unknown till now to historians, which does for the carucated districts just
what the *Inq. Com. Cant.* does for the hidated ones. Here we find the townships grouped in small blocks of from six to twenty-four 'carucatae terrae',
as a rule with almost monotonous regularity. And these blocks are further
combined in small local Hundreds, of which the very existence is unknown

to historians and antiquaries, and which are usually multiples, like the Lincolnshire Wapentake, of the six-carucate unit.[1]

Round believed that the returns of 1086 were drawn up in a manner similar to the Leicestershire Survey, 'hundred' by 'hundred' and wapentake by wapentake, but that when the Domesday Book was compiled, the wapentake headings only were retained. Moreover, he saw in the grouping of the vills into hundreds a further step in the duodecimal system of assessment.[2] This is apparent, for example, in the hundred of Waltham on the Wolds which in the Leicestershire Survey is said to contain the adjoining villages of Waltham, Stonesby and Coston. Here is the Domesday assessment for these three villages:

	Folio	Carucates	Bovates
Waltham	233	16	4
	235	2	4
Stonesby	235	8	0
Coston	233	9	0
	Total	36	0

The apparently irregular assessment of these vills thus becomes clearly duodecimal. On the basis of the assessments given in the Leicestershire Survey itself, many of these hundreds show a total approximating to a duodecimal figure, but of the thirty-two hundreds, only seven are rated at even duodecimal totals. On the evidence of the Survey, Sir Frank Stenton has compiled a list of Domesday vills whose individual assessment, 'distorted in Domesday Book from whatever cause, has been restored by the compilers of the Leicestershire Survey' to a duodecimal total.[3] Thus, it is clear that the evidence for the duodecimal system of assessment is far stronger than a casual examination of the Domesday folios of the county might lead us to suppose.

Although the assessment of the Leicestershire holdings is mainly in terms of carucates and bovates, some holdings are assessed in hides and virgates. The hide is used in eighteen entries in the Leicestershire folios, and in one entry in the Northamptonshire folios (for Little Bowden, now

[1] J. H. Round, *Feudal England*, p. 80. Note that these 'hundreds' which go to make up a wapentake are not the same as the hundreds of the southern counties.

[2] J. H. Round, *ibid.* p. 204.

[3] *V.C.H. Leicestershire*, I, p. 280.

in Leicestershire).[1] Of these nineteen places, some six are scattered over the north of the county while most of the others form a concentration south-east of Leicester city, stretching to Saddington; Burbage is an isolated entry on the south-west border of the county (Fig. 113). The relationship between the hide and the carucate is expressed in four entries:

> Kilby (236): 2 parts of one hide, that is (*id est*) 12 carucates of land.
> Bromkinsthorpe (237): There are 2 parts of one hide, that is 12 carucates of land.
> Burbage (231): There is one hide and the fourth part of one hide. There are 22½ carucates of land.
> Melton Mowbray (235 b): There are 7 hides and one carucate of land and one bovate. In each hide there are 14½ carucates of land.

The first three entries infer that a hide is equivalent to 18 carucates. The entry for Melton Mowbray is more explicit but gives a different relationship. J. H. Round maintained that 18 carucates to one hide was the normal relationship in Leicestershire, and that the Melton Mowbray entry was a special definition of a hide which departed from this usual relationship.[2]

Hides, carucates and bovates are sometimes used together to express the assessment of a holding, e.g. in the entry for Melton Mowbray above. Carucates and virgates are also used together; thus at Illston there were '2 carucates of land and one virgate' (231). The virgate is occasionally used alone; at Slawston there was 'one virgate of waste land' (234). But hides and virgates never appear together in the Leicestershire folios.[3] It seems that the Leicestershire virgate represented a quarter of a carucate or 2 bovates. Entries such as that for Barlestone—'3 carucates of land less one virgate' (232b)—cannot conceivably imply that the Leicestershire virgate represented a quarter of a hide. The ten holdings where virgates

---

[1] The places concerned are as follows: Saddington, Dishley, Shepshed (230b); Knighton, Burbage (231); Aylestone, Knaptoft (231 b); Wigston Magna (232); Saltby (234b); Foston, Arnesby (235); Melton Mowbray (235 b); Kilby, Donington le Heath (236); Burton on the Wolds, Blaby, Bromkinsthorpe, Whetstone (237); Little Bowden (223, Northants).

[2] J. H. Round, *Feudal England*, p. 83.

[3] They are used together in the entry for Little Bowden, but this comes from the Northamptonshire folios—'the same holds 2 hides and one virgate of land and the third part of one virgate in Little Bowden' (223). Here, the virgate, unlike the Leicestershire virgate, represented a quarter of a hide.

are used in the assessment are widely scattered over the southern part of the county.[1]

It is clear that the assessment, whether in carucates or hides, was artificial in character, and bore no constant relation to the agricultural resources of a vill. In a very general way, it is true that the larger the number of carucates, the greater the resources of a vill, but there are very many exceptions even to this. The variation among a representative selection of six-carucate holdings speaks for itself:

	Plough-lands	Teams	Popu-lation	Values 'Valuit'	'Valet'
Hallaton (235 b)	8	8	26	60s.	100s.
Hose (233 b)	7	4½	16	40s.	40s.
Huncote (231 b)	6	6	33	15s.	80s.
Ibstock (237)	4	4	21	5s.	40s.
Cold Newton (235 b)	4	4	9	12s.	20s.
Orton on the Hill (233)	6	9	29	40s.	100s.
Ragdale (234 b)	6	2	4	16d.	20s.
Stonton Wyville (232 b)	4	6	20	40s.	60s.

The total assessment amounted to 2,234 carucates, $\frac{1}{6}$ bovate and 28$\frac{1}{4}$ hides, but it must be remembered that this refers to the area included in the modern county. Maitland estimated the number of carucates in the county as 2,500 (?), the question mark indicating the doubt raised by the difficulty of equating hides and carucates.[2] Sir Frank Stenton gives three alternative totals for the county:[3]

(1) 2,319 carucates, 6½ bovates and 17$\frac{11}{12}$ hides. These totals are additions of the different units, but where the text is explicit, hides have been enumerated as the requisite number of carucates.

(2) 2,534 carucates and 6½ bovates. To obtain this total, one hide has been taken to represent 12 carucates.

(3) 2,642 carucates and 2½ bovates. This total was obtained by equating one hide with 18 carucates.

[1] Virgates occur in the following places: Barlestone (232 b, 234); Illston (231, 232); Slawston (234); Willoughby Waterless (236); Little Bowden (223, Northants), together with three anonymous holdings (234 tres).
[2] F. W. Maitland, *Domesday Book and Beyond* (Cambridge, 1897), p. 400. On another page Maitland declared: 'My calculations about Leicestershire are more than usually rough, owing to the appearance of the curious "hide" or "hundred" or whatever it is' (p. 409).   [3] *V.C.H. Leicestershire*, I, p. 305.

Comparable totals from the present analysis would be:

(1) 2,382 carucates, $\frac{1}{6}$ bovate and $18\frac{2}{3}$ hides.

(2) 2,606 carucates and $\frac{1}{6}$ bovate.

(3) 2,718 carucates and $\frac{1}{6}$ bovate.

It can be seen that in trying to make exact calculations many difficulties arise. All the figures can do is to indicate the order of magnitude involved.

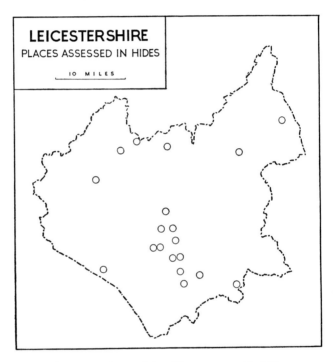

Fig. 113. Leicestershire: Places assessed in hides.

(2) *Plough-lands*

Four different types of entry are used in the Leicestershire folios to express what apparently was the plough-land. Examples of these four types are:

(a) Sharnford (231): There can be one plough (*Una caruca potest esse*).

(b) Old Dalby (235): There is land for 12 ploughs (*Terra est xii carucis*).

(c) Coston (233): In King Edward's time 10 ploughs were there (*T.R.E.
erant ibi x carucae*).

(d) Cranoe (236b): 3 ploughs were there (*Ibi iii carucae fuerunt*).[1]

In the record of any one holding, one formula only is used. Formula (a)
is infrequent, but each of the other three types is used frequently. The
distribution of the four formulae, folio by folio and fief by fief, is set out
in the table on p. 330. Maitland did hazard the suggestion that 'perhaps,
were this part of the survey explored by one having the requisite knowledge, he would teach us that the jurors of some wapentakes use the one
formula while the other is peculiar to other wapentakes'.[2] But that this
is not so can be seen, for example, when the various entries for some vills
are collected together. Two or more of the different formulae frequently
occur for a single vill. The entries for Kilworth are an example:

Fo. 234. In King Edward's time 5 ploughs were there.
Fo. 235. There is land for one and a half ploughs.
Fo. 237. 2 ploughs were there.

Maitland also suggested that the variations might be 'due to a clerk's
caprice', but, as the analysis on p. 330 shows, there is a degree of
correspondence with fiefs which may have employed 'different methods
of making returns'.[3]

Exactly what this information implies, however, is not easy to ascertain.
Taken at their face value, the statements seem to denote land fit for
ploughing in 1086, by virtue of past cultivation, contemporary conditions
or future colonisation. There is, however, a complication. It is what
Maitland called the 'horrible suspicion' that, after all, the plough-land
was as artificial an idea as the carucate.[4] In entry after entry the number of
plough-lands bears a simple ratio to that of carucates in a suspiciously
convenient manner. The examples set out in the table on p. 331 make this
abundantly clear. Moreover, the number of plough-lands is almost always
less than that of the carucates. 'It is quite possible in view of the heavy
rating of Leicestershire as a whole that the jurors in the Domesday

---

[1] There is, in fact, a further variation of formula (d). Occasionally, *ibi n carucae
fuerunt* is replaced by *ibi fuerunt n carucae*.

[2] F. W. Maitland, *op. cit.* p. 421.

[3] J. S. Moore, 'The Domesday Teamland in Leicestershire', *Eng. Hist. Rev.* (1963),
LXXVIII, p. 701.

[4] F. W. Maitland, *op. cit.* pp. 427 and 471.

Distribution of the Leicestershire Plough-land Formulae, Folio by Folio and Fief by Fief

The sign | marks the division between fiefs.

o = entry with no information; b^x = defective entry.

Folio

230   o o o o o

230b (o o o o o o o o o o o o o o o o o o o o)[1] o (o o o o o o o o o o)[2] o
      o o o o o o o o b | o o o o o o o | o

231   b o a a b b b b o o b o o b o o b b | b | b b o | o b o o o o | b b | o b o o o |

231b c o b o o | b o o o o o o o b b b b b b b b b o | b b c | c c

232   b b b o b o b b b b b b b b b b b b b b b b b b b b b

232b b b b b b b b b b b b b b b b b b b b b b b b b b b b b b b b b b b b

233   b b b b b b b b b b b | c c c o c c o o c c c c c o c c

233b o c c o o c o o c o o o c c c c o | c c o c c c c c c b b o o o b b

234   b b b | c c c c b c c o c o | b b b c c o o b b b c c c o b o o c c c

234b c c c c o b c o c c c b c c | c c c c c | b b^x b b^x b^x b^x o o b^x o b^x b^x c c

235   c o o o o o | b b b o | b b | b | b b b b b b | c o b | b b b b b | o |

235b b c c | c c c c o c | c c o (c c c c c c c)[3] d d d o d o d d d d d d o o d d
      d o | d | o |

236   d d | d | d d | d d o | d d | d d d | d | o b | d d d d o d d d d d d o d o d
      d d d

236b o d d d d d d d d d d o d d d d d d d o o o d d | b b b | d o d o d o b b o o |

237   (c c c c c c c c c c c c c c c c c c c c)[4] o o o o o | o o d d d d d d d d d d d

[1] Members of Rothley.   [2] Members of Great Bowden.
[3] Members of Melton Mowbray.   [4] Covered by one total.

Inquest may have been allowed to express the agricultural possibilities of their vills and manors in figures which bore a conscious reference to the carucates of assesssment in each case.'[1] There thus seems grounds for believing that the plough-land entry represents a less severe assessment than that of the carucate. The Northamptonshire evidence points to the same element of conventionality;[2] so does the Lincolnshire evidence.[3] In

[1] *V.C.H. Leicestershire*, I, 286.   [2] See p. 394 below.
[3] See H. C. Darby, *The Domesday Geography of Eastern England* (Cambridge, 3rd ed., 1971), p. 42.

view of these doubts, we cannot regard the plough-land as indicating the real as opposed to the rateable.[1]

Ratio of Plough-lands to Carucates on Representative Leicestershire Holdings

Ratio 1:2

	Plough-lands	Carucates
Birstall (232b)	1	2
Peatling Magna (236)	2	4
Frisby by Galby (232b)	1	2
Nailstone (232b)	½	1
Stoughton (232b)	2	4
Theddingworth (236b)	1	2

Ratio 2:3

	Plough-lands	Carucates
Burton Overy (232)	8	12
Potters Marston (231)	2	3
Oadby (232b)	1	1½
Cold Newton (235b)	4	6
Goadby (235b)	2	3
Thorpe Arnold (233)	10	15

Ratio 1:1

	Plough-lands	Carucates
Cold Overton (236)	12	12
Broughton Astley (232b)	3	3
Sproxton (235)	3	3
Shoby (231b)	11	11
Osbaston (235)	4	4
Thorpe Langton (232)	½	½

[1] For a full discussion see: (1) J. H. Round, *op. cit.* p. 90; (2) F. W. Maitland, *op. cit.* pp. 471–2; (3) F. M. Stenton in *V.C.H. Leicestershire*, I, pp. 284–6.

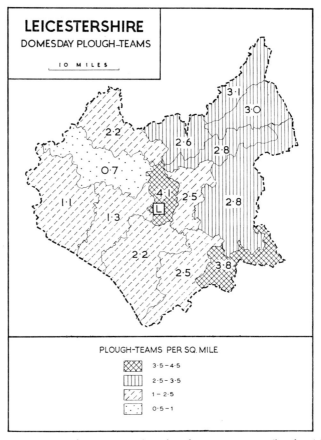

Fig. 114. Leicestershire: Domesday plough-teams in 1086 (by densities).
L indicates the Domesday borough of Leicester.

As can be seen from the table on p. 330, the information about plough-
lands is lacking in a large number of entries, and on fo. 234b there are
some seven entries where the formula appears but with a blank space
where a figure was intended to be inserted. In view of these omissions,
any total of plough-lands for the county would be misleading. Nor can
we hazard an estimate by assuming that plough-lands equalled teams on
those holdings for which no figure is available; the omissions are too
many. F. W. Maitland, in his county tables, did not enter a figure for
plough-lands in Leicestershire.[1] Of the entries for teams which record

[1] F. W. Maitland, *op. cit.* p. 401.

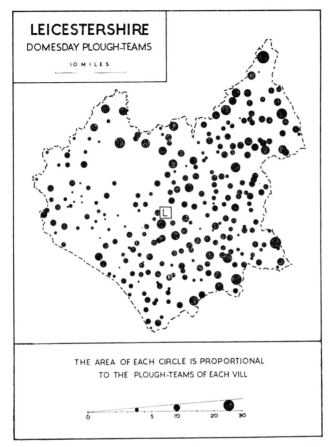

LEICESTERSHIRE
DOMESDAY PLOUGH-TEAMS
10 MILES

THE AREA OF EACH CIRCLE IS PROPORTIONAL
TO THE PLOUGH-TEAMS OF EACH VILL

Fig. 115. Leicestershire: Domesday plough-teams in 1086 (by settlements).
L indicates the Domesday borough of Leicester.

or imply plough-lands, there was a deficiency of teams in 51% and an excess in 26%.

(3) *Plough-teams*

The Leicestershire entries, like those of other counties, usually draw a distinction between the teams on the demesne and those held by the peasantry. Occasional entries are defective. On fo. 237 the name of a vill is illegible and the number of teams is partly so. Sometimes, as in the entry for Catthorpe (236), demesne teams only are recorded, but the

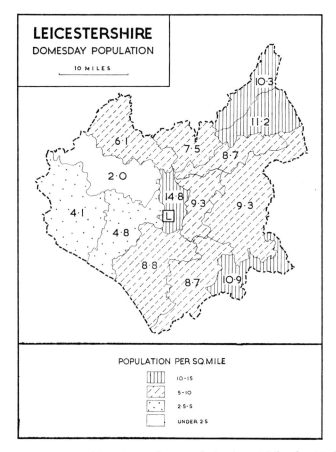

Fig. 116. Leicestershire: Domesday population in 1086 (by densities).

L indicates the Domesday borough of Leicester.

specific negation of this fact at Billesdon (235b)—*In dominio nil fuit nec est*—is unusual. But there are also entries in which the teams of the peasantry only are recorded, e.g. in that for Broughton Astley (236). There are others in which no teams at all appear, e.g. for Donnington le Heath (236).

There are many variations of the normal statement; and a few examples of these are given below:

(a) Gaddesby (236b): Half a plough was there and is [there] now (*Ibi fuit dimidia caruca et modo est*).

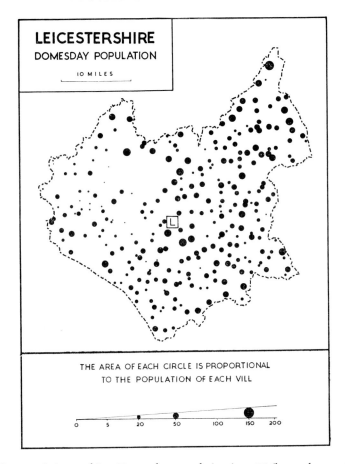

Fig. 117. Leicestershire: Domesday population in 1086 (by settlements).
L indicates the Domesday borough of Leicester.

(*b*) Cotes-de-Val (231): There can be 2 ploughs and they are there with 4 sokemen (*Duae carucae possunt esse et ibi sunt cum iiii sochemannis*).

(*c*) Bottesford (234): 7 sokemen with 2 villeins and 13 bordars having 2 ploughs between them all. Some have nothing (*vii sochemanni cum ii villanis et xiii bordariis habentes ii carucas inter omnes. Aliqui nil habent*).

(*d*) Lubbesthorpe (235): The peasantry had '2 ploughs and 5 ploughing oxen' (*ii carucas et v boves habent arantes*).

(*e*) Sharnford (236b): There are 3 bordars having 6 animals (*Ibi sunt iii bordarii habentes vi animalia*).

These last two entries are the only ones in the Leicestershire folios that specifically mention animals as opposed to plough-teams. But a similar entry comes from the Derbyshire folios relating to the Leicestershire village of Willesley (274) where 3 villeins had '5 ploughing oxen' (*v boves arantes*), a phrase found rather more frequently in the Derbyshire folios than in those of Leicestershire. For the most part, however, the record of teams is complete and fairly straightforward.

The present analysis has yielded a total of 1,881½ plough-teams for the modern county. The total for the Domesday county given by Sir Frank Stenton is 1,858½,[1] and Maitland's total is 1,817 plough-teams.[2]

(4) *Population*

The main bulk of the population was comprised in the four main categories of villeins, sokemen, bordars and serfs. In addition to these main groups were the burgesses (see p. 350 below), together with a miscellaneous group that included freemen, priests, *francigenae* and others. The details of these groups are summarised on p. 337. There are other estimates of population, by Ellis[3] and Sir Frank Stenton[4] and C. T. Smith.[5] The present estimate is not strictly comparable with any of these because it is in terms of the modern county. In any case, one thing is certain; no one who counts Domesday population can claim definitive accuracy. All that can be said for the present figures is that they indicate the order of magnitude involved. The figures are those of recorded population, and must be multiplied by some factor, say 4 or 5, in order to obtain the actual population; but this does not affect the relative density as between one area and another.[6] That is all that a map, such as Fig. 116, can roughly indicate.

It is impossible for us to say how complete were these Domesday statistics, but it does seem as if some people had been left uncounted. At Swinford (234) we are told that an unspecified number of villeins had half a team (*et villani habent dimidiam carucam*). The entry immediately

---

[1] *V.C.H. Leicestershire*, I, p. 305.   [2] F. W. Maitland, *op. cit.* p. 401.

[3] Sir Henry Ellis, *A General Introduction to Domesday Book* (London, 1833), II, p. 463. This estimate amounted to 6,772, but it included tenants-in-chief, under-tenants, burgesses and *ancillae*, and moreover was in terms of the Domesday county.

[4] *V.C.H. Leicestershire*, I, p. 305. Stenton's total for the Domesday county was 6,400.

[5] *V.C.H. Leicestershire* (London, 1955), III, pp. 129–32 and 156–62. C. T. Smith's total for the Domesday county was 6,406.

[6] But see p. 430 below for the complication of serfs.

Recorded Population of Leicestershire in 1086

A. *Rural Population*

Villeins	2,643
Sokemen	1,904
Bordars	1,362
Serfs	403
Miscellaneous	116
Total	6,428

There were also 23 bondwomen (*ancillae*) not included above.

Details of Miscellaneous Rural Population

Francigenae	33
Priests	42
Knights	14
Men (*Homines*)	10
Francigenae servientes	9
Freemen	6
Clerk (*Clericus*)	1
Deacon	1
Total	116

B. *Urban Population*

Villeins and bordars are also included in the table above.

> LEICESTER 65 burgesses; 318 *domus*; 3 villeins; 12 bordars; 1 priest; the 'men' of Countess Judith; 4 *domus vastae*.

following this speaks of Ulf and his men (*cum suis hominibus*) with a team at Walcote. For Rearsby (234b), the usual agricultural resources are set out, but there is no mention of inhabitants; and there is a similar omission in an entry for Stonton Wyville (236b). It is impossible to be certain about the significance of these omissions, but they do suggest unrecorded inhabitants.

Freemen, as such, were rare, but sokemen accounted for nearly one-third of the total population. In the north-east, they amounted to over 50% of the recorded inhabitants, while over the eastern half of the county generally, they amounted to more than 20%. Villeins and bordars constituted the main bulk of the population; villeins amounting to well over a third, and bordars to about one-fifth. Serfs constituted just over 6% and were recorded regularly in entry after entry; their percentage varied from over 11% in the north-east to less than 2% in the west. Usually, serfs were recorded with the demesne holding, but very occasionally they were linked with the other inhabitants, e.g. at Burbage (231) '20 villeins with 2 bordars and 2 serfs have 8 ploughs'. The total number of *ancillae* was 23; they are always enumerated with the demesne holding, and appear to have been the feminine counterpart of the serfs.

The miscellaneous category is a varied one. Priests were recorded fairly regularly and usually singly; *francigenae* (Frenchmen or foreigners) were recorded less regularly but in larger groups. Priests and Frenchmen seem to have partaken of the agricultural activities of the community as both classes appear to hold ploughs. Thus, at Ashby de la Zouch '8 villeins with a priest and 6 sokemen and 4 bordars have 10 ploughs' (233), and at Barkby '7 villeins with 3 bordars and 10 sokemen and 4 Frenchmen have 10 ploughs' (233b). Unnamed knights, too, appeared infrequently and in the same setting, e.g. at Alton Grange (233). These groups constituted almost all the miscellaneous population except for a deacon (*diaconus*) at Market Bosworth (233), a clerk (*clericus*) at Wigston Magna (232) and 9 French serjeants (*francigenae servientes*) at Wymeswold (232b). Finally, there were simply 'men' (*homines*), viz. at Arnesby (235), Cotesbach (232b), Swinford (234), Theddingworth (235b) and Thorpe Langton (232). No thegns were recorded for 1086, but there were 16 T.R.E.

(5) *Values*

The value of an estate is normally given for two dates, for 1086 and for some earlier date. The familiar formulae are either *Valuit n libras, Modo m libras* or *Valuit et valet n libras*, and in many entries the details descend to pence. At first sight we might naturally suppose that the earlier value refers to conditions in 1066, but, as we shall see below, this may not be so. A number of entries omit any mention of the earlier values, e.g. that for Sysonby (235). On the other hand, there are a few entries that omit the 1086 value. Furthermore, there are also a few entries that omit any mention of value, e.g. that for Ratcliffe on the Wreak (234b). Finally, there are a number of other variations in the form of the value entry,

and these are set out below. It will be noticed that the first two of these
entries provide values for three dates:

(1) The fief of Earl Hugh (237): The whole was and is worth £40. When the
earl received it, it was worth £10 (*Totum Valuit et valet xl libras. Quando
recepit comes valebat x libras*).

(2) Donington le Heath (236): It was worth 20s. Now 2s. [Nigel] received
it waste (*Valuit xx solidos. Modo ii solidos. Vastam recepit*).

(3) Members of Melton Mowbray (235 b): The whole was worth when he
received it £4. 10s. Now £15. 10s. (*Valuit totum quando recepit iiii libras
et x solidos. Modo xv libras et x solidos*).

(4) Husbands Bosworth (236): It was worth 6d. when he received it. Now
20s. (*Valuit vi denarios quando recepit. Modo xx solidos*).

(5) Bottesford (234): The whole was worth £6 when they received it. Now
£16 (*Totum valuit vi libras quando receperunt. Modo xvi libras*).

(6) Burbage (231): It was worth 2s. when the abbey received it. Now £4.
(*Valuit ii solidos quando abbatia recepit. Modo iiii libras*).

In by far the greater number of entries where two values are given, the
earlier value is very low, and it is clear that the subsequent period saw
a great increase. In order to explain this exceptionally low 'valuit',
F. W. Maitland hazarded the idea that it referred not to 1066 but to
'some time of disorder that followed the Conquest'.[1] And Sir Frank
Stenton has suggested that this time might well be that of the Conqueror's
march from Warwick to Nottingham in 1068.[2] The variations in the form
of the value entry, as set out above, have been regarded as supporting
evidence for the idea that the Leicestershire 'valuit' speaks not of 1066
but of, say, 1068. The low intermediate value of Earl Hugh's fief, the
intermediate wasting of Donington le Heath, and the low values of the
other holdings at the time they 'were received', have been taken as
indicative of the wasting in 1068, and, what is more, it has been suggested
that the low 'valuits' of other entries refer also to the same period. On
this assumption, the entries relating to Earl Hugh's fief and to Donington
le Heath are the only two occasions on which the Leicestershire folios
give a value for 1066.[3]

[1] F. W. Maitland, *op. cit.* p. 469.
[2] F. M. Stenton in *V.C.H. Leicestershire*, I, pp. 282–4.
[3] See p. 347 below for the bearing of the waste entries upon the point. For a
discussion of the problem and its relation to that of waste, see also D. Holly, 'The
Domesday Geography of Leicestershire', *Trans. Leics. Arch. Soc.* (Leicester, 1939),
XX, pp. 188–92 and 198–202.

Occasionally, two or more holdings are grouped together for purposes of valuation; thus the value of a holding of 5 carucates at Saxby (233) was included in that of Stapleford (*precium ejus in Stapeford*). The eight members of the manor of Melton Mowbray were also valued together at £15. 10s. (235 b). Or again, on fo. 237, the description of holdings in some 20 vills is followed by a statement that 'the whole was and is worth £40'. There are variations from the usual type of valuation and some examples of these are set out below.

> Rothley Manor (230): This vill is worth 62s. yearly (*Haec villa valet per annum lxii solidos*).
>
> The members of Rothley Manor (230b): they render altogether £31. 8s. 1d. (*reddunt inter omnes xxxi libras et viii solidos et i denarium*).
>
> Great Bowden (230b): the peasants 'render 30s. annually' (*reddunt xxx solidos per annum*); the demesne is worth 40s. a year (*Dominium valet xl solidos per annum*).
>
> The members of Great Bowden (230b); they render 150 shillings and 18d. (*reddunt cl solidos et xviii denarios*).
>
> Birstall (232): It was worth 40s. Now 5 ounces of gold (*Valuit xl solidos. Modo v uncias auri*).
>
> Birstall (232b): It was worth 10s. Now 3 ounces of gold (*Valuit x solidos. Modo iii uncias auri*).
>
> Shepshed (230b): From this land comes £6 as rent by order of the Bishop of Bayeux for the service of the Isle of Wight (*De hac terra exeunt vi librae ad firmam precepto episcopi baiocensis pro servitio insulae de With*).

All these complications make it difficult to compare the values of some manors one with another.

Generally speaking, the greater the number of plough-teams and men on an estate, the greater its value, but it is difficult to discern any constant relationship as the following figures for five holdings, each yielding 40s. in 1086, show:

	Teams	Population	Other resources
Somerby (233b)	3	8	Meadow
Gilmorton (234)	7	29	Nil
Bromkinsthorpe (237)	6	20	Meadow
Welby (235b)	5	16	Meadow
Swepstone (233b)	8	19	Meadow

Both the teams and population of Swepstone were more than double those of Somerby yet their values were the same. Nor does the presence of 'other resources' throw any light upon the anomaly.

Conclusion

The large size and the dispersed nature of the Domesday wapentakes do not make them convenient units for the purpose of calculating densities. Some fourteen more or less artificial units have therefore been adopted. In forming them, however, variations of soil and surface features have been borne in mind as far as possible.

Of the five standard formulae, those relating to plough-teams and population are most likely to reflect something of the distribution of the wealth and prosperity throughout the county in the eleventh century. Taken together, they supplement one another to provide a general picture (Figs. 114 and 116). The essential feature of both maps is the contrast between the land to the east of the Soar valley and that to the west. The density of plough-teams in the east ranges between 2 and 4 per square mile; and that of population between 6 and 12. West of the Soar, the density of teams is only about one while that of population is less than 5. Even among the lower densities of this western region, the Charnwood area with its plough-team density of less than 0·7 per square mile and its population density of 2, stands out as a comparatively empty place. The Soar valley itself shows a wealth and prosperity much nearer to those of the east than to those of the west.

Figs. 115 and 117 are supplementary to the density maps, but it is necessary to make one reservation concerning them. As we have seen on p. 317, it is possible that some Domesday names may have covered two or more settlements. A few of the symbols should therefore appear as two or more smaller symbols, but this limitation does not affect the main pattern of the maps. Generally speaking, they confirm and amplify the information of the density maps.

WOODLAND

Types of entries

The amount of woodland on a holding in Leicestershire was normally recorded by giving its length and breadth in terms of leagues and furlongs; thus at Huncote (231 b), for example, there was 'wood half a league in length and 4 furlongs in breadth'. The woods recorded varied in size from small ones such as that at Market Bosworth (233), which was only one furlong by half a furlong, to the large *Hereswode* entered under Leicester city (230), which was 4 leagues by one league. The wood at Barrow on Soar was valued as well as measured: *Silva i leuua longa et iiii quarentenis lata quae reddit v solidos* (237). The exact significance of these linear measurements is not clear, and we cannot hope to convert them into modern acreages.[1] All we can do is to plot them diagrammatically as on Fig. 118. Composite entries for wood are not important in the Leicestershire folios. Woodland measuring 1½ by 1 furlongs is entered for Hathern and Dishley (237), and has been plotted for the former place; another entry for Dishley alone also records wood (230b).

There are also a number of entries that record wood in different ways. On a number of holdings the wood was measured in acres, and the amounts ranged from 3 acres on each of two holdings at East Norton (235, 236b) to 40 acres on one holding in Keythorpe (230b). In the entry relating to Staunton Harold both types of measurement appear together: 'Wood 5 furlongs in length and 3 furlongs in breadth, and on the other side (*ex altera parte*) 4 acres of wood' (233). The presence of spinney is recorded at two places:

Old Dalby (235): Spinney (*Spinetum*) 2 furlongs in length and one furlong in breadth.
Ashby Folville (236b): Spinney (*Spinetum*) one furlong in length and one in breadth.

Occasionally the word *nemus* is used instead of *silva*—at Keythorpe (230b) and Cold Overton (236). At Noseley (232) there were 20 acres of brushwood (*broce*), while 'pertaining to' Aylestone (237) there were another 2 acres of brushwood the *soc* of which rendered 2*d*. yearly. It is only under Ashby de la Zouch (233) that there is any indication, in the Leicestershire folios, of the use of the woodland—'Woodland one league in length and 4 furlongs in breadth sufficient (*ad*) for 100 swine'. At Lubbesthorpe (235), however, there was 'infertile woodland' (*Silva infructuosa*); it may have been unsuitable for pannage, yet its measurements were given—6 by 3 furlongs. The Derbyshire folios record that the present-day Leicestershire vill of Willesley (272b) had wood fit for pannage (*Silva pastilis*) measuring one furlong by one furlong. No under-

---

[1] See p. 437 below.

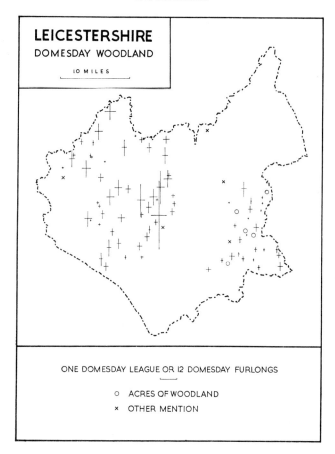

Fig. 118. Leicestershire: Domesday woodland in 1086.

Where the wood of a village is entered partly in linear dimensions and partly in some other way, only the dimensions are shown.

wood is mentioned in the Leicestershire folios, but, again under Derbyshire, the Leicestershire vill of Measham (272b) had underwood (*Silva minuta*) one furlong by one furlong.

There are three entries which depart from the usual brief statement about woodland. At Knossington (234b) there was 'wood two furlongs in length and one furlong in breadth. The fourth part of this wood belongs to a certain sokeman of the king.' Another entry under Rothley (230) indicates the division of the wood between the demesne and the villeins:

'Demesne wood (*Silva dominica*) one league in length and half a league in breadth. Wood of the villeins (*Silva villanorum*) 4 furlongs in length and 3 furlongs in breadth.' At Anstey (232) there was wood 1 league by ½ league, and 'other wood' (*alia silva*) 2 furlongs by 1 furlong.

Distribution of woodland

The woodland was disposed in two main areas—in the western half of the county and in the south-east. Elsewhere, apart from a few isolated stretches, the county seems to have been devoid of wood (Fig. 118). In the western half of the county but little wood was recorded for the Charnwood Forest area, although it was surrounded on all sides, especially to the north-west and south, by well-wooded tracts. Some of this wood may have actually been located in Charnwood itself, but we cannot be sure. The large wood recorded under Leicester city, and associated in the Survey with 'the whole sheriffdom called *Hereswode*', appears to refer to what was later to become Leicester Forest.[1] The Leicestershire folios themselves make no reference to forest anywhere in the county. In the east of the county, the high land towards the border of Rutland was the only wooded area, but the woods recorded here were much smaller than those in the west, and acres were intermingled with linear dimensions.

Comparison of Figs. 111 and 118 shows that much of this Domesday wood lay on clay soils, but it is interesting to note that two areas of fairly heavy clay soil were devoid of wood—the Vale of Belvoir and the Welland valley. The wood of these areas had presumably been cleared by the eleventh century; the rich clays had proved agriculturally attractive despite their original cover of wood.[2]

MEADOW

Types of entries

The meadow entries are comparatively straightforward. For holding after holding the same phrase is repeated monotonously—'*n* acres of meadow' (*n acrae prati*). The amount of meadow in each vill varied from one acre, e.g. at Halstead (230b), to over 100 acres. The largest total for a single vill is the 130 acres entered for Stapleford (233), but the com-

---

[1] See L. Fox and P. Russell, *Leicester Forest* (Leicester, 1948), pp. 20–1.

[2] The clearing of the Leicestershire woodland is discussed by A. R. Horwood and the Earl of Gainsborough in *The Flora of Leicestershire and Rutland* (Oxford, 1935), p. 267.

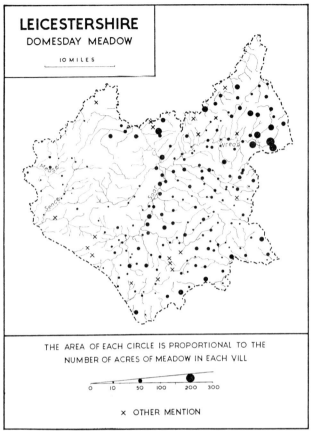

Fig. 119. Leicestershire: Domesday meadow in 1086.
Where the meadow of a village is entered partly in acres and partly in some
other way, only the acres are shown.

bined entry for Edmondthorpe and Wymondham (233) recorded 300 acres of meadow for these two vills. These combined entries are few in Leicestershire and in the entries for large manors with dependencies, e.g. Rothley (230) and Great Bowden (230b), the meadow for each member of the manor is entered separately.[1] As in the case of other counties, no attempt has been made to translate these figures into modern acreages. The Domesday acres have been treated merely as conventional units of measurement, and Fig. 119 has been plotted on that assumption.

[1] See pp. 315–16 above. The entry for Great Bowden itself omits the number of acres, and reads: *Ibi prati acrae.*

While measurement in acres is normal there are twenty-two entries in which linear measurements are given; thus, at Arnesby (235) there was meadow '4 furlongs in length and the same in breadth'. Occasionally, two holdings of meadow in the same vill are recorded differently, one in acres and the other in lengths. Thus at Dalby, for example, there were 16 acres of meadow (234b and 236) and also meadow (231) 6 furlongs in length and in breadth (*in longitudine et in latitudine*).

Distribution of meadowland

The contrast between the east and the west of the county is apparent on Fig. 119. In the east of the county the meadow is widespread. Concentrations are apparent along the valleys of the Soar, the upper Wreak and the Welland. The valleys of the smaller streams, too, are marked by substantial amounts, especially the valleys of those streams draining the higher land of the north-east. Along the south-east border of the county, on the other hand, amounts are consistently small. It was in the southeastern area, as we have already seen, that wood was comparatively plentiful. West of the Soar, meadow appears much less frequently; the Charnwood area is almost devoid of meadow, but outside Charnwood there are small concentrations in the valleys of the left-bank tributaries of the Upper Soar and in the valleys of the Anker and Mease.

WASTE

Eight villages in Leicestershire were wholly waste in 1086, and another 17 were partly waste. Prior to 1086 there were 3 villages wholly waste and 5 partly waste. Only for Ravenstone (235) was waste recorded for two dates—*Vasta fuit et est*. Another holding at Ravenstone was waste in 1086 (278), so that the village as a whole was wholly waste in 1086, and partly waste at some earlier date. The entries recording waste holdings are of three kinds, and here are some examples of them:

(*a*) Land waste in 1086:
    Fleckney (234b): *Vasta est et tamen valet xii denarios.*
    Oakthorpe (278): *Vasta est. T.R.E. valuit v solidos. Modo iiii denarios.*
    Slawston (234): *i virgata terrae vastae. Valuit iiii denarios et valet.*
    Withcote (235): *Haec vasta est. Ibi ii acrae prati. Silva i quarentena et dimidia longa et una quarentena lata. Valet xii denarios.*

(b) Land waste at some earlier date:

Weston (235): *Valet lxx solidos. Vasta fuit.*

(c) Land found or received waste (two entries only):

Shepshed (230b): *Hanc terram vastam invenit* (No T.R.E. value; 76 people with 38 teams T.R.W.).

Donnington le Heath (236): *Valuit xx solidos. Modo ii solidos. Vastam recepit.*

There were also four waste houses in the borough of Leicester itself (230).

That some vills were only partly waste, we may judge from two types of evidence. In the first place, there were those vills comprising two or more holdings of which at least one was waste. Thus of three holdings at Wyfordby, one was waste (233), and two were tilled (234b, 235b). In the second place, some single holdings were, apparently, themselves partly waste. Willesley (274) alone in Leicestershire (but described in the Derbyshire folios) seems to fall into this category. It is said to be waste, yet had men who tilled within it: *Terra i caruca. Wasta est. Ibi iii villani habent v boves arantes. T.R.E. valuit xx solidos. Modo xvi solidos.* There was another holding (272b) at Willesley which was tilled, so that, in any case, the village appears as 'partly waste' on Fig. 120.

That a holding was 'waste' did not necessarily imply that it was entirely without profit. The entries for Fleckney, Oakthorpe and Slawston, set out above, include small renders of money; that for Withcote also mentions the presence of meadow and wood. Such holdings have been regarded not as partly waste, but as entirely waste in the sense that all their arable seems to have been devastated and that they had no population. They do not necessarily appear as 'wholly waste' on Fig. 120 because some vills had fully stocked holdings in addition to their devastated land.

It seems probable that the areas waste in 1086 can be correlated with post-Conquest disturbances, and in particular, with 'the Conqueror's march from Warwick to Nottingham when he suppressed the first revolt of Edwin and Morcar towards the close of 1068'.[1] Whether the wasting of the few areas recorded as waste prior to 1086 was also due to the same cause is open to doubt.[2]

---

[1] *V.C.H. Leicestershire*, I, p. 284.

[2] For a fuller discussion of the problem of the Leicestershire waste, see (1) F. M. Stenton in *V.C.H. Leicestershire*, I, pp. 282–4; (2) D. Holly, 'The Domesday Geography of Leicestershire', *Trans. Leics. Arch. Soc.* (Leicester, 1939), XX, pp. 188–92 and 198–202. See also p. 339 above.

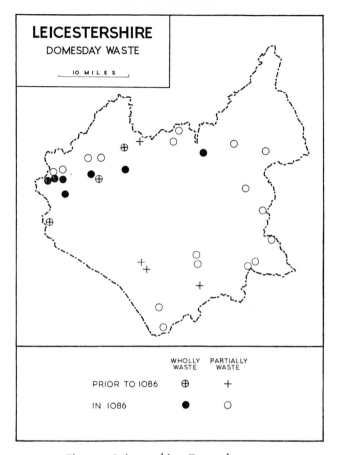

Fig. 120. Leicestershire: Domesday waste.
Ravenstone was waste both in 1086 and prior to 1086. *Windesers* (unidentified)
was waste in 1086 and does not appear on the map.

MILLS

Mills are mentioned in connection with 89 of the 296 Domesday settlements within the area covered by modern Leicestershire.[1] It is difficult to
be sure of the exact total because mills appear in two composite entries.
One mill is entered for Pickwell and Leesthorpe (235b) and two for

[1] The figure of 89 results from counting as one group the mills of a composite
entry. If it is assumed that the two mills entered for Chadwell and Wycomb were at
separate places, the total of places with mills is 90. If both Pickwell and Leesthorpe
are counted, the total becomes 91.

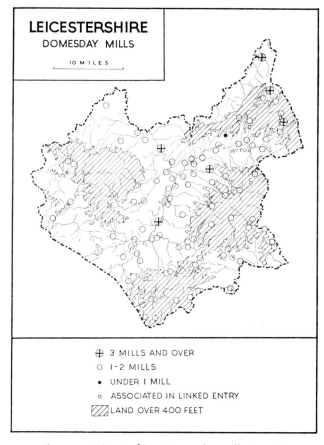

Fig. 121. Leicestershire: Domesday mills in 1086.

Chadwell and Wycomb (230b). The mills have been plotted for the first named place in each entry. The two associated places have also been indicated on Fig. 121.

In each entry the number of mills is given, and also their annual value, ranging from one mill worth only 4*d.* at Pickwell and Leesthorpe (235b) to others worth much larger sums; two mills at Sileby (233) rendered 30*s.* and one at Bruntingthorpe rendered 20*s.* (232). Only for Congerstone (233) is a mill entered without a value. Fractions of mills are occasionally recorded. In Leicester city, on the holding of the Bishop of Lincoln, 1½ mills rendered 10*s.* 8*d.* (230b), and on the holding of Countess Judith, half a mill rendered 5s. 4d. (230). It seems as if there were 2 mills there rendering a total of 16s. It is not always possible, however, to assemble the fractions in such a comprehensible manner. There was part of a mill at Frisby on the Wreake (230b), which rendered 2s., but we are given no clue to the remainder of the mill.

Domesday Mills in Leicestershire in 1086

Under 1 mill	1 settlement	3 mills	3 settlements
1 mill	59 settlements	4 mills	1 settlement
2 mills	23 settlements	7 mills	2 settlements

One of the groups of 7 mills was at Knipton, in the north-east of the county, and it was worth 18s. 4d. We can also assume there were 7 mills at Bottesford (although only 6½ are mentioned) again in the north-east but in the Vale of Belvoir; this group was worth 45s. 6d. Leicester itself is poorly represented by the 2 mills mentioned above.

Fig. 121 shows that the majority of the mills in the county were situated in the central lowlands—along the Wreak, the Soar and their associated streams. The eastern half of the county generally was better served by mills than the western half. In the west, there were a few mills around Charnwood Forest, but the small number reflected the poverty of the area.

URBAN LIFE

The Domesday description of the county begins with an account of the city (*civitas*) of Leicester (230) which is the only place in the county for which burgesses are mentioned; there is also a subsidiary entry on fo. 230b which describes the Bishop of Lincoln's holding in the city. We are told at once in the main entry that in King Edward's time the city had rendered '£30 by tale of 20 to the ounce and 15 sesters of honey'. If the King went with his army by land, the city sent 12 burgesses with him; if he went by sea, the city sent 4 horses to London to carry weapons 'or other things of which there might be need'. By 1086 the rent 'of the same city and shire' had risen to £42. 10s. For a hawk, the city rendered the large amount of £10 and, for a 'sumpter horse', 20s.

One interesting feature that Leicestershire shares with some other counties is that many of the burgesses (*burgenses*) and houses (*domus*) of Leicester in 1086 are recorded as belonging to rural manors in the villages

around. Some of these burgesses are entered under Arnesby (235) and
Poultney (231). No houses are entered under rural manors, but in the
account of Leicester itself, group after group of houses, and also some
burgesses, are said to belong to a variety of manors. The full list of
47 burgesses and 97 houses is as follows:

Entered under rural manors

Arnesby	1 burgess
Poultney	9 burgesses
Total	10 burgesses

Entered under Leicester

Anstey	24 burgesses
Sileby	13 burgesses
Total	37 burgesses

	houses		houses
Tur Langton	2	Bruntingthorpe	1
Barrow	10	Desford	2
Kegworth	6	*Legham*	3
Loughborough	1	Thurlaston	1
Ingarsby	3	Thurcaston	1
Belgrave	10	Newton Harcourt	6
Broughton Astley	4	Kibworth Harcourt	3
Stoughton	9	Dalby	1
Wigston Magna	4	Pickwell	1
Enderby	7	Shepshed	4
Earl Shilton	3	Saddington	1
Birstall	10	Thorpe Acre	1
Burton Overy	3		—
		Total	97 houses

The full extent of these contributory manors is shown on Fig. 122.
Eighteen other burgesses are recorded in the city, of whom seventeen
rendered '32*d*. yearly'. The total number of burgesses thus becomes 65.
The only other residents mentioned within the borough are the 'moneyers',
who rendered '£20 yearly of 20 to the ounce'. In contrast to this meagre

record of inhabitants, buildings are enumerated in some detail. Including the 97 houses recorded above, there is a total of 318 houses, 4 waste houses and 6 churches. The 318 houses, together with the 65 burgesses, implies a population of at least 1,500 to 2,000 in 1086.

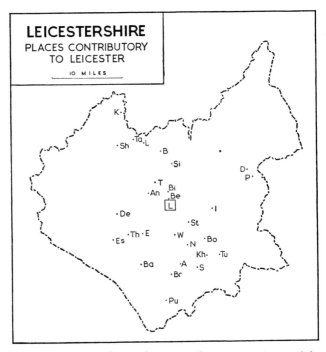

Fig. 122. Leicestershire: Places contributory to Leicester (L).

A, Arnesby; An, Anstey; B, Barrow; Ba, Broughton Astley; Be, Belgrave; Bi, Birstall; Bo, Burton Overy; Br, Bruntingthorpe; D, Dalby; De, Desford; E, Enderby; Es, Earl Shilton; I, Ingarsby; K, Kegworth; Kh, Kibworth Harcourt; L, Loughborough; N, Newton Harcourt; P, Pickwell; Pu, Poultney; S, Saddington; Sh, Shepshed; Si, Sileby; St, Stoughton; T, Thurcaston; Ta, Thorpe Acre; Th, Thurlaston; Tu, Tur Langton; W, Wigston Magna. Also the unidentified *Legham*.

Two entries indicate the agricultural element in the city. The Bishop of Lincoln (230b) held 10 carucates of land in the town. In demesne were 5 teams and 1½ mills; 'outside the wall' were 20 acres of meadow, and 3 villeins, a priest and 12 bordars had 4 ploughs. The Countess Judith (230) held half a mill in the town; 'without the borough' she had 6 carucates of land belonging to the borough, and on it she had one team and an unspecified number of men (*homines*) with 3 teams. Included with her holding, too, were 7 acres of meadow, and woodland measuring 6 furlongs by 3 furlongs.

Thus, it seems that Domesday Leicester was a flourishing city, with churches, with 'moneyers' who minted coins, with a population of perhaps 2,000 or more, and with a small agricultural element.

MISCELLANEOUS INFORMATION

But little miscellaneous information is given in the Leicestershire folios. Under Melton Mowbray (235 b) there is the tantalisingly brief reference to a market (*Mercatum*) which rendered 20s. Beehives are mentioned nowhere, yet Leicester city rendered 15 sesters of honey in 1066. No churches are recorded apart from those in Leicester city itself.

REGIONAL SUMMARY

Essentially, Leicestershire consists of eastern and western portions divided by the Soar valley. The west consists of undulating country dominated by the pre-Cambrian outcrops of Charnwood Forest. In the east, the Wreak valley divides the Wolds in the north from the rather plateau-like high land which occupies the greater part of the east and south. Much of the county, except for the higher land, is covered with boulder clay and its associated sands and gravel. The lowland strips of the Soar and the Wreak valleys are covered with alluvium and occasional patches of river gravel.

This contrast between east and west is reflected in the Domesday distributions. West of the Soar was a land of few plough-teams, sparse population, little meadow but of much woodland. The east, on the other hand, including the Soar valley, had a relatively denser population, more ploughs, meadow and mills but little woodland. Within this basic contrast there are sufficient variations, especially in the east, to warrant further subdivision, and the county as a whole may be divided into nine regions (Fig. 123).

(1) *The Vale of Belvoir*

The flat claylands of the Vale of Belvoir lie almost entirely below 200 ft. above sea-level and are drained by streams flowing away north to join

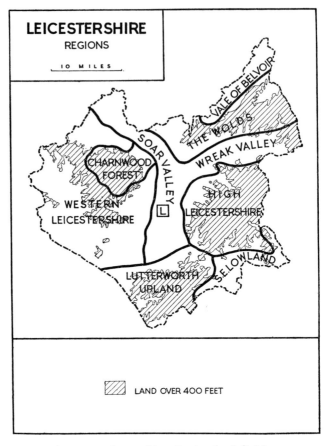

Fig. 123. Leicestershire: Regional subdivisions.
L indicates the Domesday borough of Leicester.

the river Trent. This was one of the most highly developed regions of
Domesday Leicestershire. The number of villages was small, but the
resources of each village were comparatively great. It is probable that the
large entries relating to these villages covered the resources of the missing
vills of Normanton, Muston and Plungar. The population of over 10 to
the square mile was one of the highest in the county; the corresponding
figure of 3·1 for plough-teams was also high for the county. Wood was
not recorded and presumably had been cleared by 1086. Water supply
and high fertility seem to have more than compensated for the heaviness

of the clay soil. The lack of meadow in the north-east and the absence of mills, except for the high number recorded for Bottesford, are difficult to explain.

(2) *The South-east Lowland*

The boundary of the south-eastern tip of Leicestershire is formed by the river Welland and its tributary, Eye Brook; and their valleys form a distinct unit to the south of 'High Leicestershire'. The region lies mainly between 200 and 400 ft. above sea-level. Far from the Vale of Belvoir, this region was like it in many ways. It is a district of heavy clay soils and was, again, one of the most highly developed parts of the Domesday county. The villages were not thickly distributed, but the density of population was nearly as high as 11 and that of plough-teams nearly as high as 4·0 to the square mile. Only a little wood was recorded; much had presumably been cleared. Meadow was well distributed, but mills were surprisingly few in number.

(3) *The Wolds*

This ridge of high land, mostly over 400 ft. above sea-level, runs from west to east and continues over the county boundary into Lincolnshire. The soils are much lighter than those of the neighbouring Vale of Belvoir. The widespread settlement, the population density of between 7 and about 11 per square mile, the plough-team density of about 3, together with the widespread meadow and lack of woodland, suggest a moderately high agricultural development.

(4) *High Leicestershire*

From the centre of this upland, which is over 600 ft. above sea-level, streams drain away in every direction. Boulder Clay, freely interspersed with sand and gravel patches, covers much of the region. In spite of its height, its agricultural development was of a high order; this was especially true of the western edge where the region sloped down to the Soar valley. The distribution of settlements was uniform and fairly dense; the population density was just over 9, and the plough-team density about 2·5 to 3·0 per square mile. There was a fair amount of wood in the eastern part of the region; most villages had some meadow; and about a half had one or two mills apiece.

(5) *The Lutterworth Upland*

The surface of this triangular-shaped area is almost all over 400 ft. above
sea-level. It is drained by the river Swift (a tributary of the Warwickshire
Avon), and by various tributaries of the upper Soar. It was well settled
with about 9 people and over 2 teams per square mile. No wood was
entered for its villages, but most had small amounts of meadow and some
had mills.

(6) *The Wreak Valley*

The river Wreak flows west to join the Soar below Leicester. The upper
valley, centring on Melton Mowbray, is wide, and varies in height between
200 and 400 ft. above sea-level; west of Frisby it falls below 200 ft.
Probably well settled by the Scandinavians, the valley formed a well-
developed agricultural region by the eleventh century. The fairly large
settlements, which followed the line of the river, were the key to the
prosperity of the valley. The population density of nearly 9 and the
plough-team density of nearly 3 per square mile indicate the resources of
the riverside settlements averaged out over the surrounding lowlands.
Devoid of wood, the valley was rich in meadow, especially in its upper
reaches, while mills were plentiful, especially towards the junction of the
valley with that of the Soar.

(7) *The Soar Valley*

This valley runs roughly from south to north. Above Leicester, the
Soar and its tributaries drain a wide area; below Leicester, the valley
itself is narrower but the slopes on either side form an integral part of
the valley as an economic unit. The whole valley forms a line of alluvium,
studded with gravel patches and cutting through the Boulder Clay which
lies on either side.

The middle and upper Soar was one of the richest parts of the county.
Settlements were numerous and were usually situated on the patches of
gravel in the valley. The density of population was high and varied
between about 9 and 15 per square mile. Rich arable, probably on the
older alluvium above flood-water, accounted for the average of the plough-
teams varying between 2 and 4 per square mile. The valley was also rich
in meadow, which was probably to be found on the alluvium still liable

to flooding. The region was well supplied with mills but woodland was absent except in the west.

The development of the middle and upper Soar contrasted strongly with that of the rest of the valley. In the lower tract, settlements were few and both meadow and mills were scarce. The population density was as low as 6 and that of plough-teams only about 2 per square mile. The few settlements are partly to be accounted for by Domesday omissions while tenurial connections, necessitating the inclusion of the poorer Charnwood slopes, inevitably lower the average prosperity. It is probable, too, that settlement and development generally was retarded by liability to flooding.

(8) *Charnwood Forest*

All of Charnwood Forest lies over 400 ft. above sea-level and some of it rises to over 900 ft. The core of high land consists of outcrops of pre-Cambrian rocks, rising from beneath a cover of Triassic marls. The whole area is agriculturally unattractive although the valleys of the streams, usually lying on the marl, are more inviting than the rugged high land. The influence of physical features and soils are reflected in the fact that this was the only really negative area in the county. There were a few settlements on the margin, chiefly in the west, but otherwise the map is blank. The population density was 2 per square mile and that of plough-teams was 0·7—by far the lowest in the county. Some wood was entered for the area, but scarcely any meadow and no mills.

(9) *Western Leicestershire*

The land surrounding Charwood Forest is an undulating region, varying in height from 200 ft. to over 400 ft. above sea-level. The streams radiating from Charnwood flow across this land to the larger rivers outside the county boundary. The soils, derived from Keuper Marl or heavy Boulder Clay, vary considerably, but are nowhere of high fertility. The area is a large one but neither the physical features and soil, nor the degree of agricultural development, warrant further subdivision.

The resources of this area in Domesday times were fairly uniform and of a low order. Most of the area, except the far west, was heavily wooded; this, together with its infertility, goes far to explain its poverty. The settlements were numerous, although more sporadically distributed than in the east of the county. But the communities were small and with little of the meadow so essential in eleventh-century economy. The population density of below 5 indicates the poverty of the region, while the average of about one plough-team per square mile further emphasises the fact. The general impression is relieved by the presence of slightly richer communities in the valleys of the Sence and Mease to the west, and in the valleys of the tributaries of the upper Soar to the south.

BIBLIOGRAPHICAL NOTE

(1) It is interesting to note that the first volume of John Nichol's *History and Antiquities of the County of Leicester* (London, 1795) included the text of the Leicestershire folios, together with an English translation. In the next century appeared *A literal extension of the Latin text, and an English translation of Domesday Book in relation to the county of Leicestershire* (Leicester, 1864); it was made 'to accompany the facsimile copy' and its editor is not named.

The standard translation is that made by F. M. Stenton in *V.C.H. Leicestershire* (London, 1907), I, pp. 306–38; this is accompanied by a valuable introduction (pp. 277–305) also by F. M. Stenton.

(2) The following is an early article on the Domesday geography of the county:

D. HOLLY, 'The Domesday Geography of Leicestershire' *Trans. Leics. Arch. Soc.* (Leicester, 1939), XX, pp. 168–202.

(3) Various relevant aspects are discussed in the following papers:

W. G. HOSKINS, 'The Anglian and Scandinavian Settlement of Leicestershire', *Trans. Leics. Arch. Soc.* (Leicester, 1935), XVIII, pp. 110–47.

W. G. HOSKINS, 'The Deserted Villages of Leicestershire', *Trans. Leics. Arch. Soc.* (Leicester, 1945), XXII, pp. 242–64.

W. G. HOSKINS, *Essays in Leicestershire History* (Liverpool, 1950), pp. 67–107.

R. H. HILTON, 'The Domesday Survey', *V.C.H. Leicestershire* (London, 1954), II, pp. 148–55.

C. T. SMITH, 'Population: The Domesday Returns', *V.C.H. Leicestershire* (London, 1955), III, pp. 129–32 and pp. 156–62.

J. S. MOORE, 'The Domesday Teamland in Leicestershire', *Eng. Hist. Rev.* (1963), LXXVIII, pp. 696–703.

(4) The Leicestershire Survey was first printed and discussed by J. H. Round in *Feudal England* (London, 1895), pp. 196–214. In the following year it was translated with a photographic copy by W. K. Boyd, 'Survey of Leicestershire' in *Trans. Leics. Archit. and Archaeol. Soc.* (Leicester, 1896), VIII, pp. 179–83. It was again translated and discussed by F. M. Stenton in *V.C.H. Leicestershire* (London, 1907), I, pp. 339–54. A later edition, incorporating new material, is C. F. Slade's *The Leicestershire Survey* (Leicester, 1956).

CHAPTER VIII

RUTLAND

BY I. B. TERRETT, B.A., PH.D.

It is clear from Domesday Book that the process by which the modern county of Rutland has evolved was not complete in 1086. The present county unit did not appear until early in the thirteenth century, but in 1086 part of the county, the *Roteland* of Domesday Book, was already distinct from the surrounding shires. This *Roteland* comprised an even

Fig. 124. Rutland: Domesday wapentakes.

smaller area than the modern county, including as it did, only the wapen-takes of Alstoe (*Alfnodestou*) and Martinsley (*Martinesleie*), which are described on fo. 293b (Fig. 124). These two Domesday wapentakes came later to be represented by the three hundreds of Oakham Soke, Alstoe and Martinsley. The remainder of the county comprised the hundred of Witchley (*Wiceslea*) which was an integral part of Domesday Northamp-tonshire, and which is accordingly described in the folios for the latter county (219–29). This Domesday Witchley came later to be represented by the hundreds of East and Wrangdike.

But while the southern part of Rutland was included in the adjacent county of Northampton it is also clear from the Domesday record that *Roteland* itself was intimately connected with the shires of both Lincoln and Nottingham. Thus we are told that half of Alstoe wapentake was 'in Thurgarton wapentake', and half 'in Broxtow wapentake', both in Nottinghamshire; moreover both the *Roteland* wapentakes belonged 'to the sheriffdom of Nottingham for purposes of the king's geld' (293 b).

Eight of the twelve Domesday settlements of Alstoe wapentake are described both under *Roteland* (293 b) and again under Lincolnshire, and an interesting series of duplicate entries has resulted, the study of which is illuminating. The chief points of divergence in these parallel entries are tabulated below:

	Rutland entry	Lincolnshire entry
Whissendine	fo. 293 b. No meadow 2 pre-Conquest manors 6 bordars. Value: T.R.E. £8; T.R.W. £13	fo. 367. Meadow 10 f. by 8 f. 1 pre-Conquest manor 7 bordars. No values entered
Exton	fos. 293 b and 367, entries are identical	
Thistleton[1]	fos. 293 b and 367, entries are identical	
Thistleton[1]	fo. 293 b. Value: T.R.E. 20s.; T.R.W. 60s.	fo. 358 b. Valued with South Witham
Market Overton and Stretton	fo. 293 b. Stretton described as a berewick. Wood *per loca*. Value: T.R.W. £20	fo. 366 b. Stretton not stated to be a berewick. No *per loca* added. Value: T.R.W. £40
Burley	fo. 293 b. 8 bordars. Wood *per loca*	fo. 355 b. 7 bordars. No *per loca* added
Ashwell	fo. 293 b. 3 bordars	fo. 349 b. 2 bordars
Whitwell	fo. 293 b. Value: T.R.E. 40s. (? altered from 20s.) Wood *per loca*	fo. 367. No T.R.E. value entered. No *per loca* added

The discrepancies are of no great magnitude, but they serve to show, once more, that Domesday Book is not infallible, and that a margin of error should be allowed for in the handling of the Domesday statistics. Furthermore, the Lincolnshire account of the borough of Stamford (336b) mentions holdings in the borough pertaining, some to Hambleton in

---

[1] There are also three other entries for Thistleton that appear only in the Lincolnshire folios (358 b, 366, 367). For *alia* Thistleton, see p. 363.

Rutland, others merely to Rutland. And, in addition to these complications, there were also other arrangements that lay athwart the county boundary; the Lincolnshire folios mention land in Uffington tilled with the teams of Belmesthorpe in Rutland (366b, 376b).

Of all the Mercian shires Rutland alone is not named after its county town, and moreover, within the compass of its small area both hundreds and wapentakes, and hides and carucates are encountered. These peculiarities are further evidence of the anomalous position of eleventh-century *Roteland*. Alstoe and Martinsley are described in Domesday Book as wapentakes,[1] and are assessed in carucates, but the Northamptonshire district of Witchley although sometimes termed a wapentake is three times referred to as a hundred, and is assessed in hides. Witchley hundred was the only part of midland England north of the Welland which was assessed in hides. Sir Frank Stenton concluded that 'we can hardly hesitate to believe that in Rutland as it exists today we have a fragment of eleventh-century Northamptonshire, detached from its parent county by causes which were in operation between the close of the Confessor's reign and the date of Domesday Book'.[2] It was not therefore without due cause that Canon Taylor referred to Rutland as 'that anomalous little shire'.[3]

Four composite entries occur for Rutland, but, fortunately, the vills in each group are adjacent. For the purpose of constructing distribution maps the details of these four entries have been divided equally among the villages concerned. Two of these entries include linear measurements for wood which cannot be so divided, and which has therefore been plotted for the first-named village in each group. Mills have been treated likewise.[4] The difficulties inherent in the Domesday description of the western wapentake of Martinsley, however, are more serious. Only three places are mentioned by name—Hambleton, Oakham and Ridlington—and each of them was a large royal manor embracing a number of other unspecified vills.

One final peculiarity may be mentioned. At Thorpe (228b) we are told that 'there is one sokeman whose stock is noted above' (*pecunia supra notata est*). This use of the word stock or cattle probably crept

[1] Alstoe wapentake contained two 'hundreds', and Martinsley one 'hundred'. Each of these small 'hundreds' contained 12 carucates, as in Lincolnshire.

[2] F. M. Stenton in *V.C.H. Rutland* (London, 1908), I, p. 135.

[3] Isaac Taylor in *Domesday Studies*, ed. by P. E. Dove (London, 1888), I, p. 48.

[4] See pp. 377 and 380 below.

into the Domesday record as a scribal error when the original returns were systematised and abbreviated. It suggests that stock returns were made in the original but were eliminated in the process of summarising the information to form Domesday Book as we know it.

No mention is made of fisheries, waste, salt-pans, markets or boroughs. In fact, Rutland is the only county for which no borough is mentioned in the Domesday Book.

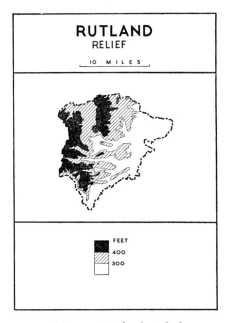

Fig. 125. Rutland: Relief.

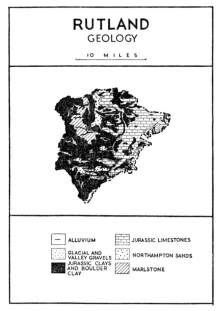

Fig. 126. Rutland: Surface geology. Based on Geological Survey One-Inch Sheet (Old Series) 64.

SETTLEMENTS AND THEIR DISTRIBUTION

The total number of Domesday settlements specifically named in modern Rutland seems to be 39. This figure, however, may not be quite accurate because there are some instances of two adjoining villages bearing the same surname today, and it is not clearly stated whether both units existed in the eleventh century. Thus the Domesday *Castretone* is now represented by Little Casterton and Great Casterton, and *Lufenham* is represented by North Luffenham and South Luffenham. As the Domesday

Book makes no distinction, each of the above place-names has been counted as one settlement. It would seem that the first mention of two settlements in each of these cases did not occur until the thirteenth century.[1] Only where there is some evidence of a second village has it been included in the total of 39. Thus folio 358b for Lincolnshire describes Thistleton and *alia* Thistleton in Rutland, implying that there may have been two distinct settlements. Curiously enough, no mention is made of *alia* Thistleton in the duplicate entries which occur in the Rutland folio 293b, nor is there any evidence of two settlements in later times.[2]

Not all the 39 Domesday names appear on the present-day parish map of Rutland. The Domesday *Alestanestorp* (Awsthorpe) appears on Speed's map of 1610 as a settlement situated midway between Cottesmore and Burley (Alstoe), but no trace of it is to be seen on the modern map. Similarly, *Smelistone* (Snelston), which is also marked on Speed's map, was formerly a village about 1½ miles south-east of Stoke Dry, 'no traces of which at present remain'.[3] *Sculetorp*, in the parish of North Luffenham, has also left no trace on the modern map. The Domesday *Belmestorp* (Belmesthorpe), however, still survives as a hamlet in the parish of Ryhall, and *Toltorp* (Tolethorpe) as the name of a country house in the parish of Casterton.

From a comparison of Fig. 127 with the present-day Ordnance Survey map, it is evident that the distribution of Domesday names in the wapentakes of Alstoe and Witchley is remarkably similar to that of the modern villages. Every modern parish name in Witchley has its counterpart in Domesday Book, and in Alstoe only three parish names are not named—Barrow, Clipsham and Pickworth. Each of the three can be traced back only to the first decade of the thirteenth century.[4] But the picture is very different in Martinsley wapentake, where no less than 19 holdings were grouped as unnamed berewicks round the large composite manors of Hambleton (7), Oakham (5) and Ridlington (7). The information for these nineteen unspecified berewicks cannot be separated from their parent vills, and this affects all the distribution maps. Unusually large symbols for plough-teams, population, wood and meadow appear on the maps for the three named places of the wapentake, and it is almost certain that these

---

[1] E. Ekwall, *Oxford Dictionary of English Place-Names* (3rd ed., Oxford, 1947), pp. 85 and 292.
[2] E. Ekwall, *ibid.* p. 444.   [3] *V.C.H. Rutland*, I, p. 141 n.
[4] E. Ekwall, *ibid.* pp. 27, 107, 348.

large totals would have appeared as many smaller units had we been given more detailed information. Of a total of twenty-three modern parish names in Martinsley, these three are the only ones which appear in the Domesday Book. Of the twenty parish names which are not mentioned, Uppingham, Wardley and Wing are known to have existed before the date of the Survey,[1] and all the others in either the twelfth or thirteenth centuries. It would therefore seem highly probable that many, if not most, of the missing names are represented by the nineteen unnamed Domesday

Fig. 127. Rutland: Domesday place-names.
The figures refer to the numbers of unnamed berewicks belonging to Hambleton, Oakham and Ridlington. The boundary of Martinsley wapentake is marked.

berewicks. The latter, however, cannot be shown on Fig. 127 as we do not know their precise location; so far as Martinsley wapentake is concerned the map thus gives us a very incomplete picture. There can be no doubt that the peculiar character of Martinsley wapentake is due to the fact that the king in 1086 held the whole area in demesne; but large discrete manors of this kind are a rarity in the Danelaw.

A further difficulty, again in Martinsley wapentake, arises in connection with the curious entry that comes at the end of the account of *Roteland* (294):

> In the above land [i.e. Martinsley wapentake] Albert the clerk has 1 bovate of land and has there one mill [of] 16*d*. The same Albert also has of the king the

[1] E. Ekwall, *op. cit.* pp. 464, 474, 499.

church of Oakham and of Hambleton and of St Peter of Stamford as much as belongs to the said churches in Hambleton, with the neighbouring lands; this is 7 bovates. In this his land there can be 8 ploughs, and nevertheless 16 teams plough there (*In ipsa eadem ejus terra possunt esse viii carucae, et tamen ibi arant xvi carucae*). He has there 4 ploughs in demesne and 18 villeins and 6 bordars having 5 ploughs.[1] In King Edward's time it was worth £8, now £10.

There is no clue here as to where precisely the various elements of this unique entry were located, and the following convention has therefore been adopted for purposes of mapping; the plough-teams and population are shown by means of open circles placed approximately in the centre of the wapentake, but the mill has not been located.[2]

Apart from the nineteen berewicks mentioned above, only one other is specifically recorded in the county. At Market Overton in Alstoe wapentake (293b), Stretton is described as a berewick, the letters *Ber'* being interlined above the place-name. For Ryhall, in the extreme north of Witchley hundred, we read that there was 'with its appendages' land for 8 ploughs; whether these 'appendages' were berewicks or not is not clear (228).

Although no generalisations can be made about Martinsley wapentake, it is apparent that in the rest of the county the distribution of settlements was widespread but by no means haphazard. In the hill and vale country of south Rutland, the Upper Lias clays of the river valleys were almost entirely avoided, and the villages were located on the lighter and more porous soils of the Northampton Sands, and on the Jurassic limestones which occupy the inter-fluvial ridges (Figs. 125 and 126). Not a single Domesday settlement was to be found close to the upper courses of the rivers Chater and Gwash; but on the higher intervening outliers of the Northampton Sands were Ridlington, Stoke Dry, Glaston and Hambleton. The main outcrop of the Northampton Sands appears in the north near Market Overton, extending southwards close to Burley, Empingham, Luffenham and Barrowden, all of which were Domesday settlements. This outcrop, though narrow, was and is, an important source of water supply, and springs are located at the junction of the Northampton Sands with the impervious Upper Lias clays. Similar spring-lines encircle the Northampton Sands outliers in the south of the county.

[1] The figures are puzzling, but, in our total count, we have added the 4 + 5 to the 16.   [2] See p. 381 below.

While the Upper Lias clays seem to have been distinctly repellent to early settlement, the clays, sandstone and marlstone of the Middle Lias seem to have been more attractive. Oakham, Ashwell, Teigh and Whissendine are all located on this formation. The Rutland villages of the Welland valley were located at some distance from the river itself, doubtless to avoid the dangers of flooding.

THE DISTRIBUTION OF PROSPERITY AND POPULATION

Some idea of the information in the Domesday folios for Rutland and of the form in which it was presented, may be obtained from the entry relating to the village of Whitwell in Alstoe wapentake (293 b). The village was held entirely by one owner, and so it is described in a single entry:

> In Whitwell Besy had one carucate of land to the geld. Land for 3 ploughs. There Herbert has of the Countess Judith one plough, and 6 villeins and 4 bordars having 2 ploughs. There a church and a priest and 20 acres of meadow, and one mill [of] 12d. Wood for pannage in places (*per loca*) 6 furlongs and 6 perches in length and 3 furlongs and 13 perches in breadth. In King Edward's time it was worth 40s.; now 40s.

This entry does not mention, for example, sokemen or demesne ploughs but it is representative enough of the county as a whole, and it contains the standard recurring formulae which are the basis for assessing the distribution of population in the county. One significant variation should be noted; in Witchley hundred, carucates are not entered, for the assessment is given in terms of hides, as in Northamptonshire.

Five basic items of information relating to most vills are enumerated: (1) carucates or hides; (2) plough-lands; (3) plough-teams; (4) population; (5) values; and the bearing of these upon regional variations in prosperity must now be considered. Reference must also be made to the measurements given for the three manors of Martinsley wapentake. In such a small area as Rutland, however, it would be unwise to overstress such differences as do occur. The true significance of the Rutland distributions may be appraised only in the light of the economic condition of the three surrounding counties.

(1) *Carucates and hides*

The normal entry for the carucated areas of Alstoe and Martinsley takes the form: 'in Teigh Godwin had 1½ carucates of land assessed to the geld' (*In Tie habebat Godwin i carucatam terrae et dimidiam ad geldum*). In the hidated district of Witchley the formula runs either, 'there are *n* hides', or 'AB holds *n* hides in C'.

(*a*) *Alstoe and Martinsley*. The total number of carucates in these two areas was 40¼, of which 27¼ were in Alstoe and 13 in Martinsley. Reference is made to the assessments of these areas in a prefatory note to the account of *Roteland* (293 b):

In Alstoe wapentake there are two hundreds. In each [there are] 12 carucates to the geld, and in each there can be 24 ploughs. This wapentake is half in Thurgarton wapentake [Notts.] and half in Broxtow wapentake [Notts.].

In Martinsley wapentake there is one hundred in which there are 12 carucates of land to the geld, and there can be 48 ploughs, saving the king's three demesne manors in which 14 teams can plough. These two wapentakes belong to the sheriffdom of Nottingham for purposes of the king's geld.

In addition to the 12 carucates mentioned above in Martinsley, Leuenot had one carucate at Oakham (293 b). There were also 3¼ additional carucates for Thistleton and *alia* Thistleton in Alstoe, entered in the Lincolnshire folios (358 b, 366, 367).

The existence of the Danish duodecimal system in the two wapentakes is quite obvious, and in this respect *Roteland* resembles the adjacent shires of Lincoln and Leicester. 'Nowhere else do we obtain so definite a clue to the system which governed' the distribution of carucates.[1] The interest of this for our purposes is that the artificial nature of the assessment is laid bare. A complete lack of correlation is evident when the assessments of the two districts, which were roughly the same in area, are compared with the numbers of plough-lands, teams and population. Martinsley with nearly 50% more teams than Alstoe was assessed at about half the number of carucates for which Alstoe was liable. The extremely low assessment was evidently not related to the economic condition of the district; it may be attributed rather to its status as a royal demesne; much of the land must have been either exempt from the assessment or else rated very leniently. It is clear, therefore, that the carucates cannot be regarded as any index of relative prosperity in 1086.

[1] F. M. Stenton in *V.C.H. Rutland*, I, p. 123.

(b) *Witchley*. The total number of hides in this wapentake or hundred was 75¾. Rutland occupies an interesting position in relation to the hide-carucate boundary; Witchley hundred was the only part of Midland England, north of the Welland, assessed in hides. This boundary followed the rivers Dove and Trent, and then Watling Street, so along the rivers Avon and Welland, but a salient included what is now Witchley hundred. It is probable that the 75¾ hides really represent 80 hides as the exact assessment; this latter figure is just half the number for which Witchley paid geld in 1067–75 according to the Northamptonshire Geld Roll.[1] Northamptonshire as a whole was subject to a greatly reduced geld in 1086, and it is probably no mere coincidence that Witchley hundred while answering for 160 hides in 1067, was only liable for 80 in 1086, and for the same nominal figure in the Pipe Roll of 1130.[2] These round figures suggest once more the artificial nature of an assessment imposed from above; they cannot be employed as an index of economic prosperity.[3]

(2) *Plough-lands*

Plough-lands are systematically entered for the villages of Rutland. The usual formula employed is simply: 'Land for *n* ploughs' (*Terra n carucis*). Occasionally the formula '*n carucae esse possunt*' is used as, for example, in the prefatory notes on fo. 293b that relate to Alstoe and Martinsley. The entry for Belmesthorpe makes no reference to plough-lands, but this was a vill dependent on Ryhall where 8 plough-lands are recorded. A total of 327¼ plough-lands is entered for the county as a whole, compared with 414¾ teams (see table on p. 369).

This excess of teams over plough-lands can be observed everywhere except in the southern half of Witchey hundred, and it was most marked in Martinsley where there were about 2½ teams for every recorded plough-land. The excess of teams over plough-lands raises the problem of the precise meaning of the plough-land in this county. To all appearances it cannot refer to potential arable, seeing that the number of plough-lands is almost everywhere less than the number of teams actually at work.

[1] *V.C.H. Rutland*, I, p. 125.

[2] J. H. Round in *V.C.H. Northamptonshire* (London, 1902), I, p. 265 n.

[3] In the Northamptonshire folios, bovates occasionally appear in conjunction with hides; four entries refer to three places in Witchley hundred: Empingham (227b), Seaton (219, 225), Tickencote (228b)—see pp. 391–2 below.

Hides, Carucates, Plough-lands, Plough-teams and Population in Rutland

	Hides	Carucates	Plough-lands	Plough-teams	Population
Alstoe	—	27¼	87¼	102⅝	321
Martinsley	—	13	56	145⅝	538
N. Witchley	41	—	88½	94	369
S. Witchley	34¾	¼	95½	72½	251
Totals	75¾	40½	327¼	414¾	1,479

The varying relationship between teams and plough-lands in individual entries is shown by the table below which gives percentages of entries:

	Excess teams	Equality	Deficient teams
Alstoe	57	21½	21½
Martinsley	100	—	—
Witchley	24	24	52

Further light is shed on the matter by the statements in the Domesday Book about the assessment and plough-lands of Martinsley and Alstoe.[1] In Alstoe in 1086 we are told that there were 24 carucates and land for 48 ploughs, but, surprisingly enough, the addition of the separate entries provides a total of no less than 84 plough-lands. 'It is no infrequent thing for Domesday arithmetic to be inconsistent with itself, and it seems plain that the compiler of the prefatory note meant to tell us that each of the two Alstoe "hundreds" contained an equal number of plough-lands, although he was wrong in his statement as to the number in question.'[2] Alstoe wapentake can in fact be divided into two 'hundreds' each containing 12 carucates and 42 plough-lands,[3] indication enough that the plough-lands are as artificial in character as the assessment itself (see p. 370). It is not possible, in view of this symmetrical arrangement of carucates and plough-lands, to give any more weight to the plough-lands of Rutland as an economic index than to the assessment. The figures for Alstoe suggest that, as in Nottinghamshire, we have in the Rutland

[1] See p. 367 above.   [2] *V.C.H. Rutland*, I, p. 122.
[3] Excluding 3¼ carucates and 3¼ plough-lands at Thistleton and *alia* Thistleton entered in the Lincolnshire folios—see p. 360 n. There was also a total of 1⅛ teams.

plough-lands the remnants of some earlier conventional assessment that had become obsolete by 1086.

Alstoe wapentake

	Carucates	Plough-lands	Plough-teams
'Hundred' i:			
Burley	2	7	6
Exton	2	12	11
Greetham	3	8	10
Whissendine	4	12	13
Whitwell	1	3	3
	12	42	43
'Hundred' ii:			
Ashwell	2	6	7
Awsthorpe	1	5	6
Cottesmore	3	12	24
Overton	$3\frac{1}{2}$	12	12
Teigh	$1\frac{1}{2}$	5	6
Thistleton	$\frac{1}{2}$	1	2
Thistleton	$\frac{1}{2}$	1	$1\frac{1}{2}$
	12	42	$58\frac{1}{2}$

The artificiality of the plough-lands is even more obvious in Martinsley, where the main entries for the three manors of the wapentake are as follows:[1]

	Carucates	Plough-lands	Plough-teams	Population
Oakham	4	16	39	158
Hambleton	4	16	45	156
Ridlington	4	16	36	200
	12	48	120	514

The Martinsley statement, referred to above, tells us that there were 12 carucates and 48 plough-lands (293b), and there is nothing unusual

---

[1] The totals that follow do not include the figures for the land of Albert the Clerk, or for the land of Leuenot at Oakham.

about this, but when we observe that on this 'land for 48 ploughs' there were over 500 persons working with 120 teams we must presume that the plough-land here is a somewhat abnormal concept. It is clear that the fours and sixteens in the table are mere abstractions unrelated to agrarian reality. At Oakham (293 b) there were 16 plough-lands and 39 teams, but even these were deemed insufficient, for we are told that 'nevertheless there can be 4 other ploughs' (*et tamen aliae iiii carucae possunt esse*). Similarly, of Albert's land in Martinsley we read, 'in this his land there can be 8 ploughs, and nevertheless 16 teams plough there' (294). It would seem likely that in this wapentake too the plough-lands are a relic of some earlier assessment.

In Witchley hundred, as we have seen, teams are not in excess of plough-lands, but even so, traces of a duodecimal system can again be observed. As Sir Frank Stenton wrote, 'the plough-lands of South Rutland, like those of the north of the modern county, were really conventional quantities, connected with an obsolete fiscal system which was based on a duodecimal system of rating and bore no necessary relation to the divisions of the soil upon which the agriculture of the period was based.'[1]

(3) Plough-teams

The Rutland entries like those for other counties draw a distinction between the teams held in demesne by the lord of the manor, and those held by the peasantry. The entry for Tinwell (221 b), for example, is typical: 'In demesne there are 2 ploughs, and 24 villeins and 11 bordars with 7 ploughs.' A few minor variations are encountered; at Seaton in Witchley (225) the teams are entered in the normal way, but we are also told that the landholder 'has only the third part of the wood and similarly of the arable land' (*de terra arabili*). This is the only mention of arable land in the county. There were 5 oxen in a team (*v boves in caruca*) at Oakham (293 b); and in the Lincolnshire folios there were 2 oxen similarly at Thistleton (367) and 3 at *alia* Thistleton (358 b). These are the only references to oxen for Rutland.

The unusual entry relating to Albert's land in Martinsley should also be noted. Here, on 8 plough-lands there were apparently 25 teams, but no evidence as to their precise location is given.[2] For this reason they are

---

[1] *V.C.H. Rutland*, I, p. 126.
[2] For the 25 teams, see p. 365 above.

shown on the distribution map by means of an open circle conventionally placed in the centre of Martinsley wapentake (Fig. 129). Of the latter district we are also told that there was land for 14 ploughs in the king's three demesne manors, although only 11 were recorded as working in the demesne of the three places (293b); and at Oakham where there were

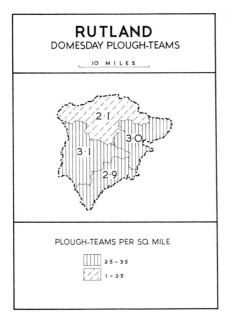

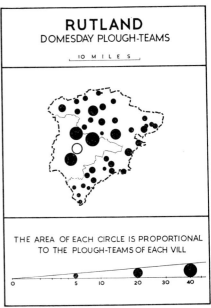

Fig. 128. Rutland: Domesday plough-teams in 1086 (by densities).

Fig. 129. Rutland: Domesday plough-teams in 1086 (by settlements).

The symbols shown in Fig. 129 for Hambleton, Oakham and Ridlington include the teams of 19 unnamed berewicks. The open circle represents the teams of Albert the Clerk, the location of which is uncertain. The boundary of Martinsley wapentake is marked; see Fig. 127.

2 demesne ploughs, there could have been four more (293b). No village in the county has less than 3 teams, and a total of 414¾ teams is recorded, a significantly large number for such a small county.

(4) *Population*

The bulk of the population consisted of the three main categories of villeins, bordars and sokemen, together with a few serfs and priests. The details of these various groups are summarised in the table below:

Recorded Population of Rutland in 1086

A. Rural Population

Villeins	1,064
Bordars	254
Sokemen	113
Serfs	34
Priests	14
Total	1,479

There were also two bondwomen (*ancillae*), not included above.

B. Urban Population
Nil

Ellis's totals for Rutland are not directly comparable with these as they refer only to Domesday *Roteland*, and also do not include the figures of the three holdings entered only in the Lincolnshire folios.[1] In addition, there were two *ancillae* recorded at Horne (228b), but these are not included in the table. It must be remembered that these figures are of recorded population and that in order to get some idea of the actual population, the figures must be multiplied by a factor—say 4 or 5— depending on one's views of the medieval family.[2]

Villeins constituted the most important element in the population, and amounted to over 70% of the total; bordars came next in importance and accounted for some 17%. The duplicate entries for the county reveal some minor discrepancies in the numbers of bordars at three manors. At Ashwell there were either 2 or 3 bordars, at Burley 7 or 8, and at Whissendine 7 or 6—but these errors are very small in comparison with the county total.[3] It is interesting to note that 105 out of the 113 sokemen were in Witchley hundred, the area surveyed under Northamptonshire, and that the 34 serfs were also in the same area. At Cottesmore (293b) the formula '3 sokemen on half a carucate of this land' is used, but this is the only instance in Rutland. The priests who numbered 14 constituted the smallest group of all.

[1] Sir Henry Ellis, *General Introduction to Domesday Book* (London, 1833), II, p. 479; see p. 360 n. above.

[2] But see p. 430 below for the complication of serfs.

[3] The latter figures have been counted here; see p. 360 above.

(5) *Values*

Values are almost always entered in terms of pounds or their equivalent in shillings. Certainly the detail never descends to pence. The entries for Alstoe and Witchley normally give details of values for the two dates 1066 and 1086, but those for Martinsley refer only to 1066 except for

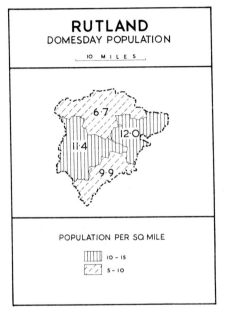

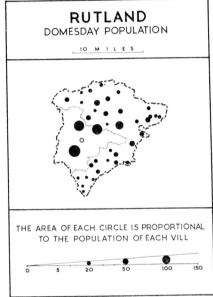

Fig. 130. Rutland: Domesday popula-   Fig. 131. Rutland: Domesday population in 1086 (by densities).   tion in 1086 (by settlements).

The symbols shown in Fig. 131 for Hambleton, Oakham and Ridlington include the population of 19 unnamed berewicks. The open circle represents the population on the land of Albert the Clerk; the location of these persons is uncertain. The boundary of Martinsley wapentake is marked; see Fig. 127.

Albert's land (294) and for part of Oakham (293b). The holdings in Alstoe were together worth £62 in 1066 and those of Martinsley £141, making a total of £203 for the two wapentakes of Domesday *Roteland*. By 1086, the total value of Alstoe had risen to £89,[1] but no details are

---

[1] The figures for Alstoe (£62 in 1066 and £89 in 1086) include those for Thistleton given only in the Lincolnshire folios—see p. 360 above.

given for Martinsley. The total values for Witchley for the same two dates were £53. 10s. and £94. 10s. respectively. We are also told that *Roteland* rendered £150 (*libras albas*) to the king (293b), in 1086. This mention of 'white' money indicates that as in some other counties, debased coins were in circulation. It is clear from the data available that no simple equation can be deduced between the numbers of teams and population on the one hand and the *valet* on the other.

The collation of the values in the duplicate entries for Thistleton sheds some interesting light on the information with which we are dealing. According to the Rutland entry the value of Thistleton was £3 in 1086 (293b), but the Lincolnshire entry states specifically that the value of Thistleton was *included* in the value for South Witham, and yet the value of the latter in 1086 was recorded as £2. 10s. (358b). Of course there may be a scribal error involved here, but it is more likely, as Sir Frank Stenton has suggested, 'that the values in the Rutland section of Domesday give us, not estimates of manorial wealth, but statements as to the contribution which each manor made towards the "firma" due from the district to the Crown...in any case the discrepancy which we are considering gives a useful warning as to the complexity of the actual facts which may underlie the specious uniformity of a series of entries in Domesday Book'.[1]

(6) *Measurements*

Reference must also be made to another type of general measurement which appears for Martinsley wapentake (293b). The relevant entries are as follows:

Oakham: The whole manor with berewicks [is] 3 leagues in length and 1 league and 8 furlongs in breadth.

Hambleton: The whole manor with berewicks [is] 3 leagues and 8 furlongs in length and 2 leagues and 8 furlongs in breadth.

Ridlington: The whole manor with berewicks [is] 3 leagues and 7 furlongs in length and 2 leagues and 2 furlongs in breadth.

What the precise significance of these dimensions may be and why they are confined to the royal manors of Martinsley must remain a matter for

[1] *V.C.H. Rutland*, I, p. 130.

conjecture.[1] They obviously imply sum totals of scattered holdings, but there is no clue to the arithmetical process by which they were computed.

Conclusion

For the purpose of calculating densities the wapentakes (Fig. 124) have been adopted as the most convenient units. Their limits are very artificial, and coincide with civil parish boundaries. Witchley hundred has been divided into two parts, north and south, thus making a total of four divisions in the county. These, if not distinct geographical regions, do provide a useful basis for distinguishing the degree of variation over the face of the county.

Of the five recurring formulae discussed above, only two, viz. plough-teams and population, may be regarded as reliable indices of the distribution of wealth and prosperity throughout the county. Neither is without uncertainty, but taken together they supplement one another to provide a general picture. But within such a small area as Rutland, we would hardly expect to encounter major geographical contrasts such as may be observed in larger counties. Further, the highly compressd form of the information for Martinsley wapentake has to be taken into account in any consideration of the western part of the county.

In Martinsley and Witchley, there were three teams (Fig. 128) and approximately eleven recorded persons per square mile (Fig. 130). In Alstoe to the north there were only two teams and about seven persons per square mile, figures which suggest a comparatively lower degree of prosperity and development. There is very little to say beyond this very general statement. The real interest of the maps is, perhaps, the comparison they afford with corresponding maps of neighbouring counties. Figs. 129 and 131 are supplementary to the density maps, but it is necessary to make one reservation about them. As we have seen on p. 362, it is possible that some Domesday names may have covered more than one settlement, e.g. the present-day villages of Great and Little Casterton are represented in the Domesday Book by only one name. A few of the symbols should therefore appear as two or more smaller symbols, but this limitation does not affect the main pattern of the maps.

[1] Measurements of a similar character occur systematically in Yorkshire, Norfolk and Suffolk, and occasionally in the counties of Chester, Derby, Lincoln, Nottingham and Oxford—but it is difficult to envisage what the figures mean.

WOODLAND

Types of entries

The Rutland folios give information about wood, underwood and spinney. The dimensions are normally expressed in terms of linear measurements, as seen in the following examples:

Oakham (293b): Wood for pannage 1 league in length and half a league in breadth (*Silva pastilis i leuua longa et dimidia leuua lata*).

Empingham (227b): Wood one furlong in length and 10 perches in breadth (*Silva i quarentena longa et x perticis lata*).

Bisbrooke (228b): Underwood 1½ furlongs in length and as much in breadth (*Silva minuta i quarentenae et dimidiae in longitudine et tantundem in latitudine*).

The exact significance of these linear measurements is not clear, and we cannot hope to convert them into modern acreages.[1] All we can do is to plot them diagrammatically as on Fig. 132. In two entries, one set of measurements is found covering two or more places. Thus at Overton and Stretton there was wood, 'fit for pannage in places', one league by a half (293b). This seems to imply, although not necessarily, some process of addition whereby the dimensions of separate tracts of wood were consolidated into one sum. On Fig. 132 the two composite entries for wood have been plotted for the first named village, e.g. for Overton in the entry above, and for Liddington (221) in the other composite entry. The associated villages in each entry have also been indicated.

The phrase 'fit for pannage in places' (*silva per loca pastilis*) is used in several entries, and there are also other variants of the more usual formula. For Hambleton, underwood 3 leagues long and 1½ leagues broad is described as '*fertilis per loca*' (293b). At Ketton in Witchley (219) there were 16 acres of *silva vilis*, i.e. wood of little value. At Thistleton (366) there were 15 acres of 'wood', and at Empingham (226, 227b) there were 16 acres in addition to other wood. Spinneys (*spineta*) are mentioned at Barrowden, Casterton, Seaton and Tixover, and are described sometimes in linear measurements and sometimes in acres. An entry for Seaton (219) makes a distinction between wood and spinney, for both are mentioned and given separate measurements (*Silva una*

----

[1] See p. 437 below.

quarentena longa et una lata. Spinetum et [sic] vi quarentenis longum et ii quarentenis latum). No mention is made of forest in the Rutland entries.

Distribution of woodland

Fig. 132 shows that the main bulk of the recorded wood was in the north and west of the county, although there were smaller scattered stretches of wood in the Welland valley area. The distribution pattern

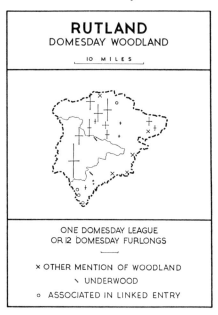

Fig. 132. Rutland: Domesday woodland in 1086.

Where the wood of a village is entered partly in linear dimensions and partly in some other way, only the dimensions are shown. Spinneys are included under 'other mention of woodland'. The boundary of Martinsley wapentake is marked; see Fig. 127.

would of course have been more exact if the information for Martinsley wapentake had been more detailed. Even so the general picture is clear enough. Some of the areas which are now wooded—Burley, Exton and Normanton parks, and Beamont Chase, relics of the ancient royal forest of Leighfield—were apparently occupied by wood at the date of the Domesday Survey. Much of the woodland was probably located on the heavy Liassic and Boulder Clays, the arable on the more attractive soils of the Marlstone and the Northampton Sands.

Types of entries

Meadow is recorded under all but three[1] of the Domesday names of Rutland, and is almost always expressed in terms of acres. The formula is simply: 'there are *n* acres of meadow' (*n acrae prati*), and the amounts thus recorded are small, varying from as little as 3 acres at Essendine and

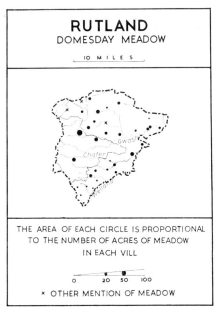

RUTLAND
DOMESDAY MEADOW

10 MILES

THE AREA OF EACH CIRCLE IS PROPORTIONAL
TO THE NUMBER OF ACRES OF MEADOW
IN EACH VILL

0   20   50   100

× OTHER MENTION OF MEADOW

Fig. 133. Rutland: Domesday meadow in 1086.

Linear measurements are indicated as 'other mention of meadow'. The boundary
of Martinsley wapentake is marked; see Fig. 127.

at Thorpe to 40 acres at Cottesmore and at Ketton. As in the case of other counties, no attempt has been made to translate these figures into modern acreages. The Domesday acres have been treated merely as conventional units of measurement, and Fig. 133 has been plotted on that assumption.

In three places the meadow is described in terms of linear measurements. At Exton (293b) it was 6 furlongs in length (*pratum vi quarentenarum in longitudine*), but we are not told how broad it was. At Teigh we read

[1] Horne, Ryhall and Thistleton.

of meadow 4 furlongs in length and 3 furlongs in breadth (293 b). Whissendine is described both in the Rutland Domesday (293 b) and under Lincolnshire (367), but meadow is only mentioned in the latter entry, the measurements being 10 by 8 furlongs.

Distribution of meadowland

Fig. 133 shows that meadow was widely distributed over the surface of the county, and while no generalisation is possible for Martinsley wapentake, it is clear that the Domesday meadow was associated with the principal streams of the area, especially the rivers Welland, Chater and Gwash. A fair proportion was located in the valleys of the higher upland country of the north and west, and some at least was to be found near the small headstreams flowing westwards into Leicestershire.

MILLS

Mills are mentioned in connection with at least 19 of the 39 Domesday names in Rutland. It is difficult to be sure of the exact total because mills appear in two composite entries. Two mills are entered for Luffenham and Sculthorpe (219); and two for Liddington and three other places (221). The mills have been plotted for the first-named place in each entry.[1] The four associated places have also been indicated on Fig. 134.

The formula commonly employed runs: 'there are n mills rendering m shillings'. With the single exception of Greetham, the values of mills are systematically entered, and the amount for a single mill varied from as little as 12d. at Whitwell (293 b) to as much as 24s. at Tickencote (228 b). In an entry for Empingham (226), one and a half mills are recorded, but there is no trace of the missing fraction.

Domesday Mills in Rutland in 1086

1 mill	10 settlements
2 mills	6 settlements
4 mills	2 settlements
12 mills	1 settlement

[1] The figure of 19 results from counting as one group the mills of a composite entry. If it is assumed that both mills in each of the two composite entries were at separate places, the total of places with mills is 21. If all six places of the composite entries are counted, the total becomes 23.

The group of 12 mills (only 11½ are recorded) was at Empingham, while Horne and Tolethorpe each had 4 mills. Ridlintgon was unique in having 2 mill sites, to which no value was attached (293 b). In Martinsley wapentake, a mill was entered for the holding of Albert the Clerk (294); this has not been shown on Fig. 134, nor included in the table on p. 380.

RUTLAND
DOMESDAY MILLS

10 MILES

⊕ 3 MILLS AND OVER
O 1–2 MILLS
S 2 SITES
o ASSOCIATED IN LINKED ENTRY
▨ LAND OVER 400 FEET

Fig. 134. Rutland: Domesday mills in 1086.
The boundary of Martinsley wapentake is marked; see Fig. 127.

Fig. 134 shows that the mills lay, for the most part in the north-east of the county. The largest number was associated with the river Gwash and its tributaries; others were located near the Welland, the Chater and other less important streams. It is notable that Martinsley wapentake, which had by far the largest number of plough-teams and the greatest population, should have only 2 mills, and 2 sites of mills.

CHURCHES

Eight churches and seven priests are mentioned in connection with four places, and a further seven priests were associated with five other places.

Seven of the eight churches are recorded in the entries relating to the three great manors of Martinsley wapentake and their appurtenant berewicks. We cannot therefore be certain as to the precise location of these churches. At Oakham (293b) there was a church and a priest, to which 4 bovates of the land of the manor belonged. This church is mentioned again on fo. 294. Three priests and three churches are mentioned at Hambleton (293b), and one of the latter was located at Hambleton itself and is subsequently recorded on fo. 294. We cannot tell where the other two were. The Hambleton churches also owned land. A further three churches and two priests were entered for Ridlington and its berewicks, but again, Domesday Book fails to give us their exact locations. The lands of each of the three great manors of Martinsley wapentake are described as 'church sokeland' (*Cherchesoch*) a reminder of the claims of the abbey of Westminster in this part of western Rutland.

The other church and priest were at Whitwell in Alstoe wapentake. In Witchley hundred no churches are recorded, but there were seven priests, and it may well be that churches did exist in this district, but as they did not possess land they do not appear in Domesday Book.

REGIONAL SUMMARY

The small size of Rutland does not give much scope for regional division. The county is situated between the Leicestershire uplands and the Fenland, and western Rutland is occupied by an extensive plateau of over 300 ft. (Fig. 125). This plateau is composed of Middle and Upper Lias Clays, separated by the Marlstone, and incised into it are a number of streams that flow eastwards to the Welland (Fig. 126). The hill and vale topography of the south of the county consists of ridges of higher ground of Northampton Sands, capped by the Lincolnshire Limestone in the east, separated by wide valleys where the rivers have cut down to the Middle Lias clays. Boulder Clay is widespread, especially in the north of the county.

The best soils are derived from the Marlstone of the Lias, the worst from the Estuarine Clay of the Great Oolite. Between these extremes are (*a*) the Liassic and Boulder Clays which produce intractable heavy soils, and (*b*) the lighter and thinner soils of the limestones and Northampton Sands.

The most densely settled part of the county was the hill and vale country of the south and the Welland valley, with about 3 teams and 10 or so recorded people per square mile. To the north, the upland of Alstoe was somewhat less prosperous, with about 2 teams and 7 people per square mile. Over the county as a whole, most places had wood and meadow entered for them, but mills were less frequent.

BIBLIOGRAPHICAL NOTE

The earliest translation of the Rutland section of the Domesday Book is that of William Bawdwen, in his *Dom Boc: A Translation of the Record called Domesday* (Doncaster, 1809). This covers the counties of Derby, Nottingham, Rutland, Lincoln, York and parts of the counties of Lancaster, Westmorland and Cumberland. The volume contains an introduction, glossary and index.

A more recent translation is that by C. G. Smith, *A Translation of that portion of Domesday Book which relates to Lincolnshire and Rutlandshire* (London, 1870). This volume also contains an introduction, glossary and index.

The standard translation now available is that prepared by Sir Frank Stenton in *V.C.H. Rutland* (London, 1908), I, pp. 138–42. This contains a series of footnotes and identifications and is preceded by a valuable introduction (pp. 121–36). The *Index to the V.C.H. Rutland* (London, 1936) contains a list of the Domesday place-names.

CHAPTER IX

NORTHAMPTONSHIRE

BY I. B. TERRETT, B.A., PH.D.

The Domesday county of Northamptonshire was somewhat larger than the modern county; it included the whole of Witchley hundred, which is now part of Rutland. In addition to the Rutland vills the folios for Northamptonshire also describe a number of other places that are not in the modern county. A number of these are immediately on the boundary; others are some distance away, a few as far afield as Staffordshire, which does not march with Northamptonshire. A list of these extraneous places, twenty-nine in all, excluding the Rutland vills, is set out in the table on p. 385. There are on the other hand, no vills that have been transferred to Northamptonshire from other counties, but the four Northampton villages of Hargrave, Luddington, Lutton and Thurning are described partly in the Huntingdonshire folios and partly in those for Northampton-shire. Similarly, the Northampton villages of Newton Bromswold, Rushden and Stanwick are described partly in the Bedfordshire folios and partly in those for Northamptonshire.[1] Furthermore, the Lincolnshire account of the borough of Stamford (336b) mentions six wards, 'five in Lincolnshire and the sixth in Northamptonshire which is beyond the bridge', but this sixth ward is nowhere described in the Northamptonshire folios, and there is no way of telling what it contained.

The value and relevance of certain auxiliary records should be noted. Most important are the Northamptonshire Geld Roll[2] (*ante* 1076) and the twelfth-century Northamptonshire Survey.[3] From a geographical point of view the value of these documents is limited in that they only give hidage, but on the other hand their collation with Domesday Book

---

[1] The Domesday folios for the holdings in these villages are as follows: Hargrave, 204, 206, 226; Luddington, 206, 221b; Lutton, 204b, 221b, 222; Thurning, 204, 206, 221; Newton Bromswold, 210, 220b; Rushden, 210, 212b, 225b; Stanwick, 210b, 221b.

[2] Printed in D. C. Douglas and G. W. Greenaway (eds), *English Historical Documents, 1042–1189* (London, 1953), pp. 483–6. This document was discussed by J. H. Round in *Feudal England* (London, 1895), pp. 147–56.

[3] Printed and discussed in *V.C.H. Northamptonshire* (London, 1902), I, pp. 357–89. This document was also discussed by J. H. Round, *Feudal England*, pp. 215–24.

Holdings described in the Northamptonshire Folios, but now in other Counties

Oxfordshire:

Charlton on Otmoor (224b)
Cottisford (224b)
Finmere (221)
Glympton (221)
Grimsbury (227b)
Hethe (221)
Heyford (221)
Mollington (226)
Shelswell (221)
Shipton on Cherwell (224b)
Sibford Gower (224b)
Wootton (221)
Worton (221)

Huntingdonshire:

Catworth (222)
Elton (221, 222)
Stibbington (229)
Winwick (221b, 228)

Leicestershire:

Little Bowden (223)

Bedfordshire:

Farndish (225b)
Podington (225b)

Warwickshire:

Berkswell (224)
Over (226)
Sawbridge (222b)
Stoneton (225)
Whichford (227b)
Whitacre (224)

Staffordshire:

Lapley (222b)
Marston in Church Eaton (222b)
West Bromwich (226)

frequently produces most useful and illuminating results, especially in enabling us to identify some places.

Thirteen composite entries, involving some forty places, occur within the modern county. For the purpose of constructing distribution maps, the details of these entries have been equally divided among all the villages concerned. But when wood is measured in terms of leagues and furlongs, it cannot be so divided, and has therefore been plotted for the first named village in each entry. Mills have been treated likewise.[1]

SETTLEMENTS AND THEIR DISTRIBUTION

The total number of separate places mentioned in the Domesday Book for the area now included in the modern county of Northamptonshire seems

[1] See pp. 404 and 412.

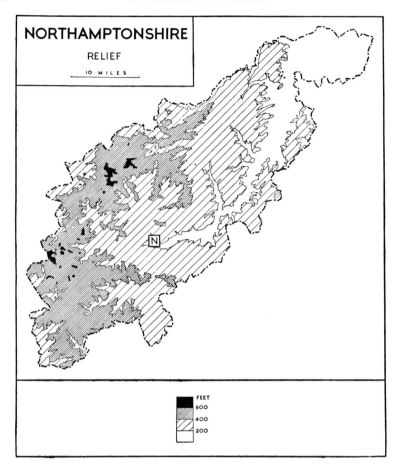

Fig. 135. Northamptonshire: Relief.
N indicates the Domesday borough of Northampton.

to be 326, including the borough of Northampton itself. This figure, however, may not be quite accurate because there are some instances of two or more adjoining villages bearing the same name today, and it is not always clear that both units existed in the eleventh century. There is no indication, say, that Great and Little Billing existed as separate villages in the eleventh century. They may well have been separate, but the relevant Domesday entries refer merely to Billing. The same applies to

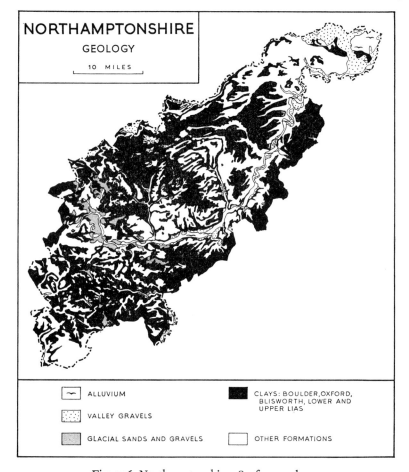

NORTHAMPTONSHIRE
GEOLOGY
10 MILES

⌐ ALLUVIUM

VALLEY GRAVELS

GLACIAL SANDS AND GRAVELS

CLAYS: BOULDER, OXFORD,
BLISWORTH, LOWER AND
UPPER LIAS

OTHER FORMATIONS

Fig. 136. Northamptonshire: Surface geology.

Based on Geological Survey One-Inch Sheets (New Series) 157, 170, 171, 185, 186, 201–3 and 218; and Quarter-Inch Sheet 12; and D. P. Bickmore and M. A. Shaw (eds), *Atlas of Britain and Northern Ireland* (Oxford, 1963), p. 54.

Great and Little Oxendon, and to Great and Little Oakley; the distinction between the respective units does not appear until the twelfth century.[1] Only where there is some evidence of a second village has it been included in the total of 326; Upper and Nether Heyford, for example, appear as

[1] J. E. B. Gover, A. Mawer and F. M. Stenton, *The Place-Names of Northampton-shire* (Cambridge, 1933), pp. 132, 118, 170.

Heiforde (220) and *altera Haiford* (223). Similarly, Great and Little
Weldon are represented in Domesday Book as *Weledene* (225) and *parva
Weledone* (224b). On fo. 226, an entry relating to Courteenhall says that
'the soc belongs to another Courteenhall' (*alio Cortenhalo*), but there is
no trace of two villages here in later times.

The similarity of Domesday names in Northamptonshire has often
led to confusion and error. Thus *Dodintone* appears in five entries; three
of these are in Wymersley hundred and refer to Denton. The other two
entries (219b and 228), Round asserted 'undoubtedly refer to Duddington (or Doddington), on the Welland, in the extreme north of the county'.[1]
This hardly seems to be so. The one-hide entry on fo. 219b, it is true,
does refer to Duddington on the Welland in Willybrook hundred[2] for
it has an unmistakable hundred rubrication. But *Dodintone* on fo. 228
occurs between Earls Barton on the one hand, and Wilby and Mears
Ashby on the other, and almost certainly refers to Great Doddington in
Hamfordshoe hundred; moreover, the Domesday Book itself says that
Great Doddington, Wilby and Mears Ashby belong to Earls Barton—all
of which places are adjacent. There are no less than 18 Domesday names
in the county containing the element -*torp*, and their identification is a
matter of some difficulty. *Westorp* (227b), we are told, is in the hundred
of *Gravesende*, and this entry probably refers to Westhorp in the parish
of Byfield. Sometimes the collation of the Domesday entries with the
corresponding ones in the twelfth-century Survey is useful, as in the case
of *Alidetorp* (224) which is said in the Survey to be held by Fucher'
Malesoures, and so can be identified with Thorpe Malsor.

Not all the 326 names appear on the present-day parish map of
Northamptonshire. Some are represented by hamlets, by houses or
farmsteads, or even by the names of topographical features. The Domesday *Neubote* (225b) survives as the hamlet of Nobottle in the parish of
Great Brington, and *Hala* (222) as Halefield Lodge in Apethorpe.
Similarly *Sewelle* (225) appears as Seawell Farm in Blakesley, while the
Domesday *Cildecote* (223b) in Cold Ashby appears on the 1825 edition
of the Ordnance Survey One-Inch map as Chilcotes Cover. In addition
there are a number of other Domesday names that have left no trace of
either their name or site; such has been the fate of *Celverdescote* (223)

---

[1] *V.C.H. Northamptonshire*, I, p. 272.

[2] It is also interesting to note that Duddington on the Welland was assessed at one
hide in the twelfth-century Survey of Northamptonshire.

which was somewhere near Everdon Magna judging from the context of the name in the twelfth-century Survey. *Calme*[1] (222) in *Stotfald*, *Brime*[2] (227) in *Elboldestou*, and *Hantone* (220b, 226b) in *Neveslund* have also completely disappeared.

On the other hand, there are about forty modern villages that are not mentioned in the Domesday Book. More than half of these modern names are located in the extreme north of the county, particularly in the Soke of Peterborough. Many of them are first mentioned in the twelfth, thirteenth, or fourteenth centuries, e.g. Etton in 1125, Thornhaugh in 1189, Borough Fen in 1307. Over the rest of the county the post-Domesday settlements are more sporadic, e.g. Claycoton 1175 and Shutlanger 1162. The names of some appear for the first time in the twelfth-century Northamptonshire Survey, e.g. Ringstead and Woodford. The parish of Newnham, on the other hand, is mentioned as early as 1021–3,[3] and yet it does not appear in the Domesday Book. Details about these places, if they existed in 1086, must have been included in the totals of other villages.[4]

But despite these complications, the fact remains that the distribution of Domesday names (Fig. 137) is remarkably similar to that of the modern villages. The only areas with unusually small numbers of settlements were the ill-drained fenland of the Soke of Peterborough, and parts of Rockingham Forest in the north, and Whittlewood Forest in the south. The rest of the county appears to have been closely settled. No well-marked village lines are apparent, as the geological boundaries are sinuous in character, but three main types of location have been described.[5] These, in brief, are: (*a*) hill-top or dry-point villages, e.g. Cold Ashby; (*b*) valley-bottom villages, e.g. Luddington in the Brook; (*c*) spring-line villages. The latter may be divided into two categories: (i) those at the junction of Northampton Sands and the Upper Lias, e.g. Walgrave; and (ii) those

---

[1] Part of Clipston Lordship, *V.C.H. Northamptonshire*, I, p. 318 n.

[2] *Brime* is assessed at $2\frac{4}{10}$ hides in Domesday Book and Culworth at 2 hides and 4 'small virgates' in the twelfth-century Survey. Round was certain that *Brime* was in the parish of Culworth in *Elboldestou* hundred; see *V.C.H. Northamptonshire*, I, p. 344 n.

[3] J. M. Kemble, *Codex Diplomaticus*, No. 736 (London, 1839–48).

[4] The dates in this paragraph are from J. E. B. Gover, A. Mawer and F. M. Stenton, *op. cit.*

[5] S. H. Beaver, *Northamptonshire* (London, 1943), p. 364, being Part 58 of *The Land of Britain*, ed. by L. Dudley Stamp.

located at the base of the Great Oolite Limestone, e.g. Croughton. The great majority of Domesday settlements in the county can be classed as spring-line in type.

Fig. 137. Northamptonshire: Domesday place-names.
N indicates the Domesday borough of Northampton.

THE DISTRIBUTION OF PROSPERITY AND POPULATION

Some idea of the information in the Domesday folios for Northampton-shire, and of the form in which it was presented, may be obtained from the entries for the village of Thrapston situated in the valley of the Nene. The village is described in two entries:

Fo. 220b. Of the same Bishop, Odelin holds 3 virgates of land in Thrapston. There is land for 2 ploughs. In demesne there is one [plough] and 2 serfs; and one villein with 4 bordars have one plough. It was worth 12*d.*; now 10*s.* Burred held [it] freely.

Fo. 228. Oger holds of the King 2½ hides in Thrapston. There is land for 5 ploughs. In demesne there are 2 ploughs, with one serf; and 7 villeins and 5 bordars have one plough; and 4 sokemen with one plough. There a mill yielding (*de*) 20*s.*, and 12 acres of meadow. Wood 6 furlongs in length and as many in breadth. It was and is worth £3.

These two entries do not include all the items of information that are given for some other places; there is no mention, for example, of churches or waste. But the entries for Thrapston are representative enough, and they do contain the standard recurring items that seem to provide a basis for assessing the distribution of population in the county. These standard entries are five in number: (1) hides; (2) plough-lands; (3) plough-teams; (4) population; and (5) values. The bearing of these five items of information upon regional variations in prosperity must now be considered.

(1) *Hides*

The normal entry for Northamptonshire states the number of hides and virgates that belonged to a landholder thus: 'Gilo holds 2 hides in Wappenham' (*Gilo tenet ii hidas in Wapenham*); sometimes the record simply runs: 'there *n* hides' (*ibi n hidae*). Three entries relating to holdings in Brafield (222b), Brampton (223) and Welton (224) contain references to acres, and in each case the amount of land involved is 5 acres. Some other peculiarities should also be noted. One of the holdings at Houghton in Wymersley hundred (226) was assessed at one hide and half a virgate, and 2 carucates of land. This solitary mention of carucates may possibly be a scribal error; the context suggests that the passage should read '2 acres', not '2 carucates'. Bovates are occasionally mentioned, e.g. a holding in Moulton (219b) answered for 1½ hides and one bovate of land; and more curious still is an entry for Boughton near Northampton (225b) which speaks of '3 virgates of land less one bovate'. This use of bovates in an area assessed in hides suggests a survival of Danish influence, and Tait asserts that, as in Cheshire, these bovates are 'evidence of Scandinavian influence upon the subdivision of the fiscal hide'.[1]

---

[1] J. Tait, *The Domesday Survey of Cheshire* (Manchester, 1916), Chetham Society, N.S., vol. 75, p. 13.

Altogether, there are some fifteen entries relating to eleven places, in which bovates are employed in the assessment.[1]

Evidence of a five-hide unit can be observed in many counties, but it is rarely employed in Northamptonshire. One of the distinctive features in this county, however, was the frequent use of a unit of 4 hides.[2] Furthermore, in the extreme north of the county the assessment was clearly duodecimal in character. The significance of these figures can only be realised when both assessment and plough-lands are considered together.[3]

Whether the four-hide unit, or a duodecimal element, can be detected or not, it is clear that the assessment was largely artificial in character and bore no constant ratio to the agricultural resources of a vill. The variation among a representative selection of four-hide vills speaks for itself:

	Teams	Population
Abington (229)	3	18
Badby (222b)	10	28
Bugbrooke (223)	13	48
Kislingbury (223b, 227b)	8	44
Marston St Lawrence (224b)	10	45
Wilby (228)	7	7

There was always a large element of artificiality about an assessment imposed from above, and the hidage cannot be taken to provide any index of the relative prosperity of different parts of the county.

---

[1] The complete list of bovates in the Northamptonshire assessment is as follows:
Boughton (225b): 3 virgates less 1 bovate.
Dingley (223): 1½ hides and 1½ bovates.
Dingley (225): 1 hide less 1½ bovates.
West Farndon (223): 1½ hides and 1 bovate.
Harleston (223): ½ bovate.
Hanging Houghton (226b): 2 virgates and 1 bovate.
Lamport (222): 1 virgate and 1 bovate.
Lamport (228): 1 bovate.
Moulton (219b): 1½ hides and 1 bovate.
Newton (227b, 228b): three holdings each assessed at 3 virgates and 1⅓ bovates.
Spratton (225b): 1 virgate and 1 bovate.
Wadenhoe (220b): 2½ hides and 1 bovate.
Walton Grounds (223b): ½ hide and 1 bovate.
For an additional four entries for three places now in Rutland (but described in the Northamptonshire folios), see p. 368 n.

[2] J. H. Round, 'The Hidation of Northamptonshire', *Eng. Hist. Rev.* (London, 1900), XV, p. 78.   [3] See p. 394 below.

The total assessment amounted to 1,180 hides, $1\frac{23}{90}$ virgates less 5 acres, and 2 carucates, $6\frac{1}{4}$ bovates. This is considerably lower than Maitland's figure of 1,356,[1] but the latter included the whole of Witchley hundred, now in Rutland, and also the twenty-nine other places not now in the county, which together accounted for $201\frac{11}{60}$ hides. If this amount is subtracted from Maitland's total, the results are substantially the same.[2] The Domesday assessment was much reduced as compared with both the County Hidage (3,200 hides), and that of the Northamptonshire Geld Roll (2,664); the artificial character of this reduction will be apparent when the plough-lands are considered.[3] The precise reason for the drastic reduction in hidage must remain a matter for conjecture. Maitland was of the view that the former assessments were much too high,[4] while Round related the question to the record of wasting in the Northamptonshire Geld Roll, which 'pointed to some terrible devastation such as is actually recorded in the English Chronicle under 1065'.[5]

(2) *Plough-lands*

Plough-lands are systematically entered for the villages of Northamptonshire. The formula employed is simply 'there is land for n ploughs' (*terra est n carucis*). For Adstone (222b) we read that 'half a plough can be employed there' (*Ibi dimidia caruca potest esse*)—but this is a variation not found elsewhere. In several entries 'land for n oxen' is recorded. Information for the year 1066, in addition to the usual plough-land figure, is given for only four places.[6] The entry for Wittering (221b) may serve as an example—*Terra est xvi carucis. T.R.E. fuerunt ibi xxx*; there were only 15 teams in 1086. The phrase *Terra est* is followed by a blank in three entries—for Denford (220b), Greens Norton (219b) and Raunds (220b); these entries account for 53 teams. Plough-lands are not mentioned in a further 20 entries with $63\frac{1}{2}$ teams.

A total of $2,524\frac{3}{8}$ plough-lands is recorded for the area covered by the modern county. If, however, we assume that plough-lands equalled teams

---

[1] F. W. Maitland, *Domesday Book and Beyond* (Cambridge, 1897), pp. 400–3.
[2] Maitland thought that his total was 'a little too large' (*ibid.* p. 457 n.), but this hardly seems to be so.
[3] See p. 394 below.   [4] F. W. Maitland, *ibid.* p. 457.
[5] J. H. Round, *V.C.H. Northamptonshire*, I, p. 260.
[6] Glinton (221), Southorpe (221b), Werrington (221), Wittering (221b). T.R.E. teams, but not plough-lands, are also entered for Burton Latimer (226b).

in these entries where a figure for the former is not given, 116½ teams must be added, and the total becomes 2,640⅞ plough-lands.[1]

Many of the figures for plough-lands seem to bear a conventional relationship with those for hides. These artificial ratios fall into three categories, examples of which are given below:

Some Representative Northamptonshire Holdings

		Hides	Plough-lands	Plough-teams
A.	Newbottle (224b)	6	15	7½
	Pattishall (226b)	8	20	14
	Kilsby (222b)	2	5	5
	Stoke Bruerne (228)	4	10	6
	Boddington (224)	1	2½	2½
B.	Abington (229)	4	8	3
	Ecton (225)	4	8	8
	Whiston (222)	1½	3	3½
	Hardwick (229)	1	2	3
C.	Duddington (219b)	1	8	4
	Oundle (221)	6	9	12
	Pilsgate (221)	6	6	12
	Tansor (220)	6	18	16

In section A of the table, which is concerned mainly with the south-west of the county, the ratio of hides to plough-lands is 2:5. The normal vill here seems to be one of 5 or 10 plough-lands, and this is clearly an artificial arrangement. Round has explained that the 2:5 ratio really represents a reduction of 60% in the assessment: 'the so-called plough-lands of the Northamptonshire Domesday are not plough-lands at all, but represent the old assessment before this great reduction. That is to say, that when a vill is entered as assessed at four "hides" and as containing 10 plough-lands, the combination really means that its assessment has been reduced from ten units to four.'[2] The rigid and conventional nature of this 2:5 system as applied to individual holdings is best seen where the assessment was given in terms of complicated fractions, e.g.:

---

[1] Maitland's total (*op. cit.* p. 400) of 2,931 included 29 vills not now in the county, together with the vills of Witchley hundred in Rutland. Maitland also made allowance for the missing plough-lands.

[2] J. H. Round, *V.C.H. Northamptonshire*, I, p. 264.

	Hides	Plough-lands
Grafton Regis (224)	$\frac{4}{5}$	2
Aynho (227)	$3\frac{1}{5}$	8
Evenley (226b)	$1\frac{3}{5}$	4
Croughton (227)	$\frac{1}{10}$	$\frac{1}{4}$
Syresham (223b)	$\frac{1}{2}$	$1\frac{1}{4}$

Round has explained[1] that two-fifths of a hide was the normal assessment of a plough-land in this part of the county—hence the strange complexity of fractions. Often, for example, the phrase 'half the fifth part of a hide' is employed. This curious quantity is described in the twelfth-century Survey as a 'small virgate', i.e. one-tenth of a hide instead of the usual virgate of one quarter of a hide.

In section B of the table the ratio of hides to plough-lands is 1:2, and this ratio is typical of many vills to the north of the Nene, in the central part of the county. The implication here, as before, is that the plough-lands represent an obsolete assessment, except that the reduction was only 50%.

In section C of the table, which refers to the north of the county, the reduction in the assessment was not uniform, but there are clear signs of a duodecimal arrangement underlying both assessment and plough-lands. This Danish system, typical of Lincolnshire and Leicestershire, was most obviously and intensely developed in the Soke of Peterborough; and it should be noted that a duodecimal arrangement may underlie the figures for some groups of vills, which when taken individually do not appear to show it. It is clear that here, too, the plough-lands were an arbitrary and highly artificial quantity.

Of course there are many entries in the county which cannot be fitted into any one of the three main systems just described, and no rigid territorial division of the county can be made on the basis of the distribution of the three types of plough-land assessment.

One result of the artificial nature of the plough-land figures is that on some holdings there is an excess of teams or apparent 'overstocking'; four examples of these are contained in the table on p. 394. For what it is worth, there was an excess of teams in 12% of the entries which record

[1] J. H. Round, 'The Hidation of Northamptonshire', *Eng. Hist. Rev.* (London, 1900), XV, pp. 78–86. For the 'small virgate' see also p. 389 n. above. For a further discussion, based upon a detailed analysis of the assessment, see C. Hart, *The Hidation of Northamptonshire* (Leicester, 1970).

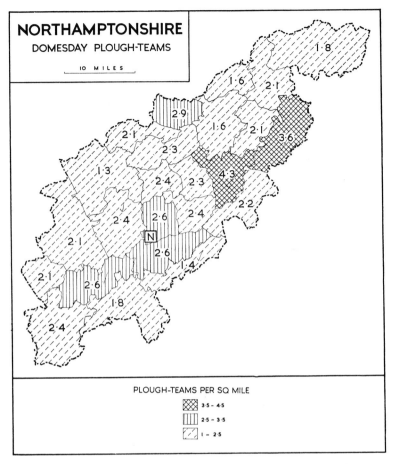

Fig. 138. Northamptonshire: Domesday plough-teams in 1086 (by densities).
N indicates the Domesday borough of Northampton.

both plough-lands and teams, and a deficiency in 38%. Occasionally, attention is drawn to the fact that a holding is overstocked. This is so in four entries in the account of Robert de Buci's fief (225–225 b). That for Brampton Ash reads: *Terra est i carucae. Tamen sunt in dominio ii carucae.* Similar formula are employed for the other three entries relating to Braybrooke, Rushton and Spratton.

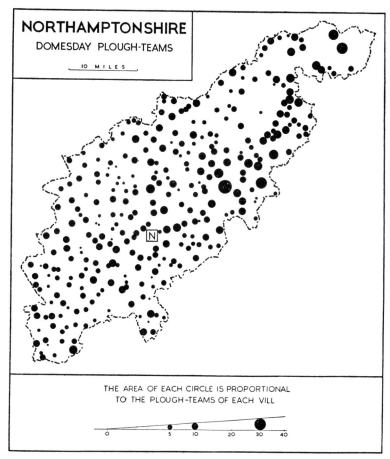

NORTHAMPTONSHIRE
DOMESDAY PLOUGH-TEAMS

10 MILES

THE AREA OF EACH CIRCLE IS PROPORTIONAL
TO THE PLOUGH-TEAMS OF EACH VILL

0   5   10   20   30   40

Fig. 139. Northamptonshire: Domesday plough-teams in 1086 (by settlements).
N indicates the Domesday borough of Northampton.

(3) *Plough-teams*

The Northamptonshire entries, like those for other counties, usually draw
a distinction between the teams held in demesne by the lord of the manor,
and those held by the peasantry. The formula employed for Courteenhall
(225 b) is typical: 'In demesne there are 2 ploughs with one serf; and
12 villeins, with one bordar and the priest, have 7 ploughs.' Occasionally
holdings with oxen are mentioned, e.g. at Billing (223). At Elkington (227)
we read: 'There is land for 3 oxen; 2 bordars who plough have these

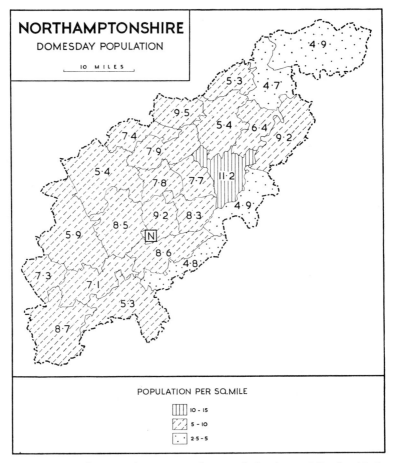

Fig. 140. Northamptonshire: Domesday population in 1086 (by densities).
N indicates the Domesday borough of Northampton.

there' (*Terra est iii boum* [*sic*]. *Hos habent ibi ii bordarii arantes*). Another
unusual entry is that for Preston Capes (224), where we are told that
'in demesne there is a moiety of a plough' (*In dominio est medietas
carucae*).

A total of 2,252⅝ teams is recorded for the area included in the modern
county. Maitland's total of 2,422 relates to the Domesday county, and
thus includes Witchley hundred, now in Rutland, as well as other vills
not now in the county.[1]

[1] F. W. Maitland, *op. cit.* p. 401.

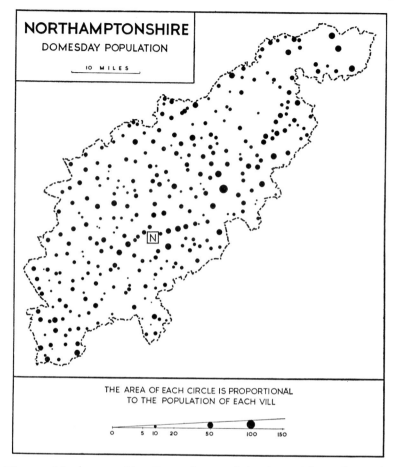

THE AREA OF EACH CIRCLE IS PROPORTIONAL
TO THE POPULATION OF EACH VILL

Fig. 141. Northamptonshire: Domesday population in 1086 (by settlements).
N indicates the Domesday borough of Northampton.

(4) *Population*

The main bulk of the population was comprised in the four categories of villeins, bordars, sokemen and serfs. In addition to these were the burgesses (at Northampton) and priests, together with a fairly small miscellaneous group that included knights (*milites*), Frenchmen (*francigenae*), smiths (*fabri*), 'men' (*homines*), serjeants (*servientes*), freemen (*liberi homines*) and a reeve (*prepositus*). The details of these groups are summarised on p. 401. The figures here set out are considerably lower than those of Ellis.[1] His total of 8,441 included tenants-in-chief, under-tenants, *ancillae*, and burgesses, together with the population recorded for many vills now in Rutland and other adjacent counties. A deduction of about 1,500 must be made from Ellis's total if the figures are to be compared. From the nature of Domesday Book it is unlikely that two people will arrive at identical results, and the statistics on p. 401 make no claim to definitive accuracy. Still they do indicate the order of magnitude involved, and the relative size of the different groups of the population. It must be remembered that these are figures of recorded population, and in order to obtain some idea of the actual population the figures must be multiplied by a factor—say 4 or 5—depending on one's views of the medieval family; but this does not affect the relative density as between one area and another.[2] That is all that a map, such as Fig. 140, can roughly indicate.

It is impossible for us to say how complete were these Domesday statistics. One entry certainly seems to be defective, and this relates to Thornby (226), where, although there was a plough-team and 4 acres of meadow, no population is entered. Another entry, relating to Grafton Underwood (228), is not precise for it tells us that one plough was there 'with certain men' (*cum quibusdam hominibus*).

The largest group of the agricultural population was that of the villeins, who accounted for almost exactly 50% of the total. In the second place came the bordars who numbered 26% of the total, while the sokemen accounted for a little over 12%, and it is interesting to note that this latter figure is higher than that for the same group in any of the adjacent counties, excepting Lincolnshire and Leicestershire. Very few sokemen are recorded in the hundreds to the west of Watling Street, but the percentage rises to the north, and in the Soke of Peterborough, adjacent to Lincolnshire, 36% of the recorded population were sokemen. Northamptonshire thus lies intermediately between the highly Scandinavian counties of Lincoln and Leicester to the north, and the non-Danish counties of the midlands to the south. Place-name evidence also supports this view. The serfs accounted for only 10% of the recorded population. About two-thirds of them were in the southern half of the county, where there were few sokemen.

---

[1] Sir Henry Ellis, *A General Introduction to Domesday Book* (London, 1833), II, p. 474.

[2] But see p. 430 below for the complication of serfs.

Recorded Population of Northamptonshire in 1086

A. Rural Population

Villeins	3,452
Bordars	1,808
Sokemen	849
Serfs	680
Miscellaneous	93
Total	6,882

There were also 57 bondwomen (*ancillae*) who have not been included in the above total.

Details of Miscellaneous Rural Population

Priests	61
Knights	14
Men (*Homines*)	7
Freemen	3
Francigenae	3
Smiths	2
Servientes	2
Reeve	1
Total	93

B. Urban Population

NORTHAMPTON   87 burgesses; 209 *domus*;
35½ waste houses (*domus* or *mansiones vastae*).

The priests are discussed on p. 414, and the burgesses on pp. 414–415. Of the miscellaneous group the smiths are mentioned on p. 415; the rest call for no comment. Finally, in addition to all the above groups there were 57 female serfs (*ancillae*) in the county. No thegns were recorded for 1086, but 20 for 1066. There were also 6 freemen in 1066, all on an unnamed holding in Higham Ferrers hundred (222b).

(5) *Values*

The value of an estate is normally given for 1066 and 1086, but sometimes for *quando recepit* and 1086 (e.g. Brackley, 224), and very rarely for all three dates (e.g. Wellingborough, 222b). A few entries give only the 1086 value, e.g. several on fo. 222b. Values are almost always stated in terms of pounds or their equivalent in shillings. Pence rarely appear except in such sums as 13s. 4d. for Dingley (225), 10s. 8d. for Brampton Ash (225) and 21s. 4d. for Scaldwell (228b). The two latter are a mark and two marks respectively on the reckoning of 16d. to the *ora*. They may therefore be but round figures like 20s. and 10s.

The 'values' are usually entered as plain statements of money, but there are occasional entries which indicate other methods of reckoning. We are specifically told that the values of some estates were reckoned by the tale or number, when coins were accepted at their nominal value. But in view of the circulation of debased coins, some values were reckoned by weight, and were also tested by assay; the latter was known as 'white' money. Some of the variations are indicated below:

> Northampton (219): *Ad canes xlii libras albas de xx in ora.*
> King's Sutton and Whitfield (219 b): *Modo xxxii libras de xx in ora.*
> Tansor (220): *T.R.E. reddebat xx libras ad numerum.*

The most significant feature of the Domesday values for the county is the recovery which took place between 1066 and 1086. With the exception of some 20 or so holdings, the 1086 value was the higher. The low figures for 1066 can be attributed to the ravages of the invading armies described in the Anglo-Saxon Chronicle for the year 1065, and it is clear, as Round wrote, that there was 'a general recovery in values as the manors were stocked afresh'.[1] The values of many holdings were at least doubled, and in the Nene and Cherwell valleys and in the Soke the increase in value was sometimes as much as fourfold, as is shown by the examples at the top of p. 403. At Aldwincle (222) some recovery is indicated by the increase in value from 20s. to 30s., but the Commissioners evidently envisaged still further improvement in the future for they wrote 'if it were well worked it would be worth 100s.' (*Valuit xx solidos. Modo xxx solidos. Si bene exerceretur c solidos valet.*) Occasional holdings were recorded as waste prior to

---

[1] J. H. Round, 'The Hidation of Northamptonshire', *Eng. Hist. Rev.* (London, 1900), XV, p. 78. See pp. 409–11 below for a discussion of 'waste'.

	Value 1066	Value 1086
Wittering (221 b)	£3	£11
Pilsgate (221)	20s.	£4
Castor (221)	20s.	50s.
Irthlingborough (221 b)	£3	£6
Knuston (227 b)	5s.	20s.

1086, e.g. Cranford (220b) and Rockingham (220), but that these had also recovered is shown by the fact that they were assigned values for 1086.

A reduction in value was recorded for about 20 holdings, many of which were in the extreme south of the county. Moreton Pinkney (227), for example, was worth £8 in 1066 but only £4 in 1086—a depreciation of 50%.

Generally speaking, the greater the number of plough-teams and men on an estate, the higher its value, but it is impossible to discern any constant relationships as the following figures for four holdings, each yielding £2 in 1086, show:

	Teams	Population	Other resources
Astwell (227)	5	17	Mill, meadow, wood
Clapton (222)	5	25	Meadow
Desborough (226)	3½	22	Mill, wood
Flore (223 b)	2	11	Mill, meadow
Haselbech (223)	9	19	Nil

It is true that the variations in the arable, as between one manor and another, did not necessarily reflect the variations in total resources, but even taking the other resources into account the figures are not easy to explain.

Conclusion

The general basis for calculating densities is that of the hundreds, but many adjustments have been made in detail. The twenty-four districts that result, although not giving as perfect a regional division as a geographer could wish for, do serve as a useful basis for distinguishing the degree of variation over the face of the county.

Of the five recurring standard formulae discussed above, only two may be regarded as reliable indices of the distribution of wealth and prosperity throughout the county. These are the statements about plough-teams and population. Neither is without some degree of uncertainty, but taken together they provide a general picture of land utilisation in the eleventh century. When the two distributions are compared, certain features stand out as common to both (Figs. 138 and 140). The highest densities were to be found in the Nene valley—up to over 4 teams and to about 11 recorded people per square mile. The poorest areas, on the other hand, were in the north and in the south-east. In the former area, the Rockingham Forest district and the Soke of Peterborough had plough-team densities of about 2 or under and population densities of about 5 per square mile. Similar low densities were to be found in the Whittlewood and Salcey districts along the south-eastern border of the county.

Figs. 139 and 141 are supplementary to the density maps, but it is necessary to make one reservation about them. As we have seen on p. 386, it is possible that some Domesday names may have covered two or more settlements, e.g. the present-day villages of Great and Little Houghton are represented in the Domesday Book by only one name. A few of the symbols should therefore appear as two or more smaller symbols, but this limitation does not affect the main pattern of the maps. Generally speaking, they confirm and amplify the information of the density maps.

WOODLAND

Types of entries

Woodland is described in the Northamptonshire folios in one of two ways. Generally a linear method of description is employed; less frequently the extent is given in terms of acres. The following entries may be regarded as typical of the two methods:

> Grafton Underwood (225 b): *Silva i leuua longa et iiii quarentenis lata.*
> Nobottle (225 b): *vi acrae silvae.*

The formula occasionally varied in detail. At Pytchley (229), for example, the wood was 3 furlongs *in longitudine et latitudine*; this has been plotted as 3 furlongs by 3. The exact significance of the linear entries is far from clear, and we cannot hope to convert them into modern acreages.[1] All we can do is to plot them diagrammatically as on Fig. 142.

Occasionally, one set of measurements is given in a combined entry

[1] See p. 437 below.

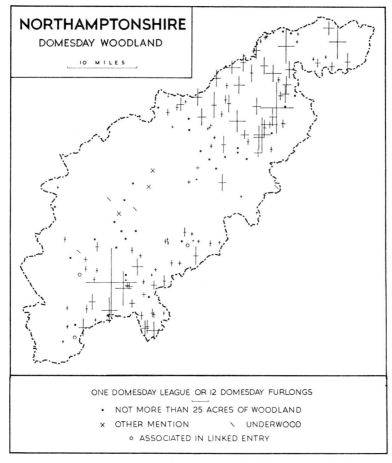

Fig. 142. Northamptonshire: Domesday woodland in 1086.
Where the wood of a village is entered partly in linear dimensions and partly
in some other way, only the dimensions are shown.

covering a number of villages. This seems to imply, although not necessarily, some process of addition whereby the dimensions of separate
tracts of woodland were consolidated into one sum. On Fig. 142 the
measurements in these combined entries have been plotted for the first-
named village only. The associated villages in all such composite entries
have also been indicated.

Wood is measured in acres for about 30 vills; the total is usually below

10 acres and never rises above 25. For some vills underwood (*silva minuta*) appears, sometimes as well as wood. There is mention of spinney (*spinetum*) at Ashley (225), Althorpe (223) and for a holding in Gravesend hundred (226b). The entry for Ashley mentions not only spinney, but wood 3 furlongs by 8 perches, and also 4 acres of wood in another place (*in alio loco*).

A few minor peculiarities may be noted in connection with the linear measurements. The occasional use of perches, e.g. at Paulerspury, suggests that some care was taken to make the wood entries as accurate as possible; here, there was 'wood 6 furlongs long and 4 furlongs and 2 perches wide' (226). The wood at Aldwincle (222) was worth 15*s*. when it bore mast (*Valet xv solidos cum oneratur*), and that at Greens Norton (219b) was 'worth 60*s*. when it bore mast, and for the honey 4*s*.'. At Oundle (221), the wood was worth 20*s*, when it bore mast; and at Fotheringhay (228) we are told that the wood was worth 10*s*. when it bore mast and the king did not hunt in it (*Silva i leuua longa et ix quarentenis lata cum oneratur et rex in ea non venatur valet x solidos*). At Cosgrove (223) there were 2 furlongs of underwood but no other measurement is added.[1] Other interesting entries relate to Lamport (226b) where an 'ash spinney' (*fraxinetum*) one furlong by one furlong is mentioned, and to Denton and Whiston (222b) where the wood was described as 'not for pannage' (*non pastilis*). At Gretton (219b) and at Corby (219b), wood is recorded in the normal linear way, and in addition we are informed that in each place, 'many things are wanting to this manor which in King Edward's time belonged to it in wood and ironworks and other matters'. This mention of wood and ironworks in the same sentence may possibly imply some connection between the two.

At Brixworth (219b) there is a solitary reference to forest. Wood and forest are mentioned, but no measurements are given. The entry says: 'To this manor pertains a wood which used to render 100*s*. a year. This is now in the king's forest (*foresta*)'. No other references to forest are to be found in the Northamptonshire folios.

Distribution of woodland

Fig. 142 shows that the recorded wood was located in two main areas, one in the north and one along the south-eastern border of the county.

---

[1] For such single dimensions see p. 437 below. There were no single dimensions for wood (as opposed to underwood) entered for Northamptonshire.

The wooded area in the north corresponded in a general way with the later Forest of Rockingham and its margins, extending to the Nene valley. That in the south-east likewise corresponded with the later Forests of Whittlewood and Salcey and their margins. Elsewhere there were only sporadic and small amounts recorded in terms of acres or in a miscellaneous fashion.

MEADOW

Types of entries

The extent of Domesday meadowland in Northamptonshire is almost always expressed in terms of acres. The formula is simply: 'there are *n* acres of meadow' (*n acrae prati*), and the amounts recorded vary from as little as one acre to as much as 107 acres. Sometimes a round figure like 20 or 30 acres is entered, and this suggests that merely an estimate of the area was made, but the figures usually give the impression of actual amounts, e.g. 3, 7, 16 acres. The meadow of composite entries has been divided equally among the component vills for purposes of mapping. As in the case of other counties, no attempt has been made to translate these figures into modern acreages. The Domesday acres have been treated merely as conventional units of measurement, and Fig. 143 has been plotted on that assumption.

A few variations from the usual formula may be mentioned. A linear method of description is employed for four places and no acres are recorded; the details are as follows:

Byfield (224b): 1 league by 7 furlongs.
Harringworth (228): 5 furlongs by 2 furlongs.
Oakley (228b): 4 furlongs by 3 perches.
Thorpe in Peterborough (221): 3 furlongs by 1 furlong.

In three other vills the meadow is described in terms of revenue; the entries are all on fo. 219b:

Fawsley: *De prato ii solidi exeunt.*
Hardingstone: *De pratis et pasturis lxvi denarii.*
King's Sutton: *De pratis xx solidi.*

No other meadow is entered for either Fawsley or King's Sutton, but at Hardingstone there was an additional 63 acres on two other holdings (219b and 228b). One other variation is recorded for Stanwick (210b), where there was 'meadow for 2 plough-teams' (*Pratum ii carucis*) and for Rushden (210) where there was meadow for 6 oxen; both entries are in the Bedfordshire folios.

Distribution of meadowland

The main feature of the distribution (Fig. 143) was the concentration of meadow in the valleys of the Nene and Ise. There were a few places with over 100 acres, e.g. Kettering and Billing, and a number with over 50 acres, but on the whole the totals for individual places were small. There was relatively little meadow in the area of the Soke, in the area of the Northampton uplands, or in the south of the county, and no well-marked distribution patterns emerge. Only very small amounts were located in the valleys of the Welland and the Cherwell—normally less than 20 acres per vill. The totals for the county as a whole are surprisingly low, as compared with say Lincolnshire, or Nottinghamshire, and the totals for the western part of the county were, as might be expected, especially small.

FISHERIES

Fisheries (*piscariae*) are not specifically recorded in the Northampton folios, but at seven places there were mills which rendered eels, so that there must have been fisheries at these places. The details are as follows:

	Number of mills	Render	
		Money	Eels
Little Addington (222)	1	12d.	200
Ashton (221 b)	2	40s.	325
Denford (220 b)	2	50s. 8d.	250
Oundle (221)	1	20s.	250
Raunds (220 b)	1	34s. 8d.	100
Wadenhoe (220 b)	1	13s. 4d.	65
Warmington (221 b)	1	40s.	325

Fig. 144 shows that these fisheries were located in the north-east of the county, in the Nene valley. Interestingly enough, no reference is made to fisheries in the fenland area of the Soke of Peterborough.

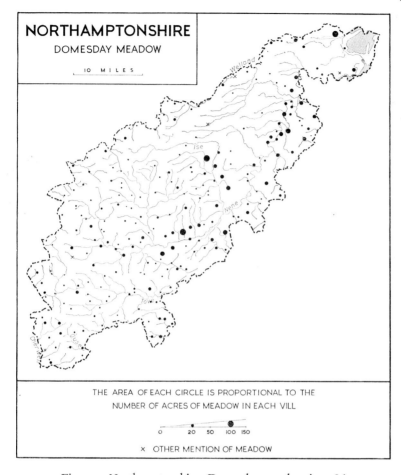

Fig. 143. Northamptonshire: Domesday meadow in 1086.

Where the meadow of a village is shown partly in acres and partly in some other way, only the acres are shown. The area of alluvium and peat is indicated. Rivers passing through this area are not shown.

WASTE

Waste does not figure prominently in the entries for Northamptonshire, although a study of the Domesday values and of the Anglo-Saxon Chronicle leads one to the view that the county had suffered considerable devastation prior to the Conquest. The latter document informs us that in 1065 :

Morkere went south with all the shire and with Nottinghamshire, and Derbyshire, and Lincolnshire, until he came to Northampton. But the northern men did much harm about Northampton...inasmuch as they slew men and burned houses and corn, and took all the cattle which they might come at, that was many thousand; and many hundred men they took and led north with them; so that the shire, and the other shires which are nigh were for many years the worse.[1]

The tale of havoc is also implicit in the Northamptonshire Geld Roll.[2] About one-third of the county was recorded as waste in this document. The low Domesday values of 1066 also suggest widespread devastation. It is surprising therefore that references to waste are so infrequent in the Domesday record for the county; as Round says, 'the devastation...had been well repaired at the time of Domesday; but we obtain a glimpse of it in the Rockingham entry: *Wasta erat quando rex W. jussit ibi castellum fieri. Modo valet xxvi solidos* (220)'.[3] Apart from this there are only three references to waste prior to 1086; they are as follows:

Cranford (220b): *Valet x solidos. Vasta fuit.*
Woodford (222): *Valet x solidos. Totum manerium wastum fuit cum acceperunt.*
In *Stotfald* hundred (226): *Vasta fuit et est.*

Cranford and Woodford were not completely wasted; there are other entries, relating to these places, which do not mention waste. The Domesday Book does not tell us precisely where the waste land in *Stotfald* hundred was located.

In 1086, one place (Foxley, 223b) was completely waste, and thirteen other places contained waste holdings. The formula used in all these entries is simply *vasta est*. Although waste, Foxley was said to be worth 5*s*. So was a holding at Charlton described in the preceding entry:

Charlton: *Vasta est. Tamen valet et valuit v solidos.*
Foxley: *Vasta est. Tamen valet v solidos.*

We have regarded these holdings as completely waste in the sense that they had no teams or population; their values were presumably derived from some other resource such as wood, though none is recorded. Each was said to be waste in 1086, but the Charlton formula (*valet et valuit*)

[1] B. Thorpe (ed.), *The Anglo-Saxon Chronicle* (London, 1861).
[2] The date of the document is 'probably 1072–78'—D. C. Douglas and G. W. Greenaway (eds) *op. cit.* p. 483.
[3] J. H. Round, *Feudal England* (London, 1895), p. 149.

makes us ask whether its holding was also waste at some earlier date. And if for Charlton, could this be true for Foxley? Moreover, we are told nothing about the pre-1086 condition of all the other holdings in the county said to be *vasta est*. Could they, too, have been waste before 1086? An unusual entry is that relating to Nortoft Grange (224) where we

Fig. 144. Northamptonshire: Domesday fisheries in 1086.

read: 'a church pertains to this land with one virgate of land in Guils-borough, and the site of a mill, with the third part of one virgate, in Hollowell. These are waste.' The distribution of the waste holdings is shown on Fig. 159.

Finally, in the borough of Northampton (219) some fourteen waste houses are mentioned in connection with the king's demesne; and, in addition to these, various landowners together had a further 21½ waste houses in the town, making a total of 35½.

MILLS

Mills are mentioned in connection with at least 155 of the 326 Domesday settlements within the area covered by modern Northamptonshire. It is difficult to be sure of the exact total because mills sometimes appear in composite entries covering a number of places. Thus one mill of 10s. is entered for Halse, Syresham and Brackley (224); two mills of 15s. are likewise entered for Greens Norton, Blakesley and Adstone (219b). For each entry the mill or mills have been plotted for the first named place.[1] The associated villages in all such entries have also been indicated on Fig. 145.

In each entry the number of mills is given, and also their annual value, ranging from a mill rendering 8d. (Easton Maudit, 226b) to one rendering 40s. and 325 eels (Warmington, 221b). Renders of both money and eels appear in a number of entries, the details of which are set out on p. 408. One mill does not appear to have been working, that at Hollowell (224), where a mill site is mentioned, and this had no value attributed to it.

Some mills were under divided ownership, and it looks as if two villages sometimes shared a mill. There may well have been some connection between the fractions entered for the adjacent villages of Rushton and Newton. The former (225b and 228b) had a mill and a half, and the latter (228b *bis*) also a mill and a half. But, on the other hand, many fractions cannot be assembled in this convenient manner. A 'part' of a mill was recorded for Easton Neston (227b) and other parts are mentioned at Flore (227b) and at Thenford (228). A total of 2½ mills was also recorded at Southorpe (221b). These places are widely separated and there is no apparent connection between the various fractions.

Domesday Mills in Northamptonshire in 1086

1 mill	104 settlements
2 mills	39 settlements
3 mills	9 settlements
4 mills	3 settlements

[1] The figure of 155 results from counting as one group the mills of a composite entry. When more than one mill is recorded in such an entry, and it is to be assumed that each was at a separate place, the total of places with mills is 157. If all the places named in the relevant composite entries are counted, the total becomes 160.

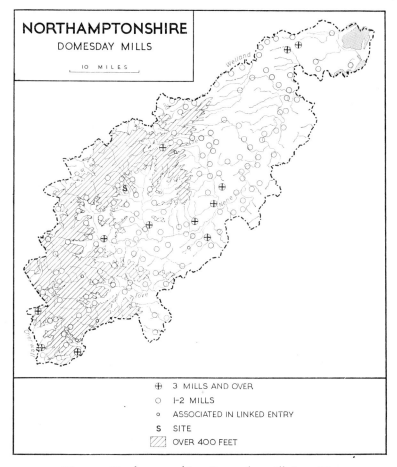

NORTHAMPTONSHIRE
DOMESDAY MILLS

10   MILES

⊕   3 MILLS AND OVER
○   1-2 MILLS
∘   ASSOCIATED IN LINKED ENTRY
S   SITE
▨   OVER 400 FEET

Fig. 145. Northamptonshire: Domesday mills in 1086.
The area of peat and alluvium is indicated. Rivers passing through
this area are not shown.

The groups of four mills were at Finedon (220, 220b) and at Harrington (222b) both on the Ise, and at Evenley (223b, 226b *bis*) on the Ouse.

It is clear from Fig. 145 that the Domesday mills were closely associated with the principal rivers of the county, and the great majority were situated below the 400 ft. contour. Well-marked linear distributions are characteristic of the Welland, Nene, Ise, Tove and Cherwell valleys, and of many tributary valleys. The only districts with few mills were Rockingham Forest and parts of the upland country in the west.

CHURCHES

The entries relating to churches are only three in number:

(1) The same [Robert] holds 2 hides in Nortoft...a church pertains to this land with one virgate of land in Guilsborough (224).

(2) Earl Aubrey held of the king 2 hides in Halse, and 2 hides in Syresham and one hide in Brackley, with a church and a mill of 10s. (224).

(3) Godwin holds of Walter 2 hides belonging to the church of Pattishall in Cold Higham (226b).

There were some sixty-one priests recorded in connection with sixty villages. Of the places with churches, Pattishall and Guilsborough apparently had no priest, but there was one at Halse. The only place with two priests was Cold Higham. One of the peculiarities of the Northamptonshire folios is that the priests are classed with the villeins and bordars and are not associated with churches. The small number of churches mentioned has been commented upon by a number of writers. Ellis, for example, observed that no injunction was laid on the jurors to make a return of churches, and said that 'unexceptionable evidence' had been adduced of the existence of several churches 'in Northamptonshire, which certainly are not noticed in the Survey'.[1]

URBAN LIFE

The Northamptonshire folios begin by giving an account of the borough of Northampton (219), which was the only place in the county where burgesses are recorded. We are told first that 'in King Edward's time there were in *Northantone*, in the king's demesne, 60 burgesses, having as many dwellings (*mansiones*). Of these [dwellings], 14 are now waste, 47 are left (*sic*).' But although the reference to waste suggests that the town had been subjected to some devastation during the Conquest, the process of rebuilding and extension had already begun in 1086. In addition to the 47 houses just mentioned there were 'in the new borough 40 burgesses (*in novo burgo xl burgenses*) in King William's demesne'. The Commissioners also recorded 209 houses (*domus*) and 21½ waste houses belonging to various landholders making a total of 296 houses and burgesses and 35½ waste houses. Among the 209 houses mentioned above,

[1] Sir Henry Ellis, *op. cit.* I, p. 287.

town houses belonging to a rural manor are mentioned only once; there were 21 houses 'pertaining to Stoke Bruerne' (219). The total population of the borough must therefore have been at least 1,500. The mention of the 'new borough' is interesting, and, as at Nottingham, the implication is that prior to 1070 there was an older English borough, beside which a new French borough was coming into existence after the Conquest.

No reference is made to the agricultural element in the borough; there were no plough-teams, wood, meadow, or any other of the normal rural features that formed part of some Domesday towns. Nor is any mention made of commercial activity, although it is probable that the town was a centre of trade.

MISCELLANEOUS INFORMATION

Ironworks and smiths

There are two entries relating to ironworks:

(1) Corby (219b): Many things are wanting to this manor which in King Edward's time belonged to it in wood and ironworks and other matters (*Multa desunt huic manerio quae T.R.E. ibi adjacebant in silva et ferrariis et aliis causis*).

(2) Gretton (219b): Very many things are wanting to this manor which in King Edward's time were appendant to it as well in wood and ironworks, as in other proceeds (*reditibus*).

Three other entries refer to smiths:

(1) Deene (222): 2 smiths (*fabri*) render 32s.

(2) Greens Norton (219b): The smiths used to render £7 in King Edward's time.

(3) Towcester (219b): The smiths used to render 100s., now [they render] nothing.

It is impossible to obtain any idea of the scope of the activities implied by these entries. There is no evidence in Domesday Book that local ore was worked, but it is interesting to note that all the places mentioned were located in well-wooded area, where presumably the supply of charcoal was abundant. It is probable that the smiths were in fact ironworkers, judging from the considerable sums of money which they rendered.

Markets

Further slight evidence of commercial activity in the county is afforded by the mention of markets at Higham Ferrers (225b), King's Sutton

(219b) and Oundle (221). The first two of these rendered 20s. per annum, that at Oundle 25s. The word *mercatum* is used at Higham Ferrers and Oundle, and *forum* at King's Sutton; the *forum* was doubtless a fair or market. It is curious that no mention of a market is made for Northampton itself.

Other references

A few solitary references may finally be noted. Rockingham (220) was waste when King William ordered a castle (*castellum*) to be built there. At Pytchley (222) there was a 'demesne building' (*dominicum aedificium*); and at Brafield on the Green (222b) we read of 'one house (*domus*) pertaining to Whiston, with 5 acres of land'. The bishop of Coutances claimed 1½ virgates at Isham (226b) in addition to '3 small gardens' (*hortuli*). A solitary reference to pasture (*pastura*) occurs on fo. 219b for Hardingstone, but no measurements are given (*De pratis et pasturis lxvi denarii*). Mention is also made of stock in an entry relating to Weston Favell (227b), where there was 'one villein with 3 beasts' (*habens iii animalia*) which we have counted as ploughing oxen.

REGIONAL SUMMARY

The fundamental contrast in the county is between the Nene valley in the east and the Northampton Heights in the west, which rise in places to over 700 ft. above sea-level. Much of the surface of valley and upland alike is covered by clays—Liassic, Oolitic, Oxford and Boulder. A division into six provides a convenient basis for considering the geography of the county in the eleventh century (Fig. 146).

(1) *The Nene Valley*

The Nene valley lies almost entirely below the 400 ft. contour. It includes not only considerable areas of heavy Oxford and other clays, but also strips of valley gravel and substantial stretches of Oolitic formations with lighter soils. It was closely settled, and the densities of teams (up to over 4 per square mile) and of population (up to over 11) were relatively high; here, too, in the valley was the county town of Northampton. The Nene and the Ise themselves were bordered by substantial amounts of meadow; there were many mills along them, and also some fishing

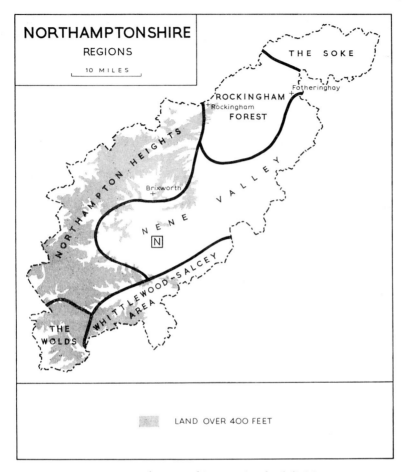

Fig. 146. Northamptonshire: Regional subdivisions.
N indicates the Domesday borough of Northampton.

activity along the Nene. There was but little wood apart from that bordering Rockingham Forest.

(2) *The Northampton Heights*

The greater part of this country lies over 400 ft. above sea-level, and in places it reaches to over 700 ft. It is covered largely by heavy Liassic clays, often associated with Boulder Clay. Most of the Domesday villages

DDG

avoided the clay valley bottoms, and may be described as 'hill-top' in character. It was an area of only moderate prosperity. The densities of teams ranged from 1·3 to 2·9 and those of population from 5·4 to 9·5 per square mile. A number of villages were without mills and there was only a little meadow and wood.

(3) *The Wolds*

Much of their surface lies over 400 ft. above sea-level and is covered by relatively light Oolitic soils. It was an area of moderate prosperity with a density of 2·4 per square mile for teams and 8·7 for population. The majority of villages had mills and a little meadow along the Ouse and other streams tributary to the Cherwell. Very few villages, on the other hand, had any wood.

(4) *The Whittlewood–Salcey area*

There was no mention or indication of the later forests of Whittlewood and Salcey in the Domesday account of Northamptonshire, but in 1086 the district certainly contained much wood and was a relatively poor area. The density of teams was under 2 per square mile and that of population about 5. Some villages had small quantities of meadow and a number also had mills.

(5) *Rockingham Forest*

There is no mention of Rockingham Forest as such in the Domesday account of Northamptonshire. But the entry for Fotheringhay (228) refers to the king's hunting, and that for Brixworth (219b) speaks of wood which formerly rendered 100s. but which was 'in the king's forest' in 1086. This may well refer to Rockingham Forest although Brixworth is some distance from the forest area as marked on Fig. 146. In post-Domesday times, Rockingham Forest extended beyond Brixworth as far south as the Nene. Whatever its extent in 1086, the northern part of the later forest contained much wood and was but sparsely settled—with about 2 teams and about 4 recorded people per square mile. Mills were frequent, but only a few villages had meadow, mainly in small amounts along the Welland. Nearby, at Rockingham itself (220), the king had built a castle.

(6) The Soke of Peterborough

In this area bordering the Fenland, villages were few in number, but some were quite large, e.g. Werrington with no less than 28 teams and 57 recorded inhabitants. Densities were low and somewhat similar to those of the forest areas—1·8 for teams and 4·9 for population. The majority of villages had mills and meadow and almost every village had some wood entered for it, occasionally in substantial amounts.

BIBLIOGRAPHICAL NOTE

(1) An early transcript of the Northamptonshire section of the Domesday Book is that of John Morton, in his 'Transcript of Domesday Book so far as it concerns Northamptonshire', printed at the end of *The Natural History of Northamptonshire*, by the same author (London, 1712).

The earliest translation is that of William Bawdwen, in the unpublished portion of his *Dom Boc: A Translation of the record called Domesday so far as it relates to the county of York, including also Lancashire, Westmoreland, Cumberland, Derby, Nottingham, Rutland and Lincoln, with an introduction, glossary and indexes* (Doncaster, 1809). The Northampton portion was translated in 1811, and now appears in British Museum MS. 27,769, compiled in 1867.

A later translation is that of S. A. Moore in *Domesday Book, the portion relating to Northamptonshire, extended and translated* (London, 1863).

The standard text now available is that of J. H. Round in the *V.C.H. Northamptonshire* (London, 1902), I, pp. 301–56. This contains a valuable series of identifications and footnotes and is preceded by one of the famous introductions by J. H. Round (pp. 257–98). It is followed by an account and translation of the Northamptonshire Survey, also by J. H. Round (pp. 357–92).

(2) The following deal with aspects of the Domesday study of the county:

J. MORTON, 'Remarks on Doomsday concerning Northamptonshire', B.M. MS. Sloane, No. 3560 (1720).

J. H. ROUND, 'The Northamptonshire Survey', *Feudal England* (London, 1895), pp. 215–24.

J. H. ROUND, 'The Hidation of Northamptonshire', *Eng. Hist. Rev.* (London, 1900), XV, pp. 78–86.

F. H. BARING, 'The Hidation of Northamptonshire in 1086', *Eng. Hist. Rev.* (London, 1902), XVII, pp. 76–83.

F. H. BARING, 'The pre-Domesday Hidation of Northamptonshire', *Eng. Hist. Rev.* (London, 1902), XVII, pp. 470–9.

C. P. BAYLEY, 'The Domesday Geography of Northamptonshire', *Jour.*

Northamptonshire Nat. Hist. Soc. and Field Club (Northampton, 1938), XXIX, pp. 1–22.

C. HART, *The Hidation of Northamptonshire* (Leicester, 1970).

(3) The Northamptonshire Geld Roll was discussed by J. H. Round in *Feudal England* (London, 1895), pp. 147–56. A translation appears in: (*a*) A. J. Robertson, *Anglo-Saxon Charters* (2nd ed. Cambridge, 1956), pp. 231–7 and 481–4; and (*b*) D. C. Douglas and G. W. Greenaway, *English Historical Documents, 1042–1189* (London, 1953), pp. 483–6.

(4) A very valuable aid to the study of the Domesday county is J. E. B. Gover, A. Mawer and F. M. Stenton's *The Place-Names of Northamptonshire* (Cambridge, 1933).

CHAPTER X

THE MIDLAND COUNTIES

BY H. C. DARBY, LITT.D., F.B.A.

The Domesday record for the Midland counties shows great diversity both in form and in content. Some of these differences arose from varying economic and social conditions. Others may reflect nothing more than the language and ideas of different sets of commissioners. In one respect, however, there was a comparatively high degree of uniformity. By far the greater part of the area was assessed in terms of hides and virgates. Only in Leicestershire and in the adjoining Rutland wapentakes of Alstoe and Martinsley were carucates and bovates to be found. To this generalisation there are a few exceptions. In the carucated shire of Leicester there were some references to 'hides', but this appears to be a peculiar use of the word; and in the hidated shires of Stafford and Northampton there are equally mysterious references to carucates and bovates. Then again, there is some mention of carucates in the country beyond the line of the Wye and of Offa's Dyke in Gloucestershire and Herefordshire; presumably, this latter area was a recent acquisition to the English realm, and had never been hidated.

The statement about plough-lands presents very baffling difficulties. For Gloucestershire, no information about plough-lands is given except in a few stray entries. But for the other counties, the following varieties of entry are encountered:

(1) For Northamptonshire, Rutland, Staffordshire and Warwickshire, the normal formula runs: 'There is land for *n* plough-teams'. This is also encountered in the later folios for Shropshire and in many folios for Leicestershire. For Leicestershire, at any rate, there seems to be a degree of correspondence with fiefs which may have employed 'different methods of making returns'.[1]

(2) For Herefordshire and Worcestershire, no plough-lands are entered, but we are frequently told that there could be other plough-teams in addition to those already at work. This variant occurs also in the earlier

---

[1] J. S. Moore, 'The Domesday Teamland in Leicestershire', *Eng. Hist. Rev.* (1963), LXXVIII, p. 701.

folios for Shropshire, and there are two somewhat similar entries in the Leicestershire folios.

(3) and (4) For Leicestershire, two other types of entry are found: 'n plough-teams were there', and 'In King Edward's time n plough-teams were there'. Both these are intermixed with other variants, and also show a degree of correspondence with fiefs.

It is tempting to believe that all four types of entry were, as Maitland and Vinogradoff argued, all giving the same information—the potential, as opposed to the actual, arable capacity of a holding. The fact that all four formulae occur, apparently interchangeably, in the Leicestershire folios lends colour to this view. Some of the minor variations in wording also support the idea. When, for example, we are told of the Leicestershire vill of Cotes-de-Val that 'there can be two ploughs and they are there' (231), we can reasonably assume that, on this holding at any rate, the land was being tilled up to capacity. The same assumption is suggested by a phrase at the end of the long account of the Worcestershire hundred of Oswaldslow (174), where we read that in none of the manors could there be more plough-teams than is stated.

On the assumption that all four formulae indicate the potential as opposed to the actual arable, we would expect the number of teams never to be more than the number of plough-lands. This is not so. On many Northamptonshire, Staffordshire, Warwickshire and Rutland holdings there was an excess of teams, and occasional excess was also to be found in Herefordshire and Shropshire.[1] At the Staffordshire village of Clifton Campville, for example, there were 13 teams but only 4 plough-lands (246b); clearly we are being told nothing about the potential arable here. It might be, of course, that the plough-lands represented some traditional estimate of conditions in 1066, and that there had been improvement, great improvement, since then. But this seems unlikely, especially in view of conditions in Staffordshire generally and, moreover, there is nothing to support such a view; the value of these holdings is usually less in 1086 than in 1066. Nor does the exceptional entry for the Herefordshire village of Preston on Wye point the way towards any general theory; here, there could be another plough on the demesne in addition to the one already there, but on the other hand, we are specifically told that the villeins had more ploughs than the land needed (181b).

[1] In eastern England there were excess teams on many Lincolnshire holdings, and on some in Cambridgeshire and Huntingdonshire.

We must also note the presence of a few sporadic entries that further complicate the issue. Thus on the Northamptonshire manor of Wittering there were 15 teams in 1086, but we are also told 'there is land for 16 teams. In King Edward's time 30 were there' (221b); four other places in the county are similarly described. Likewise at Wolverhampton in Staffordshire there seem to have been 19 teams in 1086, but again we are told 'there is land for 3 teams. In King Edward's time there were 8 teams there' (247b); the same form also appears in another Staffordshire entry for Weston, Beighterton and Brockton Grange (250b). We cannot assess the significance of such stray curiosities, nor can we hazard a guess as to how they fit into any general theory about the Domesday plough-land.

A further complication must also be noted. There seems to be a strong conventional element about the plough-land figures for some counties. This is certainly so for Leicestershire, Northamptonshire and Rutland. The plough-lands on many holdings in these counties, as in Lincolnshire, may well have been as artificial as the assessment. It was Maitland's 'horrible suspicion' that they were 'but remotely connected' with the agricultural realities of 1086.[1] For this reason alone, if for no other, we cannot use them in any attempt at reconstructing the economic geography of the Domesday Midlands. We must rely upon the plough-teams themselves.

One feature of the population statistics of the Midland counties is the very small free element, except in Leicestershire where the percentage amounted to nearly 30. The composition of the unfree population varied greatly, but easily the most important category was that of villeins, except in Worcestershire where the villeins were outnumbered by bordars. One feature of some Midland counties was the presence of a substantial miscellaneous group including such categories as radmen, oxmen, coscets and coliberts, to say nothing of Welshmen, priests and Frenchmen. Finally, another feature of the Midland counties as a whole was the high percentage of serfs along and near the Welsh border. In Gloucestershire they amounted to just a quarter of the total population, and in Shropshire, Herefordshire and Worcestershire they accounted for between 15 and 20% of the total. Some female serfs (*ancillae*) are also mentioned for every county but they are not included in the percentages set out above.

The statement about values is another variable element. Figures are most frequently given for 1066 and 1086, but a figure for an intermediate

[1] F. W. Maitland, *op. cit.* p. 427.

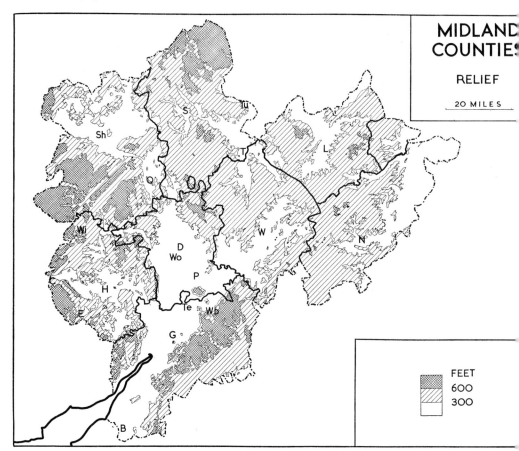

Fig. 147. Midland Counties: Relief.
Domesday boroughs are indicated by initials: B, Bristol; C, Clifford; D, Droitwich;
E, Ewias Harold; G, Gloucester; H, Hereford; L, Leicester; N, Northampton;
P, Pershore; Q, Quatford; S, Stafford; Sh, Shrewsbury; T, Tamworth; Te, Tewkes-
bury; Tu, Tutbury; W, Warwick; Wi, Wigmore; Wb, Winchcomb; Wo, Worcester.

date is sometimes given as well. There are also entries that give only one
value or even no value. All these variations are sometimes found in
entries relating to a single county. The Leicestershire entries raise particular problems, and it is possible that in them the earlier valuation refers
to 1068 and not to 1066. Renders in kind sometimes form part of the
valuation especially in Gloucestershire, Herefordshire and Worcestershire.

    The miscellaneous resources of each county are often measured differently and show great diversity. The wood entries, however, are relatively uniform, for woodland is usually measured in terms of linear dimensions, except that for Herefordshire, Shropshire and Northamptonshire other methods of measurement are important. Meadow is entered for every county except for Shropshire. Pasture is not normally mentioned except in a few sporadic entries in the folios for Gloucestershire, Herefordshire, Worcestershire, Warwickshire and Northamptonshire. Fisheries are not mentioned for Leicestershire or Rutland, although there must have been some along the rivers of those counties. Mills are regularly entered for every county. Churches, on the other hand, are only sporadically mentioned, and none at all are entered for Leicestershire and Warwickshire apart from those in the two county towns; and only two are mentioned incidentally in the Staffordshire folios. Among the miscellaneous items of the Survey, there are two that feature prominently in the folios of some Midland counties. One is the salt industry of Worcestershire, with its ramifications into other counties. The second is the 'waste' that is especially frequent in Herefordshire, Shropshire and Staffordshire. Finally, some border counties extended westward beyond their present-day limits into Wales, and we are thus able to get a glimpse of Welsh economy in the eleventh century.

Many of these differences are assembled in tabular form on pp. 452–8. These tables do not provide a complete statement of every variation in language and content within the nine Midland counties. They are intended only as a general guide to the salient features that have already been discussed in the preceding chapters. While the general framework of the Survey was the same for all counties, it is clear that the detailed recording of the information was far from uniform.

Composite maps have been made not for all the items of the Survey but only for those most relevant to an understanding of the landscape and economic geography—for settlements, plough-teams, population, woodland, forest, meadow, fisheries and waste. These maps must now be discussed separately. No attempt has been made to summarise the information relating to salt-pans in view of the general discussion of the Worcestershire salt industry on pp. 252–8.

SETTLEMENTS

Three reservations must be borne in mind when looking at a map of
Domesday names in the Midland counties. The first is the fact that
a number of Domesday names may have covered more than one settlement; the evidence of documents both before and after the date of the
Domesday Book shows that some names certainly did; others we cannot
be sure about. The so-called Leicestershire Survey, for example, which
dates from about 1129–30, mentions a number of villages which do not
appear in the Domesday Book and yet which must have existed in 1086;
their resources were presumably entered with those of neighbouring vills.
The second reservation springs from the summary nature of the information for part of Rutland and from the incomplete nature of the information for Archenfield and Ewias in south-western Herefordshire. The third
reservation arises from the fact that some Domesday names remain
unidentified and so cannot be marked on the map. Generally speaking,
this is not a serious omission, and it is always possible that subsequent
place-name investigation may clear up even these recalcitrant names. In
spite of these reservations, a map of Domesday names probably gives
a fair general picture of the intensity of settlement over the areas as a whole.

The first impression given by Fig. 148 is one of surprising uniformity;
names appear almost everywhere over the Midlands. But closer inspection
shows some areas where they are relatively few. The open area in south-
western Herefordshire, and the emptiness of much of Rutland, are, as
we have seen, due to the nature of the Domesday information for these
districts. But there are other empty areas on Fig. 148 which reflect the
absence of settlement in 1086. In Gloucestershire, the empty Forest of
Dean is clearly to be seen; so is the Clun area in the south-west of
Shropshire, and the Malvern area in the south-west of Worcestershire.
There were three empty areas in Staffordshire—the Pennine district in
the north, and Cannock Chase and Needwood Forest in the middle of
the county. Charnwood Forest in Leicestershire and the Soke of Peterborough also are conspicuously negative areas. Finally, the sprinkling of
names is fairly thin in the Arden region of Warwickshire. Other relatively
lightly settled areas may also be discerned, but these are the main ones.

The unsystematic and incomplete nature of the statistics for the
boroughs make it impossible to discuss the importance of urban life in
the Domesday Midlands. The economic activities of all nineteen boroughs

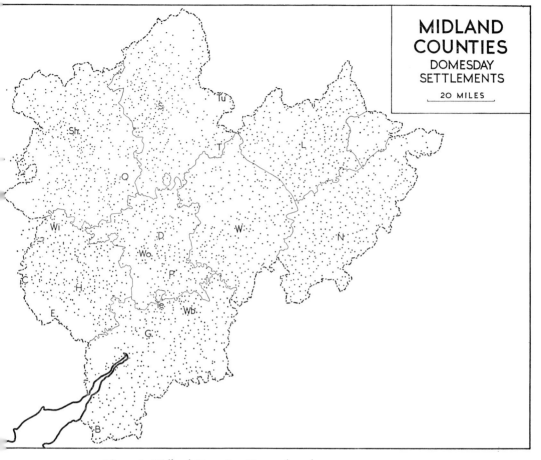

Fig. 148. Midland Counties: Domesday place-names.
Place-names to the west of the modern counties appear in the Domesday folios for Gloucestershire, Herefordshire and Shropshire. Domesday boroughs are indicated by initials: B, Bristol; C, Clifford; D, Droitwich; E, Ewias Harold; G, Gloucester; H, Hereford; L, Leicester; N, Northampton; P, Pershore; Q, Quatford; S, Stafford; Sh, Shrewsbury; T, Tamworth; Te, Tewkesbury; Tu, Tutbury; W, Warwick; Wi, Wigmore; Wb, Winchcomb; Wo, Worcester.

are wrapped in such obscurity that it is difficult to arrive at any clear idea of the relative importance of the agrarian and commercial elements in each. Markets are entered for only four of the boroughs—Hereford, Tewkesbury, Tutbury and Worcester. Seven other places also had markets. It is interesting to note that of the total of eleven markets, none is entered for Rutland, Shropshire or Warwickshire. A little information about mints and churches helps to fill out the detail given for some towns, but it is too uncertain to be of much value in obtaining a clear picture. Droitwich stands in a separate category from the other boroughs, and, with its salt industry, was an industrial as well as a commercial centre. But although we are told much about salt-making at Droitwich, it is impossible to get the industrial, commercial and agricultural elements in the economic life of the borough clearly into focus. Yet it was clearly an important centre with widespread connections extending into a number of counties (Fig. 87).

POPULATION AND PLOUGH-TEAMS

The density maps in the earlier edition did not take account of the figures for the boroughs. But in the present volume, the plough-teams and the rural population of the boroughs have been taken into consideration in the calculation of densities. The urban element itself (e.g. burgesses and houses) has been disregarded, and therefore in estimating the areas of the density units a quarter of a square mile has been deducted for each borough. As in other volumes the boundaries between the various density units inevitably have an artificial appearance because they are based upon administrative divisions.

The degree of economic development, as indicated by density of population, shows a general contrast between the relatively prosperous south and east on the one hand and the poorer north and west on the other hand. In the south and east, the density of population ranges around 10 per square mile. Over much of the area it is below, but it rises above in many localities—in the Feldon (in southern Warwickshire), in eastern Leicestershire, in Rutland and in parts of Gloucestershire and Northamptonshire. To the north and west, the density never rises above 10 per square mile, and over much of the area it is below 5. Thus the poor western part of Leicestershire around Charnwood stands in contrast with the rest of the county; the northern part of Warwickshire stands in like contrast with the Feldon, and this area of low development continues westward into Worcestershire. Staffordshire is outstandingly poor, due to the large number of waste vills and to the upland nature of much of the county. For similar reasons low densities are also found over much of Shropshire and in parts of Herefordshire.

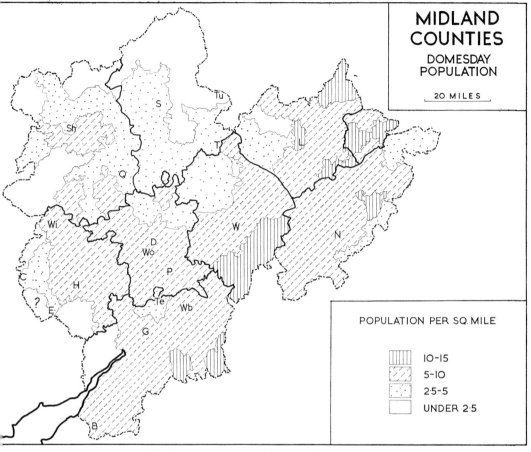

Fig. 149. Midland Counties: Domesday population in 1086.

Domesday boroughs are indicated by initials: B, Bristol; C, Clifford; D, Droitwich; E, Ewias Harold; G, Gloucester; H, Hereford; L, Leicester; N, Northampton; P, Pershore; Q, Quatford; S, Stafford; Sh, Shrewsbury; T, Tamworth; Te, Tewkesbury; Tu, Tutbury; W, Warwick; Wi, Wigmore; Wb, Winchcomb; Wo, Worcester.

It must be remembered that the densities on Fig. 149 refer not to total population but to recorded population. As Maitland said: 'Domesday Book never enables us to count heads. It states the number of the tenants of various classes, *sochemanni*, *villani*, *bordarii*, and the like, and leaves us to suppose that each of these persons is, or may be, the head of a household.'[1] Whether this be so or not, the fact remains that in order to obtain

[1] F. W. Maitland, *op. cit.* p. 17.

the actual population from the recorded population, we must multiply the latter by some factor, say 4, or perhaps 5, according to our ideas about the medieval family.[1] This, of course, does not affect the value of the population map for making comparisons between one area and another. There are, however, a number of other considerations that have to be borne in mind in interpreting the population map. And in the light of these, it may well be that the distribution of plough-teams provides a more certain index of the relative distribution of prosperity.

One reservation arises from the fact that we cannot be sure that all heads of households were counted. Thus in Gloucestershire, Leicestershire and Herefordshire there are references to landholders 'with their men' (cum suis hominibus), and to an unspecified number of villeins, and to 'certain others' (aliqui alii); this vagueness seems to indicate the presence of some unrecorded householders. A second reservation arises from the fact that serfs may stand in a different position from other categories of population. They may have been recorded as individuals and not as heads of households. Villages with many serfs may therefore appear to be relatively more populous than they really were. Maitland put the problem, but gave no answer: 'Whether we ought to suppose that only the heads of servile households are reckoned, or whether we ought to think of the servi as having no households but as living within the lord's gates and being enumerated, men, women and able-bodied children, by the head—this is a difficult question.'[2] Vinogradoff also considered the problem, and as he said, hesitated to construe the numbers of serfs as indicating individuals and not heads of households.[3]

Whatever the answer, the distribution of serfs as between one hundred and another, and between one county and the next, is uneven, so that the problem (if there be one) becomes increasingly acute as the county maps are assembled together. It is especially acute in the western parts of England where there was a very high percentage of serfs. Fig. 150 has been constructed in an attempt to meet this difficulty. The serfs have been regarded as individuals and not as heads of households. Their numbers have been divided by the arbitrary figure of four before calculating

[1] F. W. Maitland suggested 5 'for the sake of argument' (op. cit. p. 437). Russell has more recently suggested 3·5; J. C. Russell, British Medieval Population (University of New Mexico Press, Albuquerque, U.S.A., 1948), pp. 38, 52.

[2] F. W. Maitland, op. cit. p. 17.

[3] P. Vinogradoff, English Society in the Eleventh Century (Oxford, 1908), pp. 463–4: 'members of their families...must have been omitted.'

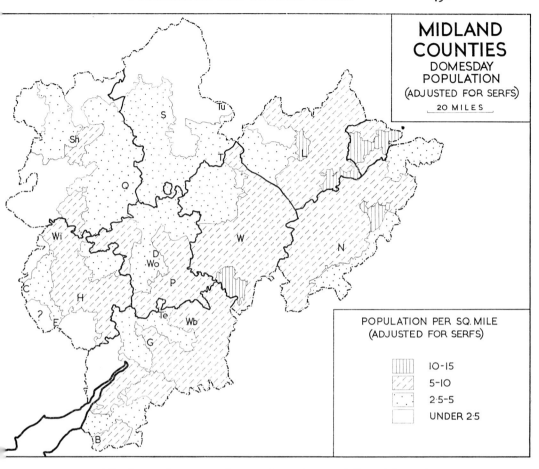

Fig. 150. Midland Counties: Domesday population in 1086 (adjusted for serfs).

On this map, the serfs have been regarded as individuals and not as heads of households. Their numbers have been divided by the arbitrary figure of four before calculating densities of population. Domesday boroughs are indicated by initials: B, Bristol; C, Clifford; D, Droitwich; E, Ewias Harold; G, Gloucester; H, Hereford; L, Leicester; N, Northampton; P, Pershore; Q, Quatford; S, Stafford; Sh, Shrewsbury; T, Tamworth; Te, Tewkesbury; Tu, Tutbury; W, Warwick; Wi, Wigmore; Wb, Winchcomb; Wo, Worcester.

densities of population per square mile. To what extent this gives a more accurate picture we cannot say. In detail the pattern of the map differs from that of Fig. 149, but the same general contrast between the south and east and the poorer north and west can be observed. What Fig. 150

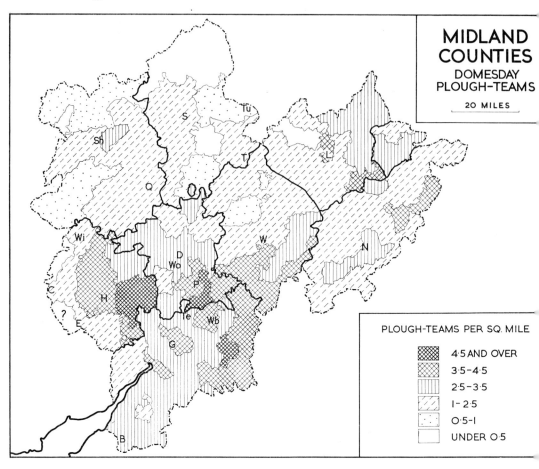

Fig. 151. Midland Counties: Domesday plough-teams in 1086.

Domesday boroughs are indicated by initials: B, Bristol; C, Clifford; D, Droitwich; E, Ewias Harold; G, Gloucester; H, Hereford; L, Leicester; N, Northampton; P, Pershore; Q, Quatford; S, Stafford; Sh, Shrewsbury; T, Tamworth; Te, Tewkesbury; Tu, Tutbury; W, Warwick; Wi, Wigmore; Wb, Winchcomb; Wo, Worcester.

does not show, however, is that while the southern and eastern areas now have fewer than 10 people per square mile, much of their surface still carried over 7·5 people per square mile. The greatest reduction is apparent in Gloucestershire for as much as one-quarter of its recorded population comprised serfs: some parts of Gloucestershire are seen to carry even less than 5 people per square mile, but nowhere does the density fall

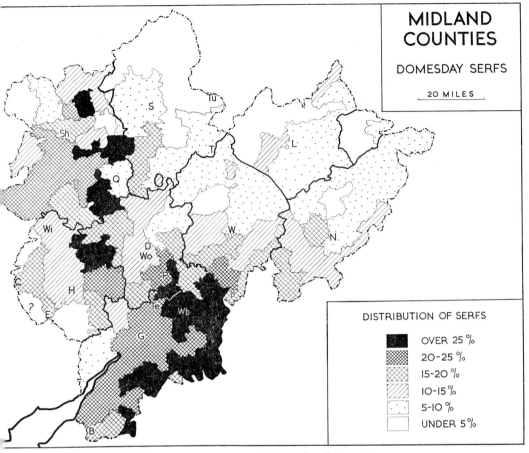

Fig. 152. Midland Counties: Distribution of serfs in 1086.

Domesday boroughs are indicated by initials: B, Bristol; C, Clifford; D, Droitwich; E, Ewias Harold; G, Gloucester; H, Hereford; L, Leicester; N, Northampton; P, Pershore; Q, Quatford; S, Stafford; Sh, Shrewsbury; T, Tamworth; Te, Tewkesbury; Tu, Tutbury; W, Warwick; Wi, Wigmore; Wb, Winchcomb; Wo, Worcester.

appreciably below 4. Over large stretches of the north and west, on the other hand, the density is well below 4 and even below 2. Thus the complications introduced by Fig. 150, while appreciable, do not invalidate the generalisation that the poorest part of the Midland counties lay to the north and west.

In a general way, the contrast is confirmed by Fig. 151 showing the distribution of plough-teams per square mile. But there is one substantial difference; Gloucestershire and the adjoining parts of Herefordshire, Worcestershire and Warwickshire stand out as areas of high development more prominently than they do on the population maps. The Severn-Avon lowland in particular appears as a very highly cultivated area. The eastern plain of Hereford and parts of the Cotswolds dip-slope are also well above the general level of prosperity.

Three other maps have been drawn to show the distribution of certain elements in the population. Fig. 152 shows the distribution of serfs as a percentage of the total recorded population for each density unit. While serfs were relatively much more numerous in the western counties than those to the east, it is interesting to note that the area with the greatest number of serfs does not lie in the most extreme west. The lands beyond the Severn have fewer serfs, both relatively and absolutely, than some other areas. Fig. 153 shows the same information plotted on the assumption that the serfs were recorded as individuals. The percentages on this second map are very much lower, but the same general pattern of distribution appears. Fig. 154 shows the distribution of the free peasantry, freemen, sokemen and radmen being reckoned together for this purpose. The radmen are mainly in the western counties, but the number is so small that their inclusion does not really affect the broad pattern of the map. The highest percentage of free peasantry is to be found in the Scandinavian areas of the eastern Midlands—in Northamptonshire, Rutland and particularly in Leicestershire. If this information be re-calculated with adjustment for serfs, many of the percentages of the free peasantry show some increase, but the differences are such that they do not appreciably modify Fig. 154, certainly not enough to justify a separate map. Supplementary to these maps are the tables on pp. 457–8 which summarise the relative importance of all the different categories of population in each county as a whole. Any discussion of the significance of these maps and tables lies far beyond the scope of this study. Moreover, their full implications cannot begin to be appreciated until they are set against similar data for the whole of England.

WOODLAND

For the most part, the woodland of the Midland counties was measured in terms of its length and breadth, and the units of measurement were the league, furlong and perch. The length of a Domesday league (*leuga*,

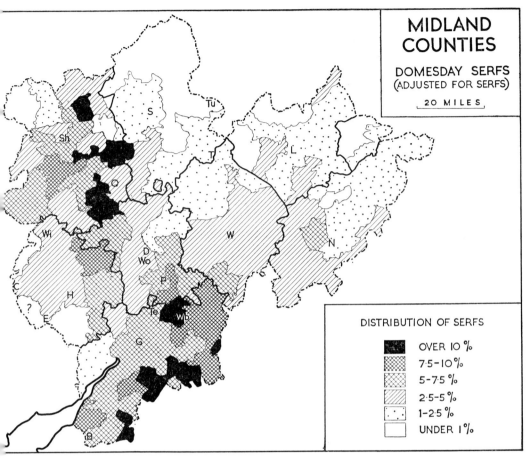

Fig. 153. Midland Counties: Distribution of serfs in 1086 (adjusted for serfs).

On this map, the serfs have been regarded as individuals and not as heads of house-holds. Their numbers have been divided by the arbitrary figure of four before calculating percentages. Domesday boroughs are indicated by initials: B, Bristol; C, Clifford; D, Droitwich; E, Ewias Harold; G, Gloucester; H, Hereford; L, Leicester; N, Northampton; P, Pershore; Q, Quatford; S, Stafford; Sh, Shrewsbury; T, Tamworth; Te, Tewkesbury; Tu, Tutbury; W, Warwick; Wi, Wigmore; Wb, Winchcomb; Wo, Worcester.

leuca or *leuua*) has been a matter for much discussion. The twelfth-century Register of Battle Abbey in Sussex states that a league comprised 12 furlongs and that a furlong comprised 40 perches. This would make a league equivalent to 1½ miles, but it must be remembered that the

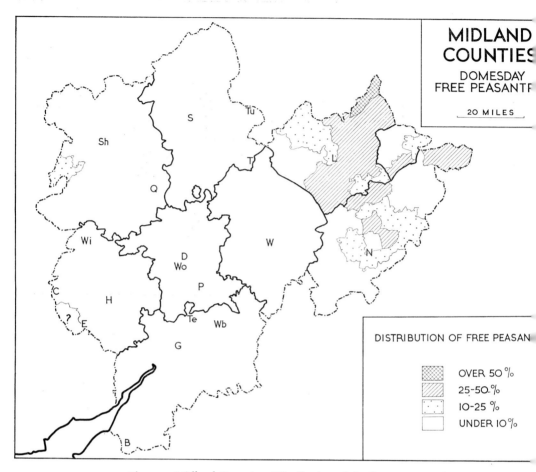

Fig. 154. Midland Counties: Distribution of the free peasantry in 1086.

The free peasantry includes freemen, sokemen and radmen. Note that along the western boundary of Shropshire, there is an area with over 10%. This high figure (16·5%) is due to the presence of radmen. The exclusion of radmen would have meant the disappearance of this shading, but would have made no other difference to the map. Domesday boroughs are indicated by initials: B, Bristol; C, Clifford; D, Droitwich; E, Ewias Harold; G, Gloucester; H, Hereford; L, Leicester; N, Northampton; P, Pershore; Q, Quatford; S, Stafford; Sh, Shrewsbury; T, Tamworth; Te, Tewkesbury; Tu, Tutbury; W, Warwick; Wi, Wigmore; Wb, Winchcomb; Wo, Worcester.

number of feet in a perch is obscure, and that the whole subject is complicated by local usage and by the existence of local or 'customary', as distinct from standard, units. Furthermore, in his Domesday study of Worcestershire, J. H. Round thought that a league might well have comprised only 4 furlongs (i.e. about half a mile). Our knowledge of the measures characterising different districts is far too slight to allow us to speak with confidence on these matters. All we can do for each county is to regard the figures as providing some rough indication of the distribution and relative size of wooded tracts.

Quite apart from the problem of the size of the units, there are other difficulties; the exact significance of this linear type of entry is not clear. Is it giving extreme diameters of irregularly shaped woods, or is it making rough estimates of mean diameters, or is it attempting to convey some other notion? We cannot tell, and we certainly cannot assume that a definite geometrical figure was in the minds of the Domesday commissioners. Nor can we hope to convert these measurements into modern acreages by any arithmetical process. It must be emphasised, therefore, that it would be rash to make any assumption about the superficial extent of woodland measured in this way. All we can safely do is to regard the dimensions as conventional units and to plot them diagrammatically as intersecting straight lines.

Occasional entries give the dimensions of wood as 'in length and breadth' or 'between length and breadth'. Other entries, particularly for Shropshire and Worcestershire, give only one dimension, and we are merely told that there was, say, one league of wood. R. W. Eyton and C. S. Taylor thought that all such measurements implied 'areal leagues' or 'areal furlongs'.[1] The evidence of the *Liber Exoniensis*, however, seems to indicate that they all are variants used when length and breadth were the same.[2] We have followed this assumption in plotting Fig. 155 but have shown single dimensions as single lines to indicate their frequency.

It occasionally looks as if these Domesday lengths and breadths represent in some way a sum total of separate tracts of woodland. This is particularly so in composite entries. Thus the composite entry that covers

[1] R. W. Eyton, *A Key to Domesday: the Dorset Survey* (London, 1878), pp. 31–5; C. S. Taylor, 'An analysis of the Domesday Survey of Gloucestershire', *Trans. Bristol and Gloucs. Archaeol. Soc.* (Bristol, 1887–9), p. 59.

[2] H. C. Darby and R. Welldon Finn (eds), *The Domesday Geography of South-west England* (Cambridge, 1967), pp. 374–5 and 382–3.

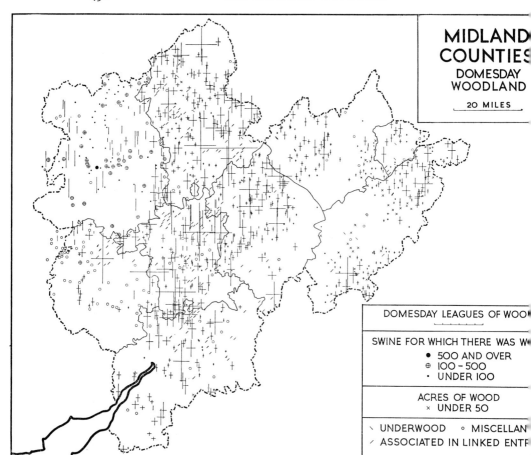

Fig. 155. Midland Counties: Domesday woodland in 1086.

Uckington and six other places in the Gloucestershire folios mentions wood 2½ leagues long by one league and 2 furlongs wide. The mere fact that one set of measurements can thus be given in a combined entry covering a number of villages (often some distance apart) seems to imply a process of addition whereby the dimensions of separate tracts of woodland have been consolidated into one total. But if this were so, it is difficult to imagine the arithmetical processes that lay behind the result. Conversely, it is impossible for us to resolve the composite measurements into their components. The measurements in these combined entries have been plotted for the first-named village only, or, in the case of a manor and its dependencies, for the manor itself. This introduces local inaccuracy, but it does not substantially affect the general picture of the distribution of woodland over a large area. The associated villages in all such composite entries have been indicated.

While linear dimensions constitute the normal method of measuring wood in the Midland counties, there is a wide variety of miscellaneous entries. We are told of small woods, of very small woods, of large woods, of woods yielding so many shillings or pence, of unproductive woods, of pasturable woods; and there are other variants and idiosyncrasies. Some entries mention brushwood, spinney and underwood. But of the various non-linear methods of indicating the presence of wood, the two most important are those giving its size in acres and those stating the number of swine it could feed. Along the eastern borders of Leicestershire, for example, measurements in terms of acres are intermingled with linear ones, and they are also frequent in Northamptonshire. Outside these two counties they are rare. In an entry for Staunton Harold, in Leicestershire, both types of measurement appear together: 'Wood 5 furlongs in length and 3 furlongs in breadth, and on the other side (*ex altera parte*) 4 acres of wood' (233). It may be that the smaller woods were measured in terms of acres, and the larger ones by giving their lengths and breadths. The only Midland county for which swine measurements predominate is Shropshire. The formula here is more explicit than for eastern England where swine measurements are usual: we are specifically told in most of the Shropshire entries that the wood was for fattening swine (*porcis incrassandis*).

It is unfortunately impossible to bring these acre and swine measures into relation with those of a linear character by reducing all three units to a common denominator. Occasionally, we are given glimpses of possible relationships between the different units, but there is no foundation for a constant and reliable ratio. The entries which suggest such a relationship are as follows:

Ashby de la Zouch (Leicestershire): Wood one league in length and 4 furlongs in breadth, sufficient for (*ad*) 100 swine (233).

Crowle (Worcestershire): Half a league of wood sufficient for (*ad*) 100 swine (176b).

Worthen (Shropshire): Wood 2 leagues long in which are 4 hays, and it is sufficient for fattening 200 swine (255b).

Leighton (now in Montgomeryshire): Wood there 2 leagues long, and it is
sufficient for fattening 200 swine (255 b).

Even were it possible to equate these entries, we could not be sure that
the relation between swine and dimensions was constant over a county,
to say nothing of over all England.

The wood entries relating to two of the nine Midland counties are of
exceptional interest because they include indications of the clearing that
was going on, and that had already transformed the face of the countryside. If, however, we expect to find many references to this clearing, we
shall be disappointed. But the cartloads of wood that were used by the
salt industry of Droitwich and that are entered in the Worcestershire
folios, point to the toll that the industrial demands of the time were
levying upon the woodlands. The details are not specific but the inferences
are clear enough. The Herefordshire evidence is even more striking. At
Much Marcle there were 58 acres 'reclaimed from the wood', and the
word 'assart' is written above 'reclaimed'; at Leominster, the profits of
the assarts in the wood were 17s. 4d.; while both at Fernhill and at
Weobley land for one plough had been reclaimed. These four stray
references are not much, but there is no reason to believe that what
happened in Herefordshire did not happen in other counties. The general
form of the Domesday entries has concealed the activity that is accidentally revealed here. But although clearing was taking place generally, the
reverse process was going on in some localities. Land devastated by
raiding or by the march of armies, soon became overgrown with thicket
and wood if allowed to remain unattended. For Herefordshire again, the
Domesday Book records plough-lands in Hezetre hundred which had
been wasted and become overgrown with wood; and elsewhere in the
county there was other land which had 'all been converted into woodland'.
Similar growth must have taken place on some of the wasted lands of
other counties, though the Domesday Book tells us nothing of it. In any
case, such regeneration of woodland was exceptional to the main trend
of the time.

In addition to woodland there was forest land about which the Domesday Book says very little. The word 'forest' is neither a botanical nor
a geographical term, but a legal one. The origin of the word has been
disputed but one view is that it implied an area outside (*foris*) the common
law, and subject to a special law that safeguarded the king's hunting.

Forest and woodland were thus not synonymous terms, for the forested areas included land that was neither wooded nor waste. Whole counties were sometimes placed under forest law. But still, a forested area usually contained a nucleus of wood, and sometimes large tracts of wooded territory. The existence and extent of these forests before 1066 is an obscure matter, but it is certain that after the Norman Conquest the forest law and forest courts of Normandy were introduced into England on a large scale. A rapid extension of forest land took place. These forests, being royal property outside the normal order, are rarely mentioned in the Domesday Book. The existence of some of them, however, is indicated by various statements to the effect that a holding, or a part of it, or sometimes its wood only, was 'in the forest', or 'in the king's forest'. A number of entries do not specifically use the word forest, but merely speak of the king's wood, or wood in the king's demesne, or in the king's enclosure, or in the hand of the king. These entries, although not specific, may well imply an extension of forest rights. For many counties 'hays', or enclosures for catching deer, are also entered, and they are particularly frequent in the Shropshire folios. A few parks are also mentioned.

The relation of these occasional Domesday references to the royal forests of the thirteenth century is shown on Fig. 156. As can be seen, the Domesday Book gives no hint of the existence of some of these forests in the eleventh century. If many did exist, it follows that Fig. 155 may not show all the wood that was present at the time, for although forests did not necessarily imply wood, tracts of wood were usually important elements in forested areas. Thus, only a small quantity of wood is marked for the Forest of Dean, but it is certain that there was much wood there in 1086. Stray Domesday hints testify to the existence of this Forest, but we can glean nothing of its extent nor of the amount of wood within it. Only about the New Forest does the Domesday Book give us any detail, and that lies outside the Midland counties.

Fig. 155 shows the distribution of woodland as recorded in the Domesday Book. How near the record was to the reality we can never say. But, in spite of many uncertainties, and difficulties, the wooded character of much of the Midland counties in the eleventh century is abundantly clear. The most densely wooded area stretched from the Arden country of north Warwickshire, northward and westward into the adjoining counties of Stafford and Worcester, although there were many open tracts within this general spread of wood. Further west, along the Welsh border, there was also a good deal of wood in Herefordshire and Shropshire; and to the south, in Gloucestershire, the Forest of Dean was certainly wooded, although that is not revealed by Fig. 155. The remainder of Gloucestershire, Vale and Cotswolds alike, carried only a scattered cover of wood. In the eastern districts of the Midlands, three main areas of wood stand out: (1) that of western Leicestershire, around and possibly in, the Charnwood region; (2) that of eastern Leicestershire, stretching into Rutland and across the Welland valley into Northamptonshire, where there was a considerable tract of wood in and about the district later occupied by Rockingham Forest; (3) that of southern Northamptonshire.

The earlier history of the Midlands had destroyed any close correlation between the most densely wooded areas and those of the heaviest soil. It is true that the main Warwick-Worcester-Stafford area was very largely covered by clays derived from Keuper Marl and glacial drift, but there were many expanses of clayland that had long been cleared. The Lower Lias clays of the Feldon, for example, carried hardly any wood in 1086; and the same was very largely true of the similar clays of the Vale of Gloucester, the Vale of Evesham and the Vale of Belvoir. Much had changed since the coming of the Anglo-Saxons, and, by 1086, the Midland countryside was well on its way to becoming the open land we know today.

MEADOW

The entries relating to meadow in all the Midland counties, except Shropshire, are normally in terms of acres. Meadow is not mentioned in the Shropshire folios, but the modern county includes seven places surveyed in the Staffordshire section of the Domesday Book, and the meadow of these is recorded in the usual Staffordshire manner. There must have been meadow in Domesday Shropshire, but, for some reason unknown to us, it was not entered in the Domesday Book as we know it. The number of acres of meadow to be found in a village ranged up to over 100 and occasionally even to over 150. But generally speaking, amounts above 50 acres were infrequent, and in Herefordshire and Staffordshire there were few places with over 25. The range of figures over all eight counties thus resembles those of Norfolk and Suffolk rather than the large quantities to be found in Lincolnshire.[1]

[1] For the problems raised by the large figures for Lincolnshire, see H. C. Darby, *The Domesday Geography of Eastern England* (Cambridge, 3rd. ed., 1971), p. 366.

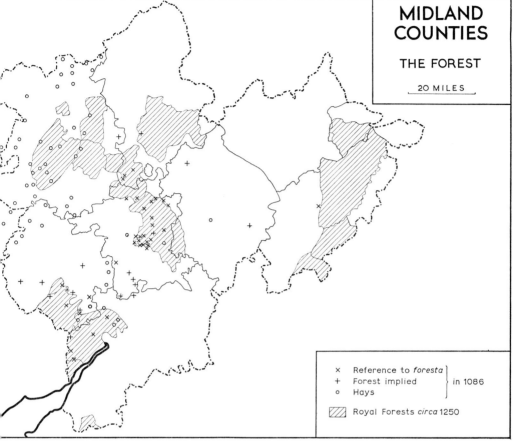

Fig. 156. Midland Counties: Domesday and later Forests.

The extent of the royal forests about A.D. 1250 is taken from M. L. Bazeley, 'The extent of the English Forest in the Thirteenth Century', *Trans. Roy. Hist. Soc.* (London, 1921), 4th series, IV, p. 148.

While acres form the normal unit of measurment, there are some exceptional entries that express the amount of meadow in other ways, e.g. in terms either of a money render or of the number of oxen the hay of a meadow could support, or by stating its length and breadth. These exceptional entries are scattered through the folios of all eight counties, but they are particularly frequent in those for Gloucestershire, Leicestershire and Herefordshire.

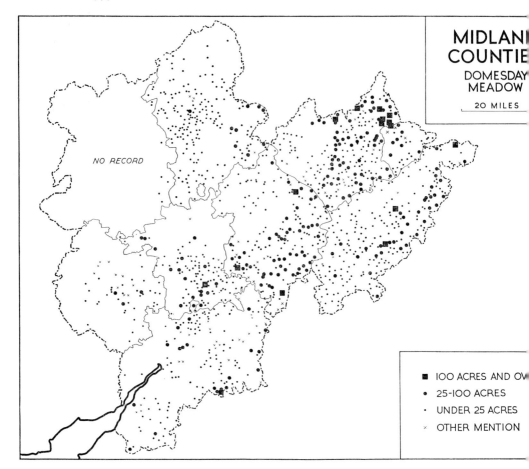

Fig. 157. Midland Counties: Domesday meadow in 1086.

Fig. 157 summarises the distribution of meadow over the eight counties, that is over all except Shropshire. In Northamptonshire, the line of meadows along the alluvial valley of the Nene stands out. In Leicestershire, there are very substantial amounts along the valleys of the Soar, the Wreak and the Eye; there is also some along the Welland in the south of the county. In Warwickshire, tracts of meadow are to be seen along the Avon and its tributaries the Leam and the Itchen, and along the Stour in the south of the county. The line of Avon meadows continues through the southern part of Worcestershire up to the point where it joins the Severn. One interesting feature of the map is the relative in-frequency of substantial entries along the Severn itself. There are some 25-acre tracts along the Gloucestershire Severn, but there is no belt of large Severn meadows through Gloucestershire and Worcestershire. Along the Trent and its tributaries in Staffordshire, and along the Wye and its tributaries in Herefordshire, only small amounts are accredited to each village—sometimes not more than 10 acres, and only occasionally more than 20 acres.

FISHERIES

Fisheries are recorded in a variety of ways. Sometimes we are merely told that there was one (or more) fishery on a holding; but more usually the render is also stated, either in money, or eels, or in both. The number of eels is given sometimes as a straightforward figure, sometimes in terms of *stiches*, the medieval stick of eels consisting of 25. Moieties or halves of fisheries are frequent, but it is not always possible to assemble the fractions of neighbouring holdings together in any comprehensible fashion. A number of entries make no mention of fisheries (*piscariae*) as such, but refer to eels in connection with mills, and these eels were presumably from the mill-stream or mill-pond. Thus at Martley in Worcestershire, there was a mill worth 8*s.* and 2 weirs (*gurgites*) which rendered 2,500 eels and 5 sticks. Obviously, the reference is to some form of fish-trap, but such detail is most exceptional. In general, the Domesday folios tell us little or nothing about the gear of fishermen or about their methods of catching fish. Eels are the only species mentioned, apart from renders of 16 salmon at Gloucester and of 6 salmon at *Turlestane* in Herefordshire.

On Fig. 158 the information has been consolidated into two categories. From the sum total of the evidence, it is clear that a fishery must have been one of those elements of wealth only sporadically mentioned in the Domesday Book. No fisheries are recorded for Leicestershire, yet it is difficult to believe that some did not exist along the Soar and the Wreak. Rutland is another county without fisheries, and only two are entered for Staffordshire. Over the rest of the Midlands, there is a marked alignment of fisheries along the Nene in Northamptonshire, and along the Severn, the Wye and their numerous tributaries in the west. But these can hardly have been all the fisheries of those rivers. Thus there were considerable stretches of the Severn for which no fisheries are recorded, but where there

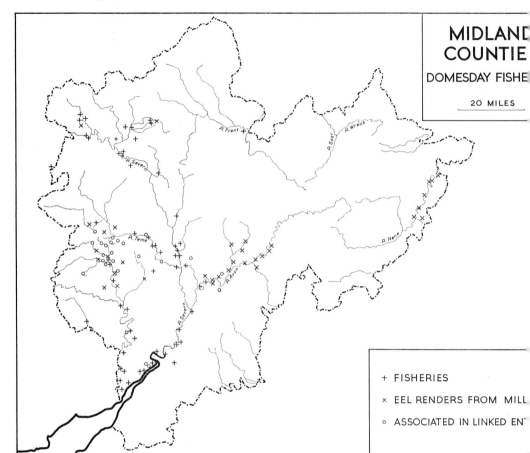

Fig. 158. Midland Counties: Domesday fisheries in 1086.

surely must have been some. They went unmentioned perhaps because they made no contribution to the profits of their respective villages, or maybe because of some idiosyncrasy of the local jurors.

WASTE

The Domesday term 'waste' implies not the natural 'waste' of mountain, marsh or heath, but land that had gone out of cultivation, mainly, it seems, as the result of deliberate devastation, but also perhaps, because of some local vicissitude that is lost to us. One feature of the Midland area in 1086 was the very considerable amount of this waste, especially in the west. It was, of course, not as great as that in the north of England, but on the other hand, it was very much greater than that of the southern and eastern counties. The Norman records make frequent reference to the destruction wrought by the armies of the king in their task of subduing revolt. The impression conveyed by these records has been summed up by Sir Frank Stenton:

> Their generalities are abundantly borne out by the evidence of Domesday Book, which shows that within the country ravaged at this time vast areas were still derelict after seventeen years. It is in Yorkshire that the desolation is most evident. But the oldest account of the harrying states that it also extended over Cheshire, Shropshire, Staffordshire, and Derbyshire, and Domesday Book proves that the devastation in those parts, though less complete than in Yorkshire, was on the same general scale. The object of the harrying was to secure that neither Mercia nor Northumbria should ever revolt again. It was the most terrible visitation that had ever fallen on any large part of England since the Danish wars of Alfred's time.[1]

Most frequently the whole of a vill still lay waste in 1086, but sometimes only one of two or more holdings comprising a vill was waste. There were a few holdings described as waste and yet accredited with some small value; presumably the arable in these was waste but the other appurtenances rendered some profit. It is possible that not all the waste was due to the march of armies. Some waste entries may reflect the hazards of farming in lean years. Others may be due to the encroachment of the king's forest. We are specifically told that *Haswic* in Staffordshire was waste for this reason—*Modo est wasta propter forestam regis*; one hide at Chacepool in the same county was also waste for the same reason. But, in general, these local vicissitudes are hidden from us.

Gloucestershire had only two waste holdings in 1086—Whittington which had also been waste in 1066 (182) and half a hide of the five-hide manor of Awre (163). There had, moreover, been a waste hide at Staunton (181), but this was said to be 'in the king's wood in 1086' and invites comparison with the Staffordshire holdings at *Haswic* and Chacepool mentioned above. No waste entries are encountered for Rutland, and relatively few for Northamptonshire, Warwickshire and Worcestershire.

---

[1] F. M. Stenton, *Anglo-Saxon England* (Oxford, 1943), pp. 596–7.

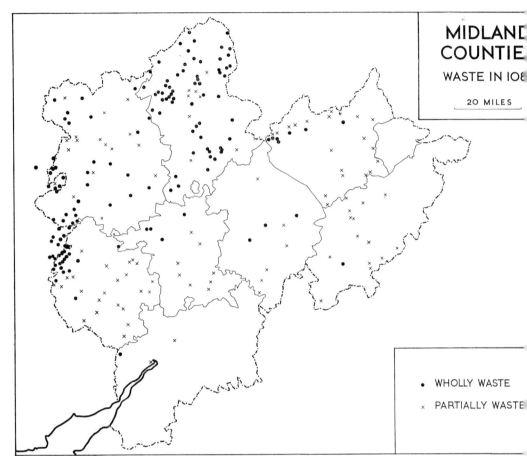

Fig. 159. Midland Counties: Domesday waste in 1086.

One of the Warwickshire entries, for Harbury, specifically ascribes the waste to the army of the king (*per exercitum regis*). This explicit statement is exceptional, but in most cases the inferences are only too clear. In Leicestershire, most of the waste holdings seem to reflect 'the Conqueror's march from Warwick to Nottingham when he suppressed the first revolt of Edwin and Morcar towards the close of 1068'.[1] The Staffordshire folios bear even more traces of the consequences of crushed revolt. Holdings at over one-fifth of the villages of the county were waste in 1086. Not all

[1] *V.C.H. Leicestershire*, I, p. 284.

this devastation must be attributed to the royal armies, for, as we have seen, *Haswic* and Chacepool were waste on account of the king's forest. The sphere of devastation extended westwards into Shropshire, and it so happens that the Shropshire entries give us some information about the pre-1086 value of many estates. There were waste holdings at maybe as many as 54 vills in 1066; this number increased possibly to as many as 128 vills in, say, 1070; and in 1086 it was still as high as 51 vills. The entry for Stoke upon Tern is representative: 'In the time of King Edward it was worth £6, and afterwards it was waste. Now it is worth £7' (256b). We may suppose that there had been similar recovery at many places in other counties, and we can occasionally discern recovery in Leicestershire and Northamptonshire.

It is probable that not all the waste of Shropshire was the result of King William's work. Even before he came, as we have seen, many villages lay waste, the result apparently of Welsh raids; and some of these places in the west and south of the county were still waste in 1086. The handiwork of the Welsh raiders could also be seen in Herefordshire. Here, especially along the western border, there were waste holdings at possibly as many as 63 vills in 1066; and by 1086 the number was still as high as 50 vills. Finally, in the Gloucestershire account of the lands beyond the Wye, we are told of four villages wasted by the Welsh King Caradoc.

Whatever its cause, Welsh raiding or Norman strategy, the sum total of the evidence set out on Fig. 159 leaves us in no doubt about the importance of devastation as an element in the economic geography of the eleventh-century Midlands. Were it possible to reconstruct a similar map for, say 1070, it is certain that the wasted villages would be more numerous and more widespread. The picture we see on Fig. 159 is that of a country-side already on its way to recovery.

SUMMARY OF DOMESDAY BOOK FOR THE MIDLAND COUNTIES

County	Assessment	Plough-lands	Plough-teams
Gloucestershire	Hides and virgates; very rarely acres 5-hide unit in well over 25% of places Carucates beyond the Wye	None except in a few stray entries	Rare mention of teams that could be added
Herefordshire	Hides and virgates; but some carucates west of Offa's Dyke Evidence of 5-hide unit	None except in a few stray entries	Frequent mention of teams that could be added
Shropshire	Hides and virgates; very rarely acres Evidence of 5-hide unit	Two types of entry: (1) Teams that could be added (earlier folios). (2) *Terra est n carucis* (later folios)	Understocking in 88% of entries
Staffordshire	Hides and virgates. Some carucates, difficult to interpret Evidence of 5-hide unit	Normally *Terra est n carucis*	Understocking in 56% of entries. Excess teams in 22%
Worcestershire	Hides and virgates Evidence of 5-hide unit in about 15% of places	None except in a few stray entries	Frequent mention of teams that could be added
Warwickshire	Hides and virgates Evidence of 5-hide unit in over 25% of places	Normally *Terra est n carucis*	Understocking in 48% of entries. Excess teams in 24%
Leicestershire	Carucates and bovates Evidence of 6-carucate unit in about 30% of places Mysterious hides at 19 places, and virgates at 5 places	Four types of entry Clear evidence fo artificiality Many omissions	Understocking in 51% of entries. Excess teams in 26%
Rutland	Carucates in Alstoe and Martinsley Hides in Witchley Both artificial Bovates for 3 places	Normally *Terra est n carucis* Clear evidence of artificiality	Varying understocking and excess teams
Northamptonshire	Hides and virgates; very rarely acres Bovates for 11 places Frequent 4-hide unit	Normally *Terra est n carucis* Clear evidence of artificiality	Understocking in 38% of entries. Excess teams in 12%

Population	Values	No. of Place-names
Main groups: villeins, serfs, bordars Varied miscellaneous group including radmen and coliberts 93 *ancillae*	Normally for 1066 and 1086 Some renders in kind. *Firma noctis* on 3 manors in 1066 and 1086.	367
Main groups: villeins, bordars, serfs Very varied miscellaneous group including oxmen, radmen and Welshmen 12 *ancillae*	Normally for 1066 and 1086; sometimes for 3 dates Some renders in kind. *Firma noctis* on one manor, apparently commuted by 1086	313
Main groups: villeins, bordars, serfs Very varied miscellaneous group including oxmen, radmen and Welshmen 8 *ancillae*	Normally for 1066 and 1086; but very often for 3 dates, when the intermediate value is usually little or nothing	437
Main groups: villeins, bordars, serfs Small miscellaneous group Only one *ancilla*	Normally only for 1086; but sporadically for 1066 as well; and, very rarely, also for an intermediate date	334
Main groups: bordars, villeins, serfs Very varied miscellaneous group including radmen and oxmen 31 *ancillae*	Normally for 1066 and 1086; sometimes for 3 dates Some renders in kind	260
Main groups: villeins, bordars, serfs Very varied miscellaneous group 4 *ancillae*	Normally for 1066 and 1086; sometimes for 3 dates One render in kind	281
Main groups: villeins, sokemen, bordars, serfs Small miscellaneous group 3 *ancillae*	Normally for 1086 and for some earlier date; the latter has been held to be 1068 and not 1066	296
Main groups: villeins, bordars, sokemen Miscellaneous group includes only serfs and priests Only 2 *ancillae*	Alstoe and Witchley: for 1066 and 1086 Martinsley: only for 1066, with two exceptions	39
Main groups: villeins, bordars, sokemen, serfs Small miscellaneous group 7 *ancillae*	Normally for 1066 and 1086; sometimes for 3 dates	326

County	Wood	Meadow	Pasture	Marsh
Gloucestershire	Normally in linear dimensions Some exceptions	Normally in acres Amounts over 20 acres rare	Only for 9 places	None
Herefordshire	Frequently in linear dimensions Many exceptions Assarts at 4 places	Normally in acres Amounts over 20 acres rare	Only for 2 places	None
Shropshire	About two-thirds in swine; one third linear. Some exceptions. Linear entries mostly give only one dimension	None in the Domesday county, but for at least 6 other places in acres	None	None
Staffordshire	Normally in linear dimensions Occasional exceptions Some underwood	Normally in acres Amounts over 20 acres rare. A few linear entries	None	None
Worcestershire	Normally in linear dimensions; some entries give only one dimension Some exceptions	Normally in acres Amounts over 50 acres rare	For at least one place	None
Warwickshire	Normally in linear dimensions Occasional exceptions	Normally in acres Amounts over 50 acres rare A few linear entries	Only for 6 places	None
Leicestershire	Normally in linear dimensions Some exceptions, including acres	Normally in acres Amounts up to over 100 acres Some linear entries	None	None
Rutland	Normally in linear dimensions Some acres Occasional underwood	Normally in acres No amounts over 40 acres A few linear entries	None	None
Northamptonshire	Normally in linear dimensions Some acres Occasional underwood	Normally in acres Amounts up to over 100 acres A few linear entries	Only for one one place	None

Fisheries	Salt-pans	Waste	Mills	Churches
For at least 16 places	For 10 places Note connection with Droitwich	For 2 places Also houses at Gloucester	For at least 126 places	For 10 places
For at least 18 places; eel renders from mills at 9 of these	For 11 places Note connection with Droitwich	For 50 places	For at least 72 places	For 13 places
For at least 18 places; eel renders from mills at 2 of these	For 3 places Note connection with Droitwich	For 51 places Also houses at Shrewsbury	For 88 places	For 22 places
For 2 places; eel renders from a mill at one of these	None	For 69 places Also houses at Stafford	For at least 50 places	For 2 places
For at least 21 places; eel renders from mills at 4 of these	Large salt-producing centre at Droitwich, described in some detail	For 12 places Also houses at Worcester	For at least 70 places	For 10 places
Eel renders from mills at 11 places	For 6 places Note connection with Droitwich	For 9 places Also houses at Warwick	For at least 94 places	For 2 places
None	None	For 25 places Also houses at Leicester	For at least 89 places	For one place
None	None	None	For at least 19 places	For 4 places
Eel renders from mills at 7 places	None	For 14 places Also houses at Northampton	For at least 155 places	For 3 places

County	Miscellaneous	Boroughs
Gloucestershire	Composite entries important for non-territorial hundreds of Berkeley, Deerhurst and Tewkesbury Forests for 6 places. Hays for 3 places Vineyard for Stonehouse Castles for Gloucester, Sharpness; and also for Chepstow Markets for Berkeley, Cirencester, Tewkesbury, Thornbury Renders of iron for Alvington, Gloucester, Pucklechurch Interesting Welsh entries	Bristol Gloucester Tewkesbury Winchcomb
Herefordshire	Forests for 13 places. Hays for 12 places Castles for Clifford, Ewias Harold, Richards Castle, Wigmore; and also for Caerleon and Monmouth One market, presumably at Hereford Render of iron for *Turlestane*, and 25 smiths elsewhere Interesting Welsh entries	Clifford Ewias Harold Hereford Wigmore
Shropshire	Many unnamed berewicks Forest for one place. Hays for 36 places Castles for Oswestry, Shrewsbury, Stanton; and also for Montgomery Interesting Welsh entries	Quatford Shrewsbury
Staffordshire	Forests for 5 places Castles for Burton, Stafford, Tutbury Market for Tutbury	Stafford Tamworth Tutbury
Worcestershire	Forests for 27 places. Hays for 3 places Vineyard for Hampton by Evesham Castle for Dudley Market for Worcester	Droitwich Pershore Worcester
Warwickshire	Forests for 2 places. Hay for one place Castle for Warwick *Ferraria* and 2 smiths for Wilnecote	Warwick
Leicestershire	Market for Melton Mowbray	Leicester
Rutland	County as such non-existent in Domesday times Linear measurements for the 3 manors with 19 unnamed berewicks in Martinsley hundred	None
Northamptonshire	Forest for one place Castle for Rockingham Markets for Higham Ferrers, King's Sutton, Oundle Smiths for 2 places	Northampton

MIDLAND COUNTIES

Summary of Rural Population in 1086

This summary includes the apparently rural element in the boroughs—
see the respective county summaries.

A. *Total Figures*

	Free-men	Soke-men	Villeins	Bordars	Cottars	Serfs	Others	Total
Gloucestershire	144*	—	3,730	1,871	7	2,052	250	8,054
Herefordshire	71*	—	1,728	1,285	19	730	617	4,450
Shropshire	191*	—	1,985	1,198	15	922	595	4,906
Staffordshire	13	6	1,669	886	—	229	40	2,843
Worcestershire	40*	—	1,653	1,732	54	721	209	4,409
Warwickshire	23*	—	3,807	1,965	6	893	148	6,842
Leicestershire	6	1,904	2,643	1,362	—	403	110	6,428
Rutland	—	113	1,064	254	—	34	14	1,479
Northamptonshire	3	849	3,452	1,808	—	680	90	6,882
Total	491	2,872	21,731	12,361	101	6,664	2,073	46,293

* Including radmen.

B. *Percentages*

	Free-men	Soke-men	Villeins	Bordars	Cottars	Serfs	Others
Gloucestershire	1·8	—	46·3	23·2	0·1	25·5	3·1
Herefordshire	1·6	—	38·8	28·8	0·5	16·4	13·9
Shropshire	3·9	—	40·5	24·4	0·3	18·8	12·1
Staffordshire	0·5	0·2	58·7	31·2	—	8·0	1·4
Worcestershire	0·9	—	37·5	39·3	1·2	16·4	4·7
Warwickshire	0·3	—	55·6	28·7	0·1	13·1	2·2
Leicestershire	0·1	29·6	41·1	21·2	—	6·3	1·7
Rutland	—	7·6	71·9	17·2	—	2·3	1·0
Northamptonshire	0·1	12·3	50·2	26·2	—	9·9	1·3
Total	1·1	6·2	46·9	26·7	0·2	14·4	4·5

MIDLAND COUNTIES

General Summary

For the various doubts associated with individual figures, see the text. The assessment, plough-lands, plough-teams and rural population of the boroughs are included in these totals, but it must be noted that the information given for the boroughs is often very fragmentary. The assessment is for 1066, and includes non-gelding hides.

	Settle-ments	Assessment	Plough-lands	Plough-teams	Rural pop.	Boroughs
Gloucestershire	367	2,422 h	—	3,823	8,054	4
Herefordshire[1]	313	1,192 h 24 c	—	2,463	4,450	4
Shropshire	437	1,403 h	3,133[2]	1,834	4,906	2
Staffordshire	334	441 h 42 c	1,317	980	2,843	3
Worcestershire	260	1,368 h	—	2,009	4,409	3
Warwickshire	281	1,503 h	—	2,276	6,842	1
Leicestershire	296	2,383 c 19 h[3]	—	1,882	6,428	1
Rutland	39	76 h 41 c	328	415	1,479	0
Northamptonshire	326	1,181 h	2,641	2,253	6,882	1
Total	2,653	—	—	17,935	46,293	19

c—carucates; h—hides.
[1] Figures incomplete; see p. 57 above.
[2] For doubts associated with this figure, see p. 124 above.
[3] For the complexity in Leicestershire, see p. 328 above.

APPENDIX II

EXTENSION AND TRANSLATION OF FRONTISPIECE

(Part of folio 165b of Domesday Book)

EXTENSION

X. TERRA SANCTI PETRI DE GLOWECESTRE

In Dudestanes hundredo. Sanctus Petrus De Glowecestre tenuit Tempore Regis Edwardi manerium Bertune cum membris adiacentibus Berneuude, Tuffelege, Mereuuent. Ibi xxii hidae una virgata minus. Ibi sunt in dominio ix carucae et xlii villani et xxi bordarii cum xlv carucis. Ibi xii servi et molinum de v solidis et cxx acrae prati et silva v quarentenis longa et iii lata. Valuit viii libras. Modo xxiiii libras. Hoc manerium quietum fuit semper a geldo et ab omni regali servitio.

Ipsa eadem ecclesia tenuit Frowecestre. *In Blacelawes hundredo.* Ibi v hidae. In dominio sunt iiii carucae et viii villani et vii bordarii cum vii carucis. Ibi iii servi et x acrae prati et Silva iii quarentenis longa et ii quarentenis lata. Valuit iii libras, modo viii libras.

Ipsa ecclesia tenet Boxewelle. *In Grimboldestowes hundredo.* Ibi v hidae. In dominio sunt ii carucae et xii villani et i Radchenistre habentes xii carucas. Ibi viii servi et molinum de v solidis. Valuit lxx solidos, modo c solidos.

In Brictuuoldesberg hundredo. Ipsa ecclesia tenet Culne. Ibi iiii hidae. In dominio sunt iii carucae et xi villani et vii bordarii cum xii carucis. Ibi iiii servi. Valuit vi libras, modo viii libras. Duo molini reddebant xxv solidos.

In Begebriges hundredo. Ipsa ecclesia tenet Aldesorde. Ibi xi hidae. In dominio sunt iii carucae et xxi villani et v bordarii et ii francigenae cum xv carucis. Ibi vi servi. Valuit c solidos. Modo viii libras.

Ipsa ecclesia tenet Bochelande. *In Wideles hundredo.* Ibi x hidae. In dominio sunt iii carucae et xxii villani et vi bordarii cum xii carucis. Ibi viii servi et x acrae prati. Valuit iii libras, modo ix libras.

In Tetboldestanes hundredo.

TRANSLATION

X. THE LAND OF SAINT PETER OF GLOUCESTER

In Dudstone hundred. Saint Peter of Gloucester held in the time of King Edward the manor of Barton with the adjoining members of Barnwood, Tuffley and Merewent. There, 22 hides less one virgate. There are in demesne 9 ploughs, and 42 villeins and 21 bordars with 45 ploughs. There, 12 serfs, and a mill yielding 5 shillings, and 120 acres of meadow, and wood 5 furlongs long and 3 broad. It was worth £8. Now £24. This manor was always free from geld and from every royal service.

The same church held Frocester *in Blacelaw hundred.* There, 5 hides. In demesne are 4 ploughs, and 8 villeins and 7 bordars with 7 ploughs. There, 3 serfs, and 10 acres of meadow, and wood 3 furlongs long and 2 furlongs broad. It was worth £3, now £8. The same church holds Boxwell *in Grumbalds Ash hundred.* There, 5 hides. In demesne are 2 ploughs, and 12 villeins and one radman having 12 ploughs. There, 8 serfs, and a mill yielding 5 shillings. It was worth 70 shillings, now 100 shillings.

In Brightwell's Barrow hundred. The same church holds Coln. There, 4 hides. In demesne are 3 ploughs, and 11 villeins and 7 bordars with 12 ploughs. There, 4 serfs. It was worth £6, now £8. Two mills used to render 25 shillings.

In Bibury hundred. The same church holds Aldsworth. There, 11 hides. In demesne are 3 ploughs, and 21 villeins and 5 bordars and 2 Frenchmen with 15 ploughs. There, 6 serfs. It was worth 100 shillings. Now £8.

The same church holds Buckland *in Wideles hundred.* There, 10 hides. In demesne are 3 ploughs, and 22 villeins and 6 bordars with 12 ploughs. There, 8 serfs, and 10 acres of meadow. It was worth £3, now £9.

In Tibaldstone hundred.

INDEX

Barley (Wo.), 226, 249
Barnacle, 278
Barnwood, 459
Barrington, 34, 42
Barrow (Ru.), 363
Barrow on Soar, 315, 317, 342, 351, 352
Barrowden, 365, 377
Barsby, 315
Barston, 277
Bartestree, 66, 70, 71, 89
Bartley, 219
Barton (St.), 168
Barton by Bristol, 40, 41, 48
Barton by Gloucester, 20, 32, 49, 459
Bascherche hundred, 116
Basford, 191, 201
Baswich, 172
Batsford, 18, 20 n.
Battle Abbey Register, 435
Baunton, 14
Bausley, 130, 161
Baveney, 117, 122
Bawdwen, William, 55, 383, 419
Baxter, F. H., 256 n.
Bayley, C. P., 419
Baysham, 61, 68
Bayton, 269
Bazeley, M. L., 443
Beadles, 72, 237, 238
Beamont Chase, 378
Bearley, 305, 306
Beaver, S. H., 389 n.
Beckford, 219, 262, 270
Beddintone, 172 n.
Bedfordshire, 32 n., 384, 385, 408
Bedminster, 2, 15, 28, 34, 35 n., 49
Bedworth, 292, 294 n.
Beehives, 266, 353; *see also* Honey
Bee-keepers, 129, 237, 238, 266; *see also* Honey
Beffcote, 167
Beighterton, 183, 194, 423
Belgrave, 322–3, 351, 352
Bellbroughton, 262
Bell Hall, 255
Bellington, 232, 244, 249, 250, 258, 259, 267, 269
Belmesthorpe, 361, 363, 368
Belton, 314

Beltrov, 62 n.
Belvoir, Vale of, 344, 350, 353–5, 442
Benehale, 119 n., 159
Beoley, 230, 244, 246, 247
Beresford, M. W., 278 n.
Berewicks, 66, 71, 77, 119, 167, 168, 192, 194, 204, 206, 215, 220 n., 223, 224, 226, 239, 244, 249, 253, 260, 261, 266, 363, 365, 372, 374, 375, 382, 456
    unnamed, 119, 165, 224, 226, 361, 363, 365
Bericote, 292
Berkeley, 4, 16, 17, 18, 20, 41, 48, 456
Berkeley hundred, 4, 5
Berkswell, 385
Bernintrev hundred, 4
Bernoldune, 62 n., 83, 88
Bescaby, 314
Bescot, 179
Besford (Sa.), 129
Besford (Wo.), 235, 238, 258, 259
Betley, 178
Betton (in Hales), 136
Beverstone, 4
Bevington, 219, 295
Bibliographical Notes, 55–6, 113–14, 161–2, 215–16, 271–2, 311–12, 358, 383, 419–20
Bibury, 16, 34, 42
Bickenhill, Hill and Middle, 278, 294
Bickford, 172 n., 192
Bickmarsh, 219, 232, 274
Bickmore, D. P., 170, 387
Biddulph, 178
Bidford, 290, 298, 305
Bilbrook, 191
Billesdon, 334
Billesley, 306
Billing, Great and Little, 386, 397, 408
Billington, 168
Bilston, 214
Binton, 300, 301, 302, 303, 304
Birdingbury, 287
Birmingham, 164
Birstall, 331, 340, 351, 352
Birtsmorton, 227
Bisbrooke, 377
Bishampton, 260, 262
Bishop's Cleeve, 27, 42